Lecture Notes in Physics

Springer-Verlag Berlin Heidelberg GmbH

Physics and Astronomy ONLINE LIBRARY

http://www.springer.de/phys/

The Editorial Policy for Proceedings

The series Lecture Notes in Physics reports new developments in physical research and teaching – quickly, informally, and at a high level. The proceedings to be considered for publication in this series should be limited to only a few areas of research, and these should be closely related to each other. The contributions should be of a high standard and should avoid lengthy redraftings of papers already published or about to be published elsewhere. As a whole, the proceedings should aim for a balanced presentation of the theme of the conference including a description of the techniques used and enough motivation for a broad readership. It should not be assumed that the published proceedings must reflect the conference in its entirety. (A listing or abstracts of papers presented at the meeting but not included in the proceedings could be added as an appendix.)
When applying for publication in the series Lecture Notes in Physics the volume's editor(s) should submit sufficient material to enable the series editors and their referees to make a fairly accurate evaluation (e.g. a complete list of speakers and titles of papers to be presented and abstracts). If, based on this information, the proceedings are (tentatively) accepted, the volume's editor(s), whose name(s) will appear on the title pages, should select the papers suitable for publication and have them refereed (as for a journal) when appropriate. As a rule discussions will not be accepted. The series editors and Springer-Verlag will normally not interfere with the detailed editing except in fairly obvious cases or on technical matters.
Final acceptance is expressed by the series editor in charge, in consultation with Springer-Verlag only after receiving the complete manuscript. It might help to send a copy of the authors' manuscripts in advance to the editor in charge to discuss possible revisions with him. As a general rule, the series editor will confirm his tentative acceptance if the final manuscript corresponds to the original concept discussed, if the quality of the contribution meets the requirements of the series, and if the final size of the manuscript does not greatly exceed the number of pages originally agreed upon. The manuscript should be forwarded to Springer-Verlag shortly after the meeting. In cases of extreme delay (more than six months after the conference) the series editors will check once more the timeliness of the papers. Therefore, the volume's editor(s) should establish strict deadlines, or collect the articles during the conference and have them revised on the spot. If a delay is unavoidable, one should encourage the authors to update their contributions if appropriate. The editors of proceedings are strongly advised to inform contributors about these points at an early stage.
The final manuscript should contain a table of contents and an informative introduction accessible also to readers not particularly familiar with the topic of the conference. The contributions should be in English. The volume's editor(s) should check the contributions for the correct use of language. At Springer-Verlag only the prefaces will be checked by a copy-editor for language and style. Grave linguistic or technical shortcomings may lead to the rejection of contributions by the series editors. A conference report should not exceed a total of 500 pages. Keeping the size within this bound should be achieved by a stricter selection of articles and not by imposing an upper limit to the length of the individual papers. Editors receive jointly 30 complimentary copies of their book. They are entitled to purchase further copies of their book at a reduced rate. As a rule no reprints of individual contributions can be supplied. No royalty is paid on Lecture Notes in Physics volumes. Commitment to publish is made by letter of interest rather than by signing a formal contract. Springer-Verlag secures the copyright for each volume.

The Production Process

The books are hardbound, and the publisher will select quality paper appropriate to the needs of the author(s). Publication time is about ten weeks. More than twenty years of experience guarantee authors the best possible service. To reach the goal of rapid publication at a low price the technique of photographic reproduction from a camera-ready manuscript was chosen. This process shifts the main responsibility for the technical quality considerably from the publisher to the authors. We therefore urge all authors and editors of proceedings to observe very carefully the essentials for the preparation of camera-ready manuscripts, which we will supply on request. This applies especially to the quality of figures and halftones submitted for publication. In addition, it might be useful to look at some of the volumes already published. As a special service, we offer free of charge LaTeX and TeX macro packages to format the text according to Springer-Verlag's quality requirements. We strongly recommend that you make use of this offer, since the result will be a book of considerably improved technical quality. To avoid mistakes and time-consuming correspondence during the production period the conference editors should request special instructions from the publisher well before the beginning of the conference. Manuscripts not meeting the technical standard of the series will have to be returned for improvement.
For further information please contact Springer-Verlag, Physics Editorial Department II, Tiergartenstrasse 17, D-69121 Heidelberg, Germany

Series homepage – http://www.springer.de/phys/books/lnpp

D. Reguera G. Platero L.L. Bonilla J.M. Rubí (Eds.)

Statistical and Dynamical Aspects of Mesoscopic Systems

Proceedings of the XVI Sitges Conference on Statistical Mechanics
Held at Sitges, Barcelona, Spain, 7-11 June 1999

Springer

Editors

David Reguera
José Miguel Rubí
Departament de Física Fonamental, Universitat de Barcelona
Diagonal 647, Barcelona 08028, Spain

Gloria Platero
Instituto de Ciencia de Materiales (CSIC)
Cantoblanco, Madrid 28049, Spain

Luis López Bonilla
Departamento de Matemáticas
Escuela Politécnica Superior
Universidad Carlos III de Madrid
Avenida de la Universidad 30, Leganés 28911, Spain

Library of Congress Cataloging-in-Publication Data applied for.

Die Deutsche Bibliothek - CIP-Einheitsaufnahme

Statistical and dynamical aspects of mesoscopic systems : proceedings of the XVI Sitges Conference on Statistical Mechanics, held at Sitges, Barcelona, Spain, 7 - 11 June 1999 / D. Reguera ... (ed.).
(Lecture notes in physics ; 547)
ISBN 978-3-662-14276-9 ISBN 978-3-540-45557-8 (eBook)
DOI 10.1007/978-3-540-45557-8

ISSN 0075-8450
ISBN 978-3-662-14276-9

Originally published by Springer-Verlag Berlin Heidelberg New York in 2000
Softcover reprint of the hardcover 1st edition 2000

Typesetting: Camera-ready by the authors/editors
Cover design: *design & production*, Heidelberg

Printed on acid-free paper
SPIN: 10720783 55/3144/du - 5 4 3 2 1 0

Preface

One of the most significant developments in physics in recent years concerns mesoscopic systems, a subfield of condensed matter physics which has achieved proper identity. The main objective of mesoscopic physics is to understand the physical properties of systems that are not as small as single atoms, but small enough that properties can differ significantly from those of a large piece of material. This field is not only of fundamental interest in its own right, but it also offers the possibility of implementing new generations of high-performance nano-scale electronic and mechanical devices. In fact, interest in this field has been initiated at the request of modern electronics which demands the development of more and more reduced structures. Understanding the unusual properties these structures possess requires collaboration between disparate disciplines. The future development of this promising field depends on finding solutions to a series of fundamental problems where, due to the inherent complexity of the devices, statistical mechanics may play a very significant role. In fact, many of the techniques utilized in the analysis and characterization of these systems have been borrowed from that discipline.

Motivated by these features, we have compiled this new edition of the Sitges Conference. We have given a general overview of the field including topics such as quantum chaos, random systems and localization, quantum dots, noise and fluctuations, mesoscopic optics, quantum computation, quantum transport in nanostructures, time-dependent phenomena, and driven tunneling, among others.

The Conference was the first of a series of two Euroconferences focusing on the topic *Nonlinear Phenomena in Classical and Quantum Systems.* It was sponsored by CEE (Euroconference) and by institutions who generously provided financial support: DGCYT of the Spanish Government, CIRIT of the Generalitat of Catalunya, the European Physical Society, Universitat de Barcelona and Universidad Carlos III de Madrid. It was distinguished by the European Physical Society as a Europhysics Conference. The city of Sitges allowed us, as usual, to use the Palau Maricel as the lecture hall.

Finally, we are also very grateful to all those who collaborated in the organization of the event, Profs. F. Guinea and F. Sols, Drs. A. Pérez-Madrid and O. Bulashenko, as well as M. González, T. Alarcón and I. Santamaría-Holek.

Barcelona, February 2000 The Editors

Contents

Part I Quantum Dots

Part II Quantum Chaos

Part VII Random Systems and Localization

Part VIII Mesoscopic Superconductors, Nanotubes and Atomic Chains

Appendix I Contributions Presented as Posters

Part I

Quantum Dots

Thermopower in Quantum Dots

K.A. Matveev

Department of Physics, Duke University, Durham, NC 27708-0305, USA

Abstract. At relatively high temperatures the electron transport in single electron transistors in the Coulomb blockade regime is dominated by the processes of sequential tunneling. However, as the temperature is lowered the cotunneling of electrons becomes the most important mechanism of transport. This does not affect significantly the general behavior of the conductance as a function of the gate voltage, which always shows a periodic sequence of sharp peaks. However, the shape of the Coulomb blockade oscillations of the *thermopower* changes qualitatively. Although the thermopower at any fixed gate voltage vanishes in the limit of zero temperature, the amplitude of the oscillations remains of the order of $1/e$.

1 Introduction

1.1 Coulomb Blockade

The phenomenon of Coulomb blockade is usually observed in devices where the electrons tunnel in and out of a small conducting grain. A simplest example of such a system is shown in Fig. 1. The small grain here is connected to a large metal electrode—the lead—by a layer of insulator, which is so thin that the electrons can tunnel through it.

When this happens, the grain acquires the charge of the electron $-e$. As a result, the grain is now surrounded by an electric field, and there is clearly some energy accumulated in this field. The energy can be found from classical electrostatics as $E_C = e^2/2C$, where C is the appropriate capacitance of the grain. Since the capacitance of small objects is small, the charging energy can be quite significant. In a typical experiment E_C/k_B is on the order of 1 Kelvin. A typical temperature in this kind of experiment is $T \sim .1$ K, i.e., $T \ll E_C$. Since it is impossible for an electron to tunnel into the grain without charging it, the electron must have the energy $E \geq E_C$ before it tunnels. At low temperatures $T \ll E_C$ the number of such electrons in the lead is negligible, and no tunneling is possible. This phenomenon is called the *Coulomb blockade* of tunneling.

How can one observe the absence of tunneling? To do this, one needs to add another metal electrode to the system—the gate, see Fig. 1. It is far enough from the grain, so that no tunneling between these two pieces of metal is possible. However by applying the voltage V_g to the gate one can change the charging energy and control the Coulomb blockade. Indeed, if we apply positive voltage to the gate, the positive charge in it will attract electron to

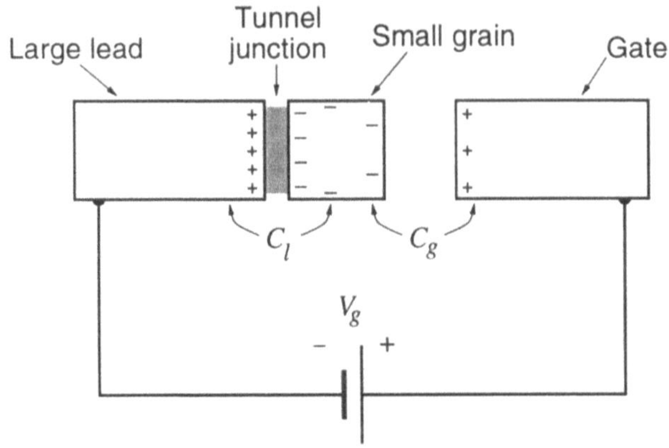

Fig. 1. A small metallic grain is coupled to the lead electrode via a tunnel junction. The electrostatic energy of the system is tuned by applying voltage V_g to the gate electrode. C_l and C_g are the capacitances between the grain and the lead and gate electrodes.

the grain and decrease the charging gap. Mathematically, this is expressed as the following dependence of the electrostatic energy on the number n of extra electrons in the grain and the gate voltage:

$$E(n, V_g) = E_C \left(n - \frac{C_g V_g}{e} \right)^2 . \tag{1}$$

To discuss the effect of the gate voltage on electron tunneling in this system, it is helpful to plot the energy (1) as a function of V_g for various values of n, see Fig. 2(a).

Clearly the energy (1) depends on V_g quadratically, so for each value of n we get a parabola centered at $C_g V_g / e = n$. If the number of electrons in the grain can change due to the possibility of tunneling through the insulating layer, the ground state of the system is given by the parabola with n being the integer nearest to $C_g V_g / e$. Thus the number of the extra electrons in the grain behaves according to Fig. 2(b). The steps of the grain charge as a function of the gate voltage were observed by Lafarge et al. (1993).

Although the measurements of the charge of a small grain are possible, it is far easier to measure *transport* properties of the systems with small metallic conductors. The most common device studied experimentally is single electron transistor shown in Fig. 3. Unlike the device in Fig. 1, there are two leads coupled to the grain by tunneling junctions. By applying bias voltage between the two leads one can study the transport of electrons through the grain. Instead of making the device based on true metallic grains and leads one can achieve the same basic setup by confining two-dimensional electrons

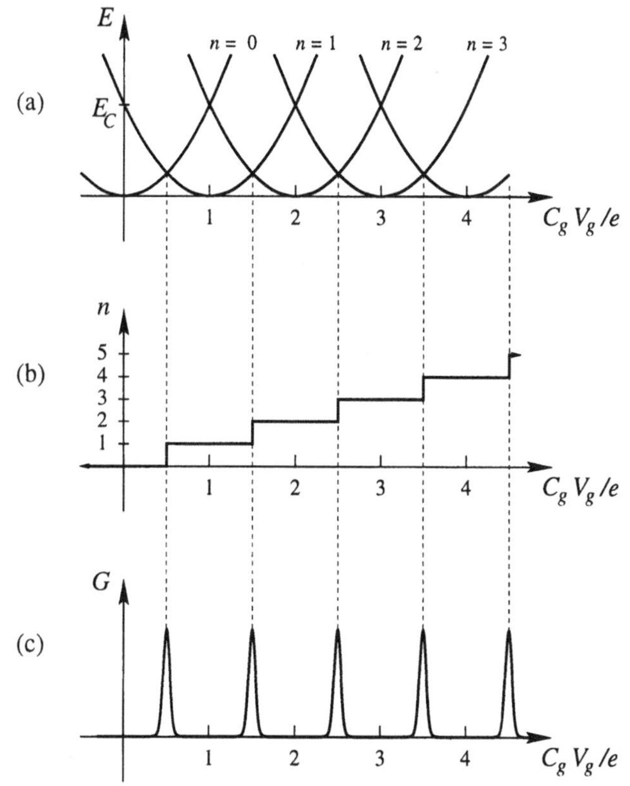

Fig. 2. (a) Electrostatic energy (1) of the system in Fig. 1 as a function of the charging energy for various values of the number of extra electrons n in the dot; (b) the number of electrons in the dot as a function of the gate voltage found by minimization of the electrostatic energy; (c) the conductance of a single electron transistor shows peaks at the points where the charge has steps.

in semiconductor heterostructures by additional gates, see, e.g., (Kastner 1993). In this case the role of the grain is played by a small isolated "puddle" of electrons—*a quantum dot.* Although there are significant differences between these experimental techniques, they will not be important for the following discussion.

An interesting behavior is observed when a small bias voltage is applied, $eV \ll T$, and the conductance G of the single electron transistor is measured as a function of the gate voltage. The experiment shows periodic peaks in the conductance as a function of V_g, see, e.g., (Kastner 1993).

The origin of the peaks is quite clear from Fig. 2(a). At the points where $C_g V_g/e = m + 1/2$, the electrostatic energy of the states with m and $m+1$ extra electrons in the grain are equal. At these values of the gate voltage an

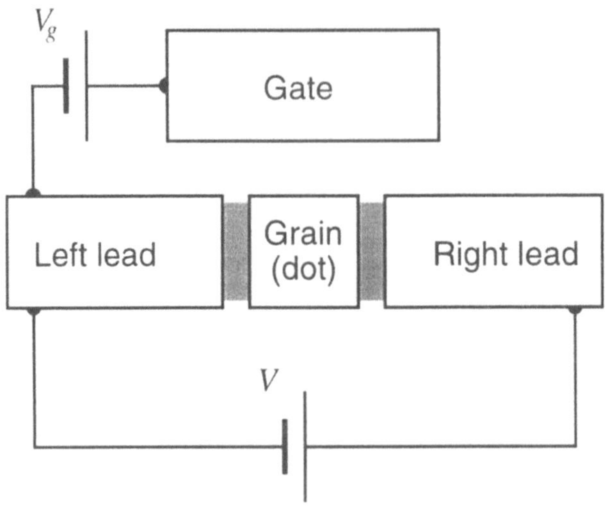

Fig. 3. Single electron transistor. The central electrode can be either a metal grain or a semiconductor quantum dot. The bias voltage V is applied between the two leads.

electron can tunnel between the grain and the leads without changing the electrostatic energy of the system. As a result the Coulomb blockade is lifted, and the transport is greatly enhanced. Thus the conductance has periodic peaks, as shown in Fig. 2(c).

1.2 Mechanisms of Transport

Apart from the positions of the peaks in conductance of a single electron transistor, it is interesting to discuss their shapes. This requires a more detailed understanding of the mechanisms of charge transfer through the grain. The relative importance of different mechanisms is determined primarily by the temperature. We will concentrate on the regime of temperatures much smaller than E_C, where the conductance does show the sharp peaks of Fig. 2(c). In this case the two most important mechanisms are sequential tunneling and cotunneling.

Sequential tunneling. This mechanism is the foundation of the so-called *orthodox model* of Coulomb blockade (Averin and Likharev 1991). In order for the current to flow from the left lead to the right one, one electron tunnels from the left lead to the dot, and another electron tunnels from the dot to the right lead. The two processes are assumed to be real transitions, so that the energy of the system is conserved at every step. The resulting peak shape was found by Glazman and Shekhter (1989):

$$G^{\rm sq} = \frac{G_l G_r}{2(G_l + G_r)} \frac{u/T}{\sinh(u/T)}. \tag{2}$$

Here G_l and G_r are the conductances of the two tunnel barriers. The energy u is the Coulomb blockade gap, which is proportional to the distance from a peak, $u = (eC_g/C)(V_g^{(n)} - V_g)$, with $V_g^{(n)} = \frac{e}{C_g}(n - \frac{1}{2})$ being the center of the n-th peak, Fig. 2(c). The important features of the sequential tunneling result (2) are:

- The peak height is

$$G_0^{\rm sq} = \frac{G_l G_r}{2(G_l + G_r)}. \tag{3}$$

  This result can be interpreted as the sum of the resistances of the two tunneling barriers. The additional factor of 1/2 results from the fact that near any given peak only two charge states n and $n+1$ are allowed, and all the tunneling events which would give rise to states with charged $n-1$ and $n+2$ are forbidden.
- Away from the center of the peak the conductance falls off exponentially, $G^{\rm sq} \propto e^{-u/T}$. The reason for this behavior is that the electron tunneling from a lead has to charge the grain, which requires for it to have the energy u above the Fermi level, see Fig. 4(a). At low temperature $T \ll u$, the probability of finding such an electron in the lead is exponentially small.

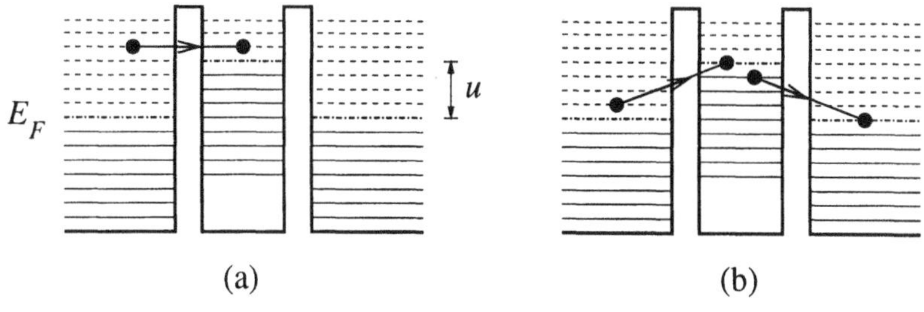

Fig. 4. Energy states of electrons in a single electron transistor. Quantum dot is shown as a small region between the barriers separating it from the left and right leads. Solid and dashed lines represent states below and above the Fermi level, respectively. Arrows illustrate the elementary tunneling processes leading to (a) sequential tunneling and (b) inelastic cotunneling.

Inelastic cotunneling. At low temperatures $T \ll E_C$, the conductance $G^{\rm sq}$ in the valleys between peaks is exponentially small. As a result, another transport mechanism—the inelastic cotunneling—becomes important. This mechanism is illustrated in Fig. 4(b). At the first stage, an electron tunnels from a state near the Fermi level in the left lead to the dot. The energy of the system increases by an amount close to u, assuming that we are not too close to the center of the peak, i.e., $u \gg T$. Since the energy is not conserved, the process does not stop here, and the state of the higher energy is only a virtual state. At the second stage, another electron tunnels from the dot to the right lead. This brings the energy back to its original value, and the tunneling process is complete.

In the linear regime, when the bias is small, $eV \ll T$, the contribution of inelastic cotunneling to the conductance was found by Averin and Nazarov (1990):

$$G^{\rm co} = \frac{\pi\hbar}{3e^2} G_l G_r \frac{T^2}{u^2}. \tag{4}$$

To compare this result with the sequential contribution, we need to estimate $G^{\rm co}$ at the center of the peak and in the valleys:

- At the center of a peak $u = 0$, and (4) formally diverges. This is because the calculation was performed under the assumption $u \gg T$, and the quasiparticle energies $\xi \sim T$ were neglected compared to u in the calculation of the energy of the virtual state. Thus the correct way to fix the singularity in (4) is by substituting $u \sim T$. Thus the peak value of $G^{\rm co}$ is

$$G_0^{\rm co} \sim \frac{\hbar}{e^2} G_l G_r. \tag{5}$$

- In the valleys, at $u \gg T$, the conductance is inversely proportional to the square of the distance from the peak u. This result is easy to understand, because the amplitude of the second-order process is inversely proportional to the energy of the virtual state $E_v \simeq u$, and the tunneling probability is square of the amplitude. The temperature dependence is T^2, because the original electron of energy $\xi \sim T$ decays into three quasiparticles, resulting in a phase space volume $W \propto \xi^2 \sim T^2$. This argument is quite analogous to the one used to evaluate the lifetime of a quasiparticle in a Fermi liquid, see, e.g, (Abrikosov 1988).

Comparison of the two mechanisms. At the center of the conductance peaks one needs to compare the results (3) and (5). We are interested in the case of weak tunneling between the quantum dot and the leads, i.e., $G_l + G_r \ll e^2/\hbar$. Then obviously the sequential tunneling mechanism gives the dominant contribution. On the other hand, the cotunneling conductance (5) decays much slower than the sequential one, (3), when the gate voltage is tuned away from the center of the peak. As a result, at $u > u_c$, where

$$u_c \sim T \ln \frac{e^2}{\hbar(G_l + G_r)}, \tag{6}$$

the cotunneling mechanism dominates the conduction. Note that this only happens if u_c is less than the distance to the center of the valleys $u = E_C$. Thus the cotunneling becomes an important mechanism of transport at low enough temperatures $T \lesssim E_C / \ln[e^2/\hbar(G_l + G_r)]$.

It is worth mentioning that this crossover occurs far from the center of the peak, i.e., $u_c \gg T$, where the conductance is already very small. Thus the presence of two transport mechanisms is not immediately obvious from looking at the data for the conductance as a function of the gate voltage.

2 Thermopower

In a number of recent experiments a different transport property of single electron transistors, the thermopower, was studied (Staring et al. 1993, Dzurak et al. 1997). We will see below that the thermopower S is very sensitive to the transport mechanism, and the crossover from sequential tunneling to cotunneling changes the behavior of $S(V_g)$ qualitatively.

2.1 Definition

To measure the thermopower, one first needs to ensure that the temperatures of the two leads T_l and T_r, are slightly different, $\Delta T = |T_l - T_r| \ll T_l$. Then, one must be able to measure the voltage V generated on the device under the condition that there is no electric current I through it. The thermopower is defined as

$$S \equiv - \lim_{\Delta T \to 0} \left. \frac{V}{\Delta T} \right|_{I=0} . \tag{7}$$

It is helpful to think about the thermopower from the point of view of the linear response theory. Most generally the current in a device is a linear function of the voltage V and temperature difference ΔT, i.e.,

$$I = G_T \Delta T + GV. \tag{8}$$

Here G is the usual conductance of the system, and the kinetic coefficient G_T describes the current response to the temperature difference. Since the definition of S calls for zero current I through he device, we can express the thermopower (7) as

$$S = \frac{G_T}{G}. \tag{9}$$

Thus, one can find the thermopower S by calculating the kinetic coefficients G_T and G.

2.2 Physical Meaning of the Thermopower

Before we proceed with the discussion of the thermopower of single electron transistors, let us try to get a better idea of the physical meaning of this quantity.

The electric current in a rather arbitrary electronic device can be presented as

$$I = -e \int [n_l(\epsilon) - n_r(\epsilon)] w(\epsilon) d\epsilon, \tag{10}$$

where ϵ is the energy of an electron measured from the Fermi level, $n_l(\epsilon)$ and $n_r(\epsilon)$ are the Fermi distribution functions corresponding to the temperatures and chemical potentials of the left and right leads, respectively. The quantity $w(\epsilon)$ represents the remaining relevant physical properties of the system, such as tunneling densities of states in the leads, transmission coefficients of the tunneling barriers, etc.

Expression (10) is quite generic: it applies not only to simple tunneling junctions, where w has the meaning of transmission coefficient, but also to many other devices, including single electron transistors. If both the electrochemical potentials and temperatures in the two leads coincide, i.e., $\mu_l - \mu_r = -eV = 0$ and $\Delta T = 0$, we have $n_l = n_r$ and the current (10) vanishes, as expected. One can then apply a small ΔT or V and discuss the kinetic coefficients G_T and G,

$$G_T = \frac{\partial I}{\partial T_l} = -e \int \frac{\epsilon}{T} \left(-\frac{\partial n}{\partial \epsilon}\right) w(\epsilon) d\epsilon, \tag{11}$$

$$G = \frac{\partial I}{\partial V} = -e \int (-e) \left(-\frac{\partial n}{\partial \epsilon}\right) w(\epsilon) d\epsilon. \tag{12}$$

Note that we have differentiated only the Fermi function n_l with respect to the temperature and chemical potential of the left lead. Although the transmission probability w may also depend on T_l and μ_l, in the linear approximation this should be neglected.

Using the results (11) and (12), we can present the thermopower (9) as

$$S = -\frac{\langle\epsilon\rangle}{eT}. \tag{13}$$

Here $\langle\epsilon\rangle$ has the meaning or the average energy of the electrons carrying the current through the system. It is defined as

$$\langle\epsilon\rangle = \frac{\int \epsilon \left(-\frac{\partial n}{\partial \epsilon}\right) w(\epsilon) d\epsilon}{\int \left(-\frac{\partial n}{\partial \epsilon}\right) w(\epsilon) d\epsilon} \tag{14}$$

We see from (13) that the thermopower S of a single electron transistor measures the average energy of electrons tunneling between the left and right leads.

It is worth mentioning that $\Pi = -\langle\epsilon\rangle/e$ is the Peltier coefficient of the device, and that the relation $S = \Pi/T$ equivalent to (13) follows from Onsager relations, see, e.g. (Abrikosov 1988).

2.3 Thermopower in the Sequential Tunneling Regime

The first experiments on the thermopower of a single electron transistor (Staring 1993) were performed at relatively high temperature, and the transport in the device was dominated by the sequential tunneling processes. The theory of thermopower in this regime was developed by Beenakker and Staring (1992). At $T \ll E_C$ their results can be easily understood from Fig. 5.

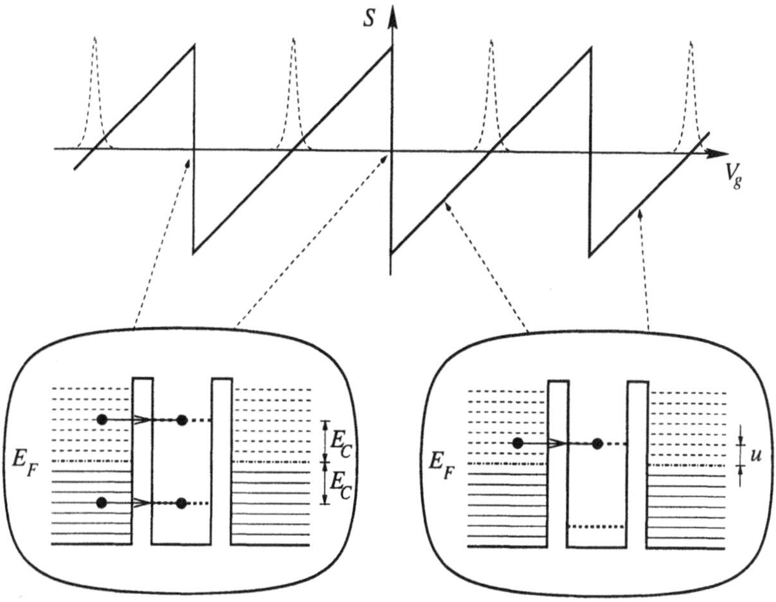

Fig. 5. The thermopower of a single electron transistor as a function of the gate voltage shows sawtooth behavior. This result was obtained within the framework of the sequential tunneling theory by Beenakker and Staring (1992). The dashed peaks correspond to the linear conductance $G(V_g)$.

We will interpret the result in terms of the average energy of tunneling electrons (13). In the centers of the valleys separating the conductance peaks the system possesses a certain symmetry: the change of the electrostatic energy when one electron is either added to or removed from the dot is the same, $u = E_C$. As a result, the two processes shown in the left insert in Fig. 5 contribute equally to the transport, and the average energy of tunneling electrons is zero. However, when the gate voltage is tuned slightly away from the centers of the valleys, one of the processes gives much greater contribution to the transport, resulting in $\langle\epsilon\rangle = \pm E_C$. Thus the thermopower shows sharp steps in the middles of the valleys of conductance. When the gate voltage is tuned away from the centers of the valleys, the change in the charging energy u varies linearly with the gate voltage. One then expects $\langle\epsilon\rangle = u$, and $S = -u/eT$. In fact, the theory (Beenakker and Staring 1992) predicts

$$S^{\rm sq} = -\frac{u}{2eT}. \tag{15}$$

The additional factor of $\frac{1}{2}$ is due to the fact that in the sequential tunneling mechanism the energy ϵ of the tunneling electron can be less than u, if there are holes in the dot at energy $\epsilon - u$. The density of electrons with energy ϵ in the lead is proportional to $e^{-\epsilon/T}$, and the density of holes at energy $\epsilon - u$ is $e^{-(u-\epsilon)/T}$. The product of these two exponentially small factors is simply $e^{-u/T}$, meaning that the tunneling probability is the same for all electrons with energies ϵ between 0 and u. The average energy of such electrons is then $\langle\epsilon\rangle = u/2$, in agreement with (15).

An important feature of the result (15) is that in the limit of low temperature, $T \to 0$, the amplitude of the thermopower oscillations $S_0^{\rm sq} = E_C/2eT$ diverges. This unusual behavior is specific to the sequential tunneling mechanism. Unlike most other cases, the transport is due to electrons which are far from the Fermi level, i.e., at energies $\epsilon \sim E_C \gg T$. Thus, according to (13) the thermopower diverges as $1/T$ at $T \to 0$.

The sawtooth behavior of the thermopower, Fig. 5, was observed experimentally by Staring et al. (1993). The finite temperature of the experiment gives rise to rounding of the "teeth" of the sawtooth dependence; the relative positions of the peaks of conductance $G(V_g)$ to the sawtooth $S(V_g)$ correspond to Fig. 5.

2.4 Cotunneling Thermopower

The problem of thermopower in single electron transistors has been recently revisited in the experiment by Dzurak et al. (1997). Although the observed behavior of $S(V_g)$ is somewhat similar to Fig. 5, there were a number of important differences:

- The jumps aligned with the *peaks* of conductance, instead of the valleys.
- The behavior of $S(V_g)$ between the jumps was not linear.
- The direction of the "teeth" was opposite to the one shown in Fig. 5.
- The amplitude of the oscillations of $S(V_g)$ was estimated to be on the order of $S_0 \sim 1/e$, i.e., much smaller than $S_0 = E_C/2eT$.

In order to understand the deviations from the theory (Beenakker and Staring 1992), one needs to take into account the fact that the temperature in this experiment was significantly lower than in (Staring et al. 1993). Indeed the ratio T/E_C in (Dzurak et al. 1997) was estimated to be on the order of 0.012, i.e., much less than 0.13 in (Staring et al. 1993). It is then natural to conjecture that the new behavior observed by Dzurak et al. (1997) is caused by cotunneling mechanism of transport, which is expected to dominate at low temperatures, Sect. 1.2. Here we review the theory of the thermopower in the regime of inelastic cotunneling (Turek and Matveev 1999).

Contrary to the case of sequential tunneling, the transport in the cotunneling regime is always due to the electrons which are within a strip of width $\sim T$ around the Fermi level. Since the cotunneling occurs in the second order of the perturbation theory, the cotunneling probability w is inversely proportional to the square of the difference of energies of virtual and initial states:

$$w(\epsilon) \propto \frac{1}{(u + \epsilon' - \epsilon)^2}. \tag{16}$$

Here ϵ is the energy of the electron in the left lead, and ϵ' is its energy after it tunnels into the dot. It is clear from (16) that at positive u the electrons above the Fermi level tunnel more effectively than those below the Fermi level. Thus one expects to find non-zero average energy (14).

One can easily estimate $\langle\epsilon\rangle$ as follows. Since the typical ϵ is of order T, one can expand (16) in small ϵ/u,

$$w(\epsilon) \propto \frac{1}{u^2}\left(1 + \frac{2\epsilon}{u}\right). \tag{17}$$

Thus although the electrons with positive ϵ do tunnel more effectively than the ones with negative ϵ, this effect is small as ϵ/u, or, for typical electrons, $\sim T/u$. Since typical electrons have energies $\epsilon \sim T$, the average energy is $\langle\epsilon\rangle \sim T^2/u$. We can now use (13) to estimate the cotunneling thermopower as $S \sim -T/eu$. A careful calculation supports this estimate and gives the numerical prefactor:

$$S^{\mathrm{co}} = -\frac{4\pi^2}{5}\frac{T}{e}\left(\frac{1}{u} + \frac{1}{u - 2E_C}\right). \tag{18}$$

The second term in (18) accounts for the processes when first an electron tunnels from the dot to the right lead, and then another electron tunnels from the left lead to the dot.

The cotunneling thermopower given by (18) diverges at $u = 0$. The origin of this behavior is the same as that of divergence in cotunneling conductance result (4), namely the calculation at $T \ll u$ neglects contributions of quasiparticle energies to the energy of the virtual state. Taking this effect into account, one can study the behavior of the thermopower at any u. This leads to the smearing of the singularities at $u \to 0$. In order to understand the correct behavior of $S(V_g)$, one should also remember that at small u the transport is dominated by sequential tunneling, Sect. 1.2. Thus both contributions have to be taken into account in calculating G and G_T in (9). The resulting thermopower (Turek and Matveev 1999) is shown schematically in Fig. 6. It is described by (18) in the valleys between the peaks of $G(V_g)$ and coincides with sawtooth (15) in the peak regions.

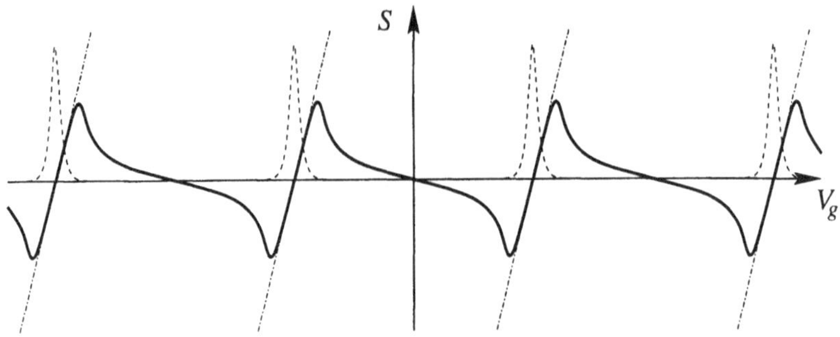

Fig. 6. Schematic view of the thermopower of a single electron transistor at low temperatures. For comparison, the conductance peaks are shown by dashed line, and the sawtooth behavior (15) is indicated by dash-dotted lines.

Note that the apparent slope of the new sawtooth is opposite to that of the original one. It is also clear that the sharpest regions are now aligned with the peaks of the conductance $G(V_g)$. To estimate the amplitude of the thermopower oscillations, one can simply notice that the maxima are at $u = u_c$, where the crossover from sequential tunneling to cotunneling occurs. Substituting (6) into the sequential tunneling result, one arrives at the estimate of the amplitude of the oscillations

$$S_0 \sim \frac{1}{e} \ln \frac{e^2}{\hbar(G_l + G_r)}. \tag{19}$$

It is interesting that although at $T \to 0$ and fixed gate voltage the thermopower vanishes in accordance with (18), the amplitude (19) is independent of the temperature.

The behavior of Fig. 6 is in qualitative agreement with the experiment (Dzurak et al. 1997). The exact amplitude of the thermopower oscillations could not be measured in the experiment due to the uncertainty in measurements of the temperatures of the leads. However, the order of magnitude estimate of the amplitude of thermopower oscillations observed in that experiment is in reasonable agreement with (19).

3 Conclusions

We discussed the thermopower of single electron transistors in the regime of low temperatures, when sequential tunneling is no longer the main mechanism of electron transport. We found that as the temperature is lowered and inelastic cotunneling starts to dominate the conduction between the peaks of Coulomb blockade, the dependence of the thermopower on the gate voltage undergoes a qualitative change. This can be easily seen by comparing figures 5 and 6. The fact that the mechanism of transport can be clearly identified by the general shape of $S(V_g)$ is new compared to the case of linear conductance $G(V_g)$, which shows periodic peaks for either mechanism.

The results reviewed in this paper were obtained under the assumption that the quantum dot is coupled weakly to the leads, i.e., the conductances of the tunneling barriers are small compared to $e^2/\hbar$. In a recent experiment by Möller et al. (1998) a different regime, in which one of the contacts is strongly coupled to the lead, $G_r \sim e^2/\hbar$, was investigated. The above theory is not applicable to this case, however one can still explore the limit of almost perfect transmission between the dot and one of the leads, when the conductance G_r approaches $e^2/\pi\hbar$. The results will be published elsewhere (Andreev and Matveev 1999).

Another limitation of this work is that we have limited it to the regime of relatively large dots or, equivalently, not too low temperatures. It is known that in the limit $T \to 0$ the transport will be dominated by *elastic* cotunneling (Averin and Nazarov 1990). This happens at temperatures below $\sqrt{E_C \Delta}$, where Δ is the quantum level spacing in the dot. Therefore, in small dots one should expect that as the temperature is lowered the thermopower will cross over from the sawtooth behavior of Fig. 5 to the inelastic cotunneling dependence of Fig. 6, and then to a new regime of elastic cotunneling, which needs to be studied in the future.

The author is grateful to A.V. Andreev, L.I. Glazman, and M. Turek for useful discussions. This work was supported by A.P. Sloan Foundation and by NSF Grant DMR-9974435.

References

Abrikosov A.A. (1988): *Fundamentals of the theory of metals* (Elsevier, Amsterdam)
Andreev A.V., Matveev K.A. (1999): *in preparation.*
Averin D.V., Likharev K.K. (1991): in *Mesoscopic Phenomena in Solids*, edited by B. Altshuler, P.A. Lee, and R.A. Webb (Elsevier, Amsterdam)
Averin D.V., Nazarov Yu.V. (1990): Phys. Rev. Lett. **65**, 2446
Beenakker C.W.J., Staring A.A.M (1992): Phys. Rev. B **46**, 9667
Dzurak A.S., Smith C.G., Barnes C.H.W., Pepper M., Martin-Moreno L., Liang C.T., Ritchie D.A., Jones G.A.C. (1997): Phys. Rev. B **55**, 10197
Glazman L.I., Shekhter R.I. (1989): J. Phys. Conden. Matter **1**, 5811
Kastner M.A. (1993): Physics Today **46**, 24
Lafarge P., Joyez P., Esteve D., Urbina C., Devoret M.H. (1993): Nature **365**, 422
Möller S., Buhmann H., Godijn S.F., Molenkamp L.W., (1998): Phys. Rev. Lett. **81**, 5197
Staring A.A.M., Molenkamp L.W., Alphenhaar B.W., van Houten H., Buyk O.J.A., Mabesoone M.A.A., Beenakker C.W.J., Foxon C.T. (1993): Europhys. Lett. **22**, 57
Turek M., Matveev K.A. (1999): *in preparation.*

Kondo Effect in Quantum Dots

L.I. Glazman[1], F.W.J. Hekking[2], and A.I. Larkin[1,3]

[1] Theoretical Physics Institute, University of Minnesota, Minneapolis, MN 55455, USA
[2] Theoretische Physik III, Ruhr-Universität Bochum, 44780 Bochum, Germany
[3] L.D. Landau Institute for Theoretical Physics, 117940 Moscow, Russia

Abstract. Kondo effect in a quantum dot is discussed. In the standard Coulomb blockade setting, tunneling between the dot and leads is weak, the number of electrons in the dot is well-defined and discrete; Kondo effect may be considered in the framework of the conventional one-level Anderson impurity model. It turns out however, that the Kondo temperature T_K in the case of weak tunneling is extremely low. In the opposite case of almost reflectionless single-mode junctions connecting the dot to the leads, the average charge of the dot is not discrete. Surprisingly, its spin may remain quantized: $s = 1/2$ or $s = 0$, depending (periodically) on the gate voltage. Such a "spin-charge separation" occurs because, unlike Anderson impurity, quantum dot carries a broad-band, dense spectrum of discrete levels. In the doublet state, Kondo effect with a significantly enhanced T_K develops.

1 Introduction

The Kondo effect is one of the most studied and best understood problems of many-body physics. Initially, the theory was developed to explain the increase of resistivity of a bulk metal with magnetic impurities at low temperatures (Kondo 1964). Soon it was realized that Kondo's mechanism works not only for electron scattering, but also for tunneling through barriers with magnetic impurities (Appelbaum 1966, Anderson 1966, Rowell 1969). A non-perturbative theory of the Kondo effect has predicted that the cross-section of scattering off a magnetic impurity in the bulk reaches the unitary limit at zero temperature (Nozières 1974). Similarly, the tunneling cross-section should approach the unitary limit at low temperature and bias (Ng and Lee 1988, Glazman and Raikh 1988) in the Kondo regime.

The Kondo problem can be discussed in the framework of Anderson's impurity model (Anderson 1961). The three parameters defining this model are: the on-site electron repulsion energy U, the one-electron on-site energy ε_0, and the level width Γ formed by hybridization of the discrete level with the states in the bulk. The non-trivial behavior of the conductance occurs if the level is singly occupied and the temperature T is below the Kondo temperature $T_K \simeq (U\Gamma)^{1/2}\exp\{\pi\varepsilon_0(\varepsilon_0+U)/2\Gamma U\}$, where $\varepsilon_0 < 0$ is measured from the Fermi level (Haldane 1979).

It is hard to vary these parameters for a magnetic impurity embedded in a host material. One has much more control over a quantum dot attached to leads by two adjustable junctions. Here, the role of the on-site repulsion U is played by the charging energy $E_C = e^2/C$, where C is the capacitance of the dot. The energy ε_0 can be tuned by varying the voltage on a gate which is capacitively coupled to the dot. In the interval

$$|\mathcal{N} - (2n+1)| < \frac{1}{2} \tag{1}$$

of the dimensionless gate voltage $\mathcal{N}$, the energy $\varepsilon_0 = E_C[(2n+1) - \mathcal{N} - 1/2] < 0$, and the number of electrons $2n+1$ on the dot is an odd integer. The level width is proportional to the sum of conductances $G = G_L + G_R$ of the left (L) and right (R) dot-lead junctions, and can be estimated as $\Gamma = (hG/8\pi^2e^2)\Delta$, where Δ is the discrete energy level spacing in the dot.

The experimental search for a tunable Kondo effect brought positive results (Goldhaber-Gordon *et al.* 1998) only recently. In retrospect it is clear, why such experiments were hard to perform. In the conventional Kondo regime, the number of electrons on the dot must be an odd integer. However, the number of electrons is quantized only if the conductance is small, $G \ll e^2/h$, and the gate voltage $\mathcal{N}$ is away from half-integer values (see, *e.g.*, Glazman and Matveev 1990, Matveev 1991). Thus, in the case of a quantum dot, the magnitude of the negative exponent in the above formula for T_K can be estimated as $|\pi\varepsilon_0(\varepsilon_0 + U)/2\Gamma U| \sim (E_C/\Delta)(e^2/hG)$. Unlike an atom, a quantum dot has a non-degenerate, dense set of discrete levels, $\Delta \ll E_C$. Therefore, the negative exponent contains a product of two large parameters, E_C/Δ and e^2/hG.

Further complication becomes evident if one compares the $\propto \ln T$ correction G_K, which is the textbook manifestation of the Kondo effect at $T \gg T_K$, with the background temperature-independent conductance G_{el} provided by the elastic co-tunneling mechanism,

$$G_K \sim G_{\mathrm{el}} \frac{\hbar G}{e^2} \left(\frac{\Delta}{E_C}\right)^2 \ln\left(\frac{E_C}{T}\right). \tag{2}$$

As one can see from Eq. (2), the Kondo correction remains small compared to the background conductance everywhere in the temperature region $T \gtrsim T_K$. The Kondo contribution G_K becomes of the order of e^2/h and therefore dominates the conductance only in the low-temperature region $T \lesssim T_K$. [The ensemble-averaged value of G_{el} at $G_L, G_R \ll e^2/\hbar$ can be estimated (Averin and Nazarov 1990) as $\langle G_{\mathrm{el}} \rangle \simeq (\hbar G_L G_R/e^2)(\Delta/E_C)$.]

To bring T_K within the reach of a modern low-temperature experiment, one may try smaller quantum dots in order to decrease E_C/Δ; this route obviously has technological limitations. Another, complementary option is to increase the junction conductances, so that $G_{1,2}$ come close to $2e^2/h$, which is the maximal conductance of a single-mode quantum point contact. Junctions in the experiment (Goldhaber-Gordon *et al.* 1998, Cronenwett *et al.* 1998, Schmid 1998) were tuned to $G \simeq (0.3-0.5)e^2/\pi\hbar$. A clear evidence for the Kondo effect was found at the gate voltages away from the very bottom of the odd-number valley, where $|\varepsilon_0|$ is relatively small. Only in this domain of gate voltages the anomalous increase of conductance $G(T)$ with lowering the temperature T was clearly observed. (The unitary limit and saturation of G signalling that $T \ll T_K$, were not reached even there.) The anomalous temperature dependence of the conductance, though, was hardly seen at $\mathcal{N} = 2n+1$, where $|\varepsilon_0|$ reaches maximum. To increase the Kondo temperature and to observe the anomaly of $G(T)$ function in these unfavorable conditions, one may try to make the junction conductances larger. However, if $G_{1,2}$ come close to $e^2/\pi\hbar$, the discreteness of the number of electrons on the dot is almost completely washed out (Matveev 1995). Exercising this option, therefore, raises a question about the nature of the Kondo effect in the absence of charge quantization. It is the main question we address in this work.

2 Main Results

We show that the spin of a quantum dot may remain quantized even if charge quantization is destroyed and the average charge $\langle N\rangle e$ is not integer. Spin-charge separation is possible because charge and spin excitations of the dot are controlled by two very different energies: E_C and Δ, respectively. The charge varies linearly with the gate voltage, $\langle N\rangle \simeq \mathcal{N}$, if at least one of the junctions is almost in the reflectionless regime, $|r_{L,R}| \ll 1$, and its conductance $G_{L,R} \equiv (2e^2/h)(1 - |r_{L,R}|^2)$ is close to the conductance quantum. We will show that the spin quantization is preserved if the reflection amplitudes $r_{L,R}$ of the junctions satisfy the condition $|r_L|^2|r_R|^2 \gtrsim \Delta/E_C$. These two constraints on $r_{L,R}$ needed for spin-charge separation are clearly compatible at $\Delta/E_C \ll 1$.

Under the condition of spin-charge separation, the spin state of the dot remains singlet or doublet, depending on $e\mathcal{N}$. If $\cos\pi\mathcal{N} < 0$, the spin state is doublet, and the Kondo effect develops at low temperatures $T \lesssim T_K$. The Kondo temperature we find is

$$T_K \simeq \Delta\sqrt{\frac{\Delta}{T_0(\mathcal{N})}}\exp\left\{-\frac{T_0(\mathcal{N})}{\Delta}\right\}; \tag{3}$$

$$T_0(\mathcal{N}) = \alpha E_C |r_L|^2 |r_R|^2 \cos^2 \pi\mathcal{N}. \tag{4}$$

In the derivation presented below, we entirely disregard the mesoscopic fluctuations. In this case, $\alpha > 0$ is some fixed numerical factor. Fluctuations would result in a statistical distribution of α, with variance $\langle(\delta\alpha)^2\rangle \sim \langle\alpha\rangle^2$. Eps. (3) and (4) demonstrate that in the case of weak backscattering in the junctions, the large parameter E_C/Δ in the Kondo temperature exponent may be compensated by a small factor $\propto |r_L|^2|r_R|^2$. This compensation, resulting from quantum charge fluctuations in a dot with a dense spectrum of discrete states, leads to an enhancement of the Kondo temperature compared with the prediction for T_K of a single-level Anderson impurity model, discussed in the Introduction. Despite the modification of the Kondo temperature, strong tunneling does not alter the universality class of the problem. The temperature dependence of the conductance at $T \lesssim T_K$ is described by a known Costi *et al.* 1994 universal function $F(T/T_K)$,

$$G_K(T/T_K, \mathcal{N}) \simeq \frac{e^2}{h}\left|\frac{r_R}{r_L}\right|^2 (\cos\pi\mathcal{N})^2 F(T/T_K), \tag{5}$$

with F(0)=1. Unlike the case of weak tunneling Ng and Lee 1988, Glazman and Raikh 1988, the conductance (5) explicitly depends on the gate voltage. Eqs. (3) – (5) were derived for an asymmetric set-up, $|r_R|^2 \ll |r_L|^2$. In the special case $|r_L| \to 1$, we can determine the energy T_0, Eq. (4), exactly;

$$T_0(\mathcal{N}) = (4e^{\mathbf{C}}/\pi)E_C|r_R|^2\cos^2\pi\mathcal{N}, \quad |r_L| \to 1, \tag{6}$$

where $\mathbf{C} = 0.5772...$ is the Euler constant. The above results, apart from the detailed dependence of T_K and G_K on $\mathcal{N}$, remain qualitatively correct at $|r_L|^2 \simeq |r_R|^2 \ll 1$. The universality of the Kondo regime is preserved as long as $T_K \ll \Delta$.

3 Bosonization for a Finite-Size Open Dot

We proceed by outlining the derivation of Eqs. (3)–(5). To see how the dense spectum of discrete levels of the dot affects the renormalization of T_K, we first consider the special case $|r_L| \to 1$ and $|r_R| \ll 1$.

In the conventional constant-interaction model, the full Hamiltonian of the system, $\hat{H} = \hat{H}_F + \hat{H}_C$, consists of the free-electron part,

$$\hat{H}_F = \int d\mathbf{r}\left[\frac{1}{2m}\nabla\psi^\dagger\nabla\psi + (-\mu + U(\mathbf{r}))\,\psi^\dagger\psi\right], \tag{7}$$

and of the charging energy

$$\hat{H}_C = \frac{E_C}{2}\left(\frac{\hat{Q}}{e} - \mathcal{N}\right)^2, \quad \frac{\hat{Q}}{e} = \int_{\text{dot}} d\mathbf{r}\psi^\dagger\psi. \tag{8}$$

Here the potential $U(\mathbf{r})$ describes the confinement of electrons to the dot and channels that form contacts to the bulk, μ is the electron chemical potential, and operator $\hat{Q}$ is the total charge of the dot. To derive Eq. (3) for the Kondo temperature, we start with a single-junction system. Following Matveev 1995, we reduce the Hamiltonian (7) – (8) to the one-dimensional (1D) form, and then use the boson representation for the electron degrees of freedom. In this representation, the free-electron term is $\hat{H}_F = \hat{H}_0 + \hat{H}_R$,

$$\hat{H}_0 = \frac{v_F}{2}\int_{-L}^{\infty} dx \sum_{\gamma=\rho,s}\left[\frac{1}{2}(\nabla\phi_\gamma)^2 + 2(\nabla\theta_\gamma)^2\right], \tag{9}$$

$$\hat{H}_R = -\frac{2}{\pi}|r_R| D\cos[2\sqrt{\pi}\theta_\rho(0)]\cos[2\sqrt{\pi}\theta_s(0)], \tag{10}$$

where v_F is the Fermi velocity of the electrons in the single-mode channel connecting the dot with the bulk, and D is the energy bandwidth for 1D fermions, which are related to the boson variables by the transformation (Haldane 1981):

$$\psi_\sigma^\dagger(x) = \hat{\eta}_\sigma\sqrt{\frac{D}{2\pi v_F}}\exp\left\{i\sqrt{\frac{\pi}{2}}\left[\phi_\rho(x) + \sigma\phi_s(x)\right]\right\}$$
$$\times\sum_{l=\pm1}\exp\left\{ilk_Fx + il\sqrt{\frac{\pi}{2}}\left[\theta_\rho(x) + \sigma\theta_s(x)\right]\right\}. \tag{11}$$

We introduced Majorana fermions $\eta_{\pm1}$ here to satisfy the commutation relations for the fermions with opposite spins, $\{\eta_{+1},\eta_{-1}\} = 0$. Anti-commutation of the electrons of the same spin ($\sigma = 1\,\text{or} - 1$), is ensured by the following commutation relations between the canonically conjugated Bose fields:

$$[\nabla\phi_\gamma(x'),\theta_\gamma(x)] = [\nabla\theta_\gamma(x'),\phi_\gamma(x)]$$
$$= -i\delta(x-x'), \quad \gamma = \rho, s. \tag{12}$$

The interaction term (charging energy) becomes also quadratic in the boson representation: $\hat{H}_C = (E_C/2)\left[2\theta_\rho(0)/\sqrt{\pi} - \mathcal{N}\right]^2$. The operators $(2e/\sqrt{\pi})\nabla\theta_\rho(x)$ and $(2/\sqrt{\pi})\nabla\theta_s(x)$ are the smooth parts of the electron charge (ρ) and spin (s) densities, respectively. The continuum of those electron states outside the dot, which are capable to pass through the junction, is mapped (Matveev 1995) onto the Bose fields defined on the half-axis $[0;\infty)$. Similarly, states within a finite-size dot are mapped onto the fields defined on the interval $[-L;0]$ with the boundary condition $\theta_{\rho,s}(-L) = 0$, which corresponds to $|r_L| = 1$. The length in this effective 1D problem is related (Matveev 1995) to the average density of states $\nu_d \equiv 1/\Delta$ in the dot by $L \simeq \pi v_F \nu_d$, and scales proportionally to the area A of the dot formed in a two-dimensional electron gas.

To the leading order in the reflection amplitude $|r_R| \ll 1$ and in the level spacing $\Delta/E_C \ll 1$, the average charge of the dot can be found by minimization of the energy $\hat{H}_C$. The charge is not quantized, and, to this order, it varies linearly with the gate voltage, $(2e/\sqrt{\pi})\langle\theta_\rho(0)\rangle = e\mathcal{N}$. Within the same approximation, the factor $\cos[2\sqrt{\pi}\theta_\rho(0)]$ in (10) at low energies $E \ll E_C$ may be replaced by its average value. This procedure yields (Matveev 1995) the effective Hamiltonian $\hat{H}_s = \hat{H}_0^s + \hat{H}_R^s$ for the spin mode,

$$\hat{H}_0^s = \frac{v_F}{2}\int_{-L}^{\infty} dx\left[\frac{1}{2}(\nabla\phi_s)^2 + 2(\nabla\theta_s)^2\right], \tag{13}$$

$$\hat{H}_R^s = -\left[\frac{4e^{\mathbf{C}}}{\pi^3}E_C D\right]^{1/2} |r_R| \cos(\pi\mathcal{N})\cos[2\sqrt{\pi}\theta_s(0)]. \tag{14}$$

This is a Hamiltonian of a one-mode, $g = 1/2$ Luttinger liquid with a barrier at $x = 0$. At $L \to \infty$ (*i.e.*, at $E \gg \Delta$) the backscattering at the barrier, described by the Hamiltonian $\hat{H}_R^s$, is known to be a relevant perturbation (Kane and Fisher 1992): even if $|r_R|$ is small, at low energy $E \to 0$ the amplitudes of transitions between the minima of the potential of (14) scale to zero. These minima are $\theta_s(0) = \sqrt{\pi}n$ if $\cos\pi\mathcal{N} > 0$, or $\theta_s(0) = \sqrt{\pi}(n+1/2)$, if $\cos\pi\mathcal{N} < 0$. The crossover from weak backscattering $|r_R(E)| \ll 1$ to weak tunneling $|t_R(E)| \ll 1$ occurs at $E \sim T_0(\mathcal{N})$, Eq. (6). To describe the low-energy $[E \lesssim T_0(\mathcal{N})]$ dynamics of the spin mode, it is convenient to project out all the states of the Luttinger liquid that are not pinned to the minima of the potential (14). Transitions between various pinned states then are described by the tunnel Hamiltonian $\hat{H}_0^s + \hat{H}_{xy} + \hat{H}_z$, where

$$\hat{H}_{xy} = -\frac{D^2}{2\pi T_0(\mathcal{N})}\cos\left\{\sqrt{\pi}[\phi_s(+0) - \phi_s(-0)]\right\} ;$$
$$\hat{H}_z = \frac{v_F^2}{2T_0(\mathcal{N})}\nabla\theta_s(-0)\nabla\theta_s(+0). \tag{15}$$

Here a discontinuity of the variable $\phi_s(x)$ at $x = 0$ is allowed, and the point $x = 0$ is excluded from the region of integration in Eq. (13). The term $\hat{H}_{xy}$, which is a sum of two operators of finite shifts for the field $\theta_s(0)$, represents hops $\theta_s(0) \to \theta_s(0) \pm \sqrt{\pi}$ between pinned states. This term is familiar from the theory of DC transport in a Luttinger liquid (Kane and Fisher 1992). However, the usual scaling argument (Kane and Fisher 1992) is insufficient for deriving the term $\hat{H}_z$ and for establishing the exact coefficients in $\hat{H}_{xy}$ and $\hat{H}_z$. We have accomplished these tasks by matching the current-current correlation function $\langle[\hat{I}_s(t), \hat{I}_s(0)]\rangle$ calculated from (15) with the proper asymptote of the exact result which we obtained starting with Eqs. (13), (14) and proceeding along the lines of Furusaki and Matveev 1995.

At $L \to \infty$ the ground state of the spin mode is infinitely degenerate, different states may be labeled by the discrete boundary values $\theta_s(0)$. At finite L, however, this degeneracy is lifted due to the energy of spatial quantization, coming from the Hamiltonian (13). If $\cos\pi\mathcal{N} > 0$, the spatial quantization entirely removes the degeneracy, and the lowest energy corresponds to $\theta_s(0) = 0$ (spin state of the dot is $s = 0$). If $\cos\pi\mathcal{N} < 0$, the spatial quantization by itself, in the absence of tunneling, would leave the ground state doubly degenerate, $\theta_s(0) = \pm\sqrt{\pi}/2$ (spin state of the dot is $s = 1/2$). Hamiltonian (15) hybridizes the spin of the dot with the continuum of spin excitations in the lead. The Kondo effect consists essentially of this hybridization, which ultimately leads to the formation of a spin singlet in the entire system. The energy scale at which the hybridization occurs, is the Kondo temperature of the problem at hand.

4 The Effective Exchange Hamiltonian

At energies $E \ll T_0(\mathcal{N})$, the spin field $\theta_s(0)$ is pinned at the point contact. Recalling that $\theta_s(-L) = 0$, we see that the spin of the dot indeed takes discrete values only, as was mentioned above. We can analyze the low-energy spin dynamics staying in the bosonized representation of a finite-size fermion system, but it is more instructive to return, following Haldane 1981, to the fermion variables. After the two parts of the Hamiltonian (15) are found, one can explicitly see that the initial $SU(2)$ symmetry of the problem is preserved. Therefore, the effective Hamiltonian which replaces (15) at low energies, corresponds to the isotropic exchange interaction,

$$\hat{H}_{\text{ex}} = J_{RR}(\tilde{D})\hat{\mathbf{S}}_R\hat{\mathbf{S}}_d. \tag{16}$$

Here $\hat{\mathbf{S}}_R = \hat{\psi}^\dagger_{\sigma_1}(\mathbf{R}_R)\mathbf{s}_{\sigma_1\sigma_2}\hat{\psi}_{\sigma_2}(\mathbf{R}_R)$ and $\hat{\mathbf{S}}_d = \hat{\chi}^\dagger_{\sigma_1}(\mathbf{R}_R)\mathbf{s}_{\sigma_1\sigma_2}\hat{\chi}_{\sigma_2}(\mathbf{R}_R)$ are the operators of spin density in the dot ($x < 0$) and in the lead ($x > 0$) respectively, at the point $\mathbf{R}_R$ of their contact; $\rho_d \equiv \nu_d/A$ and ρ_R are the corresponding average densities of states. The electron creation-annihilation operators $\psi^\dagger$ and ψ, and the Hamiltonian (16) are defined within a band of some width $\tilde{D} \ll T_0(\mathcal{N})$.

If the dot is in the spin-doublet state, the exchange constant gets renormalized at low energies. Unlike the "ordinary" Kondo model with only one localized orbital state involved, here the renormalization occurs due to virtual transitions in both the continuum spectrum of the lead and the discrete spectrum of the dot:

$$J_{RR}(E,\tilde{D}) = J_{RR}(\tilde{D}) + J^2_{RR}(\tilde{D})\int_{|E|}^{\tilde{D}} \nu_R d\xi_R \sum_{|\xi_d^{(n)}|<\tilde{D}} \frac{1}{|\xi_R| + |\xi_d^{(n)}|}, \tag{17}$$

where the integral is taken over the continuum energy spectrum in the leads. The sum over the discretized energy levels of electrons in the dot includes the term corresponding to a spin-flip within the same orbital level, $\xi_d^{(0)} = 0$. It is this term that is responsible for the logarithmic singularity in the "ordinary" Kondo effect. The rest of the terms in the sum corresponds to the virtual transitions onto (from) the partially occupied level from (to) doubly-occupied (empty) electron levels. At energies $E \gg \Delta$ the discreteness of the spectrum is not important, and the sum in Eq. (17) can be replaced by an integral,

$$J(E,\tilde{D}) = J(\tilde{D}) + J^2(\tilde{D})\int_{|E|}^{\tilde{D}} \nu_R d\xi_R \int_{-\tilde{D}}^{\tilde{D}} \frac{\nu_d d\xi_d}{|\xi_R| + |\xi_d|}, \quad E \gg \Delta. \tag{18}$$

Here, the limit $|E| \to 0$ is not singular; the relative correction to $J_{RR}(\tilde{D})$ is of the order $\tilde{D}/T_0(\mathcal{N})$ and small. The low-energy observable quantities (such as scattering amplitudes) should not depend on $\tilde{D}$. It means that $J(E,\tilde{D})$

in the domain $\tilde{D} \gg E \gg \Delta$ should be just equal to the exchange constant found from Eq. (15) upon returning to the fermion variables,

$$J_{RR}(E, \tilde{D}) = J_{RR} \equiv [\nu_d \nu_R T_0(\mathcal{N})]^{-1}, \quad \tilde{D} \gg E \gg \Delta. \tag{19}$$

Combining Eqs. (17)–(19), we find the following representation of $J_{RR}(E, \tilde{D})$, convenient at $E \ll \Delta$:

$$J_{RR}(E, \tilde{D}) = J_{RR} + J_{RR}^2 \nu_R \left(\ln \tfrac{\Delta}{E} + \Lambda\right), \tag{20}$$

$$\Lambda = \lim_{\tilde{D}\to\infty} \lim_{E\to 0} \left\{ -\ln\frac{\Delta}{E} + \int_{|E|}^{\tilde{D}} \nu_R d\xi_R \left[\sum_{|\xi_d^{(n)}|<\tilde{D}} \frac{1}{|\xi_R| + |\xi_d^{(n)}|} - \int_{-\tilde{D}}^{\tilde{D}} \frac{\nu_d d\xi_d}{|\xi_R| + |\xi_d|} \right] \right\}. \tag{21}$$

The constant Λ is of the order of unity. Equation (20) demonstrates that in the strong tunneling regime the bandwidth for the effective Kondo problem at hand is Δ rather than E_C. Once Eq. (20) is established, one can obtain the results (3) and (4) following the lines of Haldane 1979.

The Kondo effect in a single-junction system results in a specific behavior of the spin polarization. If the dot is in a singlet state, the gap for its spin polarization is $\sim \Delta$. In the doublet state, the contribution of the dot to the susceptibility at low temperature and fields, $T, \mu_B H \ll \Delta$, is identical to that of a Kondo impurity (Nozières 1974) with T_K given by Eq. (3), (4); here μ_B is the Bohr magneton for the electrons of the dot. The manifestation of the most interesting effect, the enhanced low-temperatue conductance, requires a two-junction dot geometry.

5 A Dot with Two Junctions

To consider the low-temperature conductance through a dot, we derive a Hamiltonian that generalizes Eq. (16) to the case of two junctions and acts within the energy band $|E| \leq \Delta$:

$$\begin{aligned}\hat{H}_{\text{ex}} = \Big[& J_{LL} \hat{\psi}^\dagger_{\sigma_1}(\mathbf{R}_L) \hat{\chi}^\dagger_{\sigma_3}(\mathbf{R}_L) \hat{\chi}_{\sigma_4}(\mathbf{R}_L) \hat{\psi}_{\sigma_2}(\mathbf{R}_L) \\ & + J_{RR} \hat{\psi}^\dagger_{\sigma_1}(\mathbf{R}_R) \hat{\chi}^\dagger_{\sigma_3}(\mathbf{R}_R) \hat{\chi}_{\sigma_4}(\mathbf{R}_R) \hat{\psi}_{\sigma_2}(\mathbf{R}_R) \\ & + J_{LR} \hat{\psi}^\dagger_{\sigma_1}(\mathbf{R}_L) \hat{\chi}^\dagger_{\sigma_3}(\mathbf{R}_L) \hat{\chi}_{\sigma_4}(\mathbf{R}_R) \hat{\psi}_{\sigma_2}(\mathbf{R}_R) \Big] \mathbf{s}_{\sigma_1\sigma_2} \mathbf{s}_{\sigma_3\sigma_4}. \end{aligned} \tag{22}$$

The derivation of the low-energy theory goes through stages similar to Eqs. (13), (14) and (15). We will explain first how to derive the relevant exchange constants in the least involved case of a strongly asymmetric set-up: $G_L \ll e^2/h$ and $|r_R| \ll 1$. In this case the largest constant $J_{RR} \propto G_L^0$

exists even in the limit $G_L = 0$, and is defined by Eq. (16); the smallest constant, $J_{LL} \propto G_L^2$, is unimportant in the calculation of the conductance; the intermediate constant J_{LR} is proportional to G_L. To find the proportionality coefficient, we calculate the conductance through the dot in the lowest-order perturbation theory in the Hamiltonian (22), and obtain $G(T) = (\pi^4 e^2/3h) J_{LR}^2 \rho_L \rho_R \rho_d^2 T^2$. When deriving this formula, we set also $T \gg \Delta$, which allows us now to compare $G(T)$ with the exact at $\Delta = 0$ result (Furusaki and Matveev 1995) for the conductance of the same system. The comparison yields:

$$J_{LR}^2 = 4(h/e^2) G_L \left[\pi e^{\mathrm{C}} E_C T_0(\mathcal{N}) \rho_L \rho_R \rho_d^2\right]^{-1} . \qquad (23)$$

At $T \lesssim \Delta$, only the lowest discrete level in the dot remains important. If the gate voltage is close to an odd integer, $\cos \pi\mathcal{N} < 0$, then the level is spin-degenerate. This way, the initial problem of the dot, which has a dense spectum of discrete levels, and is strongly coupled to the leads, is reduced to the problem of a single-level Kondo impurity in a tunnel junction (Ng and Lee 1988, Glazman and Raikh 1988). Using the found values of the exchange constants, and the result of Glazman and Raikh 1988 for a strongly asymmetric junction ($J_{LL} \ll J_{LR} \ll J_{RR}$), we obtain the conductance in the problem under consideration:

$$\begin{aligned} G_K(T/T_K, \mathcal{N}) &= (e^2/h)(J_{LR}/J_{RR})^2 F(T/T_K) \\ &\simeq (64/\pi^2) G_L |r_R|^2 (\cos \pi\mathcal{N})^2 F(T/T_K). \end{aligned} \qquad (24)$$

Note that Kondo conductance (24) in the strongly asymmetric set-up is significantly smaller than the conductance quantum e^2/h even at $T = 0$. The maximal value of G_K is substantially increased, if the asymmetry between the junctions is reduced, and the condition $G_L \ll e^2/h$ is lifted. To show this, we further generalize the above results to include the experimentally important case $|r_R| \ll |r_L| \ll 1$. Like in the case of a single strong junction considered above, the backscattering in the junctions becomes increasingly effective at low electron energies. Initially, at energies below E_C, the reflection amplitudes grow independently of each other (Furusaki and Matveev 1995) as $|r_{L,R}(E)| \sim |r_{L,R}|(E_C/E)^{1/4}$. Upon reducing the energy scale, the weaker junction reaches the crossover region first: at $E \sim T_1 \equiv E_C |r_L|^4$ the backscattering in this junction becomes significant, $|r_L(E)| \sim 1$.

To consider conductance at temperatures $T \ll T_1$, we can formulate now an effective Hamiltonian, which acts within the narrow energy band T_1, and describes weak reflection in the right junction, $|r_R(T_1)| \sim |r_R/r_L|$, and strong reflection in the left junction, $|r_L(T_1)| \sim 1$. Both junctions eventually cross over into the weak tunneling regime at sufficiently low temperatures. Replacing E_C by the bandwidth T_1 and $|r_R|$ by $|r_R/r_L|$ in Eq. (6), we find Eq. (4) for the new crossover temperature. Below it, the exchange Hamiltonian (22) is applicable. The largest exchange constant J_{RR} is independent of $|r_L|$ in the leading approximation; it is still defined by Eq. (16) with

$T_0(\mathcal{N})$ from Eq. (4). To find the new value of J_{LR}, we replace $E_C \to T_1$, $G_L \to (e^2/h)(1 - |r_L(T_1)|^2) \sim e^2/h$, and use Eq. (4) for $T_0(\mathcal{N})$ in Eq. (23); the result is $J_{LR}^2 \sim [E_C^2 |r_L|^6 |r_R|^2 \rho_L \rho_R \rho_d^2]^{-1}$. Substituting the exchange constants J_{RR} and J_{LR} in Eq. (24), we arrive at Eq. (5).

6 Overall Temperature Dependence of the Conductance

We finally discuss the overall temperature dependence of the conductance, see Fig. 1. In this discussion, we use the above results for the Kondo regime, and the results of Furusaki and Matveev 1995, Aleiner and Glazman 1998 for co-tunneling, generalized properly onto the case $|r_R| \ll |r_L| \ll 1$. The conductance decreases slowly (Furusaki and Matveev 1995), as the temperature is reduced from E_C to T_1. At lower tempertures, the leading mechanism of transport is inelastic co-tunneling, which yields $G \sim T/T_1$ and $G \sim T^2/T_1 T_0(\mathcal{N})$ at T above and below $T_0(\mathcal{N})$, respectively. At yet lower temperatures, the main contribution to the conductance $G(T)$ is provided by elastic co-tunneling, $G_{\text{el}} \sim (\Delta/T_1)\ln(T_1/\Delta)$. The crossover between the two co-tunneling mechanisms occurs at $T^* \sim \sqrt{T_0(\mathcal{N})\Delta \ln(T_1/\Delta)}$. It is instructive to compare G_{el} with the zero-temperature Kondo conductance (5). Taking into account the definition of T_1, we see that the Kondo mechanism dominates, if $T_0(\mathcal{N})/\Delta \gtrsim \ln(E_C|r_L|^4/\Delta)$. This condition simultaneously ensures the smallness of the Kondo temperature compared to the level spacing, so that the Kondo singlet state remains distinct.

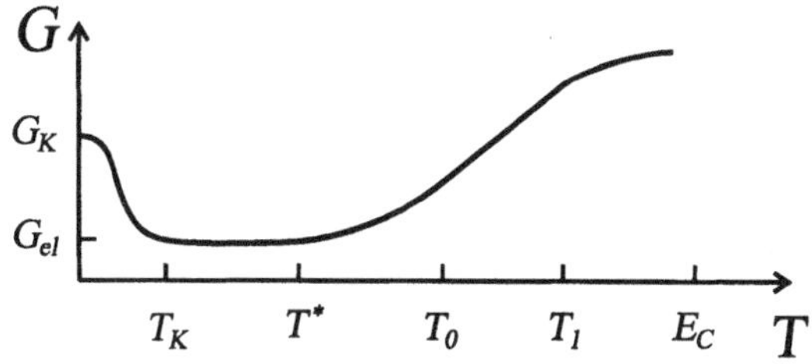

Fig. 1. The overall temperature dependence of conductance. The estimates of the crossover temperatures and the two characteristic values of the conductance, $G_K \equiv G_K(0, \mathcal{N})$ and G_{el}, are given in the text.

Upon the increase of the conductance G_L towards the value $2e^2/h$, the spin quantization of the dot eventually is destroyed. We expect that, at the same time, the oscillations in the conductance $G(\mathcal{N})$ with period $\delta\mathcal{N} = 2$ give way to mesoscopic fluctuations, which do not distinguish between the "even" and "odd" intervals of $\mathcal{N}$.

7 Conclusions

We developed a theory of the Kondo effect in a quantum dot which has a dense spectum of discrete one-particle states. It turns out that the spin of the quantum dot may remain quantized, even if the quantization of charge is destroyed by strong dot-lead tunneling. In the spin-doublet state, the Kondo effect develops at low temperature, yielding a non-monotonous temperature dependence of the conductance. We found that the Kondo temperature is significantly enhanced by charge fluctuations, compared to the standard case of weak dot-lead tunneling.

Acknowledgments

The work at the University of Minnesota was supported by NSF Grants DMR-9731756 and DMR-9812340; FH acknowledges financial support through SFB 237 of the Deutsche Forschungsgemeinschaft. The authors are grateful to A. Kaminski for help in preparation of the manuscript.

References

Aleiner, I.L. and Glazman L.I. (1998): Phys. Rev. B **57**, 9608.
Anderson, P.W. (1961): Phys. Rev. **124**, 41.
Anderson, P.W. (1966): Phys. Rev. Lett. **17**, 95.
Appelbaum, J. (1966): Phys. Rev. Lett. **17**, 91.
Averin, D.V. and Nazarov, Yu.N. (1990): Phys. Rev. Lett. **65**, 2446.
Costi, T.A., Hewson, A.C. and Zlatić, V. (1994): J. Phys. Condens. Matter **6**, 2519.
Cronenwett, S.M., Oosterkamp, T.H., and Kouwenhoven, L.P. (1998): Science **281**, 540.
Furusaki, A. and Matveev, K.A. (1995): Phys. Rev. B **52**, 16676.
Glazman, L.I. and Raikh, M.E. (1988): JETP Lett. **47**, 452.
Glazman, L.I. and Matveev, K.A. (1990): Sov. Phys. JETP **71**, 1031.
Goldhaber-Gordon, D. *et al.* (1998): Nature (London) **391**, 156; Phys. Rev. Lett. **81**, 5225.
Haldane, F.D.M. (1979): Phys. Rev. Lett. **40**, 416.
Haldane, F.D.M. (1981): J. Phys. C: Solid State Phys. **14**, 2585.
Kane, C.L. and Fisher, M.P.A. (1992): Phys. Rev. B **46**, 15233.
Kondo, J. (1964): Prog. Theor. Phys. **32**, 37.
Matveev, K.A. (1991): Sov. Phys. JETP **72**, 892.
Matveev, K.A. (1995): Phys. Rev. B **51**, 1743.
Ng, T.K. and Lee, P.A. (1988): Phys. Rev. Lett. **61**, 1768.
Nozières, P. (1974): J. Low Temp. Phys. **17**, 31.
Rowell, J.M. (1969): in *Tunneling Phenomena in Solids*, edited by E. Burstein and S. Lundquist (Plenum, New York, 1969), p. 385.
Schmid, J. (1998): Physica B **256-258**, 182.

Interpolative Method for Transport Properties of Quantum Dots in the Kondo Regime

A. Levy Yeyati, A. Martín-Rodero, and F. Flores

Departamento de Física Teórica de la Materia Condensada CV, Universidad Autonoma de Madrid, 28049 Madrid, Spain

Abstract. We present an interpolative method for describing coherent transport through an interacting quantum dot. The idea of the method is to construct an approximate electron self-energy which becomes exact both in the limits of weak and strong coupling to the leads. The validity of the approximation is first checked for the case of a single (spin-degenerate) dot level. A generalization to the multilevel case is then discussed. We present results both for the density of states and the temperature dependent linear conductance showing the transition from the Kondo to the Coulomb blockade regime.

1 Introduction

The Kondo effect constitutes a prototypical correlation effect in condensed matter physics. Although originally studied in connection to magnetic impurities in metals, there is now a renewed interest in this many-body problem fostered by the recent observation of Kondo effect in semiconducting quantum dots [1,2]. Quantum dots provide an almost ideal laboratory where the relevant parameters can be controlled, which allow to test the predictions of theoretical models.

From the theoretical side, Kondo physics in quantum dots has been mainly analyzed in the light of the so called single level Anderson model. There were predictions for the Kondo effect in quantum dots based on this model since the early 90's [3,4]. The theory predicts an enhancement of the linear conductance due to Kondo effect at very low temperatures, which is in qualitative agreement with recent experiments.

However, in most realistic situations, the single level Anderson model constitutes a crude approximation for a quantum dot. Actual semiconducting quantum dots contain a large number ($\sim$ 100) of electrons and the single-particle level separation between dot levels may be not so large compared to the level broadening, which restricts the validity of the single-level approximation. The actual situation would be more appropriately described by a multilevel model, including several instead of a single dot level. Unfortunately, there are no simple theoretical approaches to extract the electronic and transport properties from such a microscopic model.

In this paper we present results on the Kondo effect in quantum dots based on the interpolative method. The basic idea of this method is to construct an interpolative electron self-energy which becomes exact both in the limits of weak and strong coupling to the leads. These ideas were first introduced in Ref. [5] in connection to the single-level Anderson model and have, since then, been adapted by several authors to different problems involving strongly correlated electrons. In this way, the method has been used to study the Hubbard model [6], the non-equilibrium Anderson model [4], the metal-insulator transition in infinite dimensions [7], to incorporate correlation effects into band-structure calculations [8], the ac-Kondo effect in quantum dots [9] and finally extended by the present authors to analyze the multilevel Anderson model [10].

The paper will be organized as follows: In section 2 we present the interpolative method. We first discuss the single level case, showing the accuracy of the method with the help of a simple exactly solvable model. We then consider the multilevel situation. In section 3 we present results which illustrate the behavior of the conductance with temperature in a multilevel situation.

2 The interpolative method

For describing a multilevel quantum dot (QD) we consider a model Hamiltonian $H = H_{\text{dot}} + H_{\text{leads}} + H_T$ where $H_{\text{dot}} = \sum_m \epsilon_m \hat{d}^\dagger_m \hat{d}_m + U \sum_{l>m} \hat{n}_m \hat{n}_l$ corresponds to the uncoupled QD ($\hat{n}_m = \hat{d}^\dagger_m \hat{d}_m$); $H_{\text{leads}} = \sum_{k \in L,R} \epsilon_k \hat{c}^\dagger_k \hat{c}_k$ to the uncoupled leads, and $H_T = \sum_{m,k\in L,R} t_{m,k} \hat{d}^\dagger_m \hat{c}_k + h.c.$ describes the coupling between the dot and the leads. The labels m and l in H denote the different dot levels including spin quantum numbers. The number of dot levels will be denoted by M (i.e. $1 \leq m, l \leq M$). We adopt the usual simplifying assumption of having the same electron-electron interaction U between any pair of dot states.

The main objective of our method is to determine the dot retarded Green functions $G_m(\tau) = -i\theta(\tau) < [\hat{d}_m(\tau), \hat{d}^\dagger_m(0)]_+ >$ from which the different level charges and the dot linear conductance can be obtained. In the frequency representation we can write G_m as:

$$G_m(\omega) = \frac{1}{\omega - \epsilon^{HF}_m - \Sigma_m(\omega) - \Gamma_{m,L}(\omega) - \Gamma_{m,R}(\omega)}, \tag{1}$$

where $\epsilon^{HF}_m = \epsilon_m + U \sum_{l \neq m} n_l$ is the Hartree-Fock level (we adopt the notation n_l for the mean charge on level l) and $\Gamma_{m,L}, \Gamma_{m,R}$ are tunneling rates coupling the dot to the leads, given by $\Gamma_{m,L(R)}(\omega) = \sum_{k\in L(R)} t^2_{m,k}/(\omega - \epsilon_k + i0^+)$. We shall neglect indirect coupling between dot levels through the leads (non-diagonal elements $\Gamma_{m,m',L(R)}$) and adopt the usual approximation of considering $\Gamma_{m,L(R)}$ as a pure imaginary constant independent of the energy.

The self-energy $\Sigma_m(\omega)$ takes into account electron correlation effects beyond the Hartree approximation. The idea of the present approximation is to determine an interpolative self-energy which yields the correct exact results both in the $\Gamma/U \to 0$ limit (atomic limit) and in the opposite $U/\Gamma \to 0$ limit.

2.1 The single-level case

Let us first discuss how to proceed for the simple single-level case. In this case $m = 1, 2$, the two indexes corresponding to up and down spin orientations. These will be denoted by σ and $\bar{\sigma}$. In the atomic limit G_σ can be obtained using the equation of motion technique [11] as

$$G_\sigma^{(at)}(\omega) = \frac{1 - n_{\bar{\sigma}}}{\omega - \epsilon + i0^+} + \frac{n_{\bar{\sigma}}}{\omega - \epsilon - U + i0^+} \tag{2}$$

This expression can be formally written in the usual Fermi liquid form, i.e. $G_\sigma^{(at)}(\omega) = [\omega - \epsilon - U n_{\bar{\sigma}} - \Sigma_\sigma^{(at)}(\omega)]^{-1}$ by introducing the "atomic" self-energy

$$\Sigma_\sigma^{(at)}(\omega) = \frac{U^2 n_{\bar{\sigma}}(1 - n_{\bar{\sigma}})}{\omega - \epsilon - U(1 - n_{\bar{\sigma}}) + i0^+}$$

In the opposite limit, $U/\Gamma \to 0$, the electron self-energy can be calculated by second-order perturbation theory in U, which yields

$$\Sigma_\sigma^{(2)}(\omega) = U^2 \int_{-\infty}^{\infty} d\epsilon_1 \int_{-\infty}^{\infty} d\epsilon_2 \int_{-\infty}^{\infty} d\epsilon_3 \, \frac{\tilde{\rho}_\sigma(\epsilon_1)\tilde{\rho}_{\bar{\sigma}}(\epsilon_2)\tilde{\rho}_{\bar{\sigma}}(\epsilon_3)}{\omega + \epsilon_2 - \epsilon_1 - \epsilon_3 + i0^+} \times$$
$$[f(\epsilon_1)f(\epsilon_3)(1 - f(\epsilon_2)) + (1 - f(\epsilon_1))(1 - f(\epsilon_3)) f(\epsilon_2)], \tag{3}$$

where $f(\omega)$ is the Fermi distribution function and $\tilde{\rho}_\sigma(\omega) = \Gamma/\pi((\omega - \tilde{\epsilon}_\sigma)^2 + \Gamma^2)$ is the local density of states for an effective level $\tilde{\epsilon}_{\bar{\sigma}}$, which will be determined in order to fulfill exact Fermi liquid properties at zero temperature.

It is important to stress the following simple property of $\Sigma^{(2)}$:

$$\lim_{\Gamma \to 0} \Sigma_\sigma^{(2)}(\omega) = U^2 \frac{\tilde{n}_{\bar{\sigma}}(1 - \tilde{n}_{\bar{\sigma}})}{\omega - \tilde{\epsilon}_\sigma + i0^+}$$

Thus, when extrapolated to the atomic limit $\Sigma^{(2)}$ has the same functional form as $\Sigma^{(at)}$. This property suggests that one can smoothly interpolate between the two limits. The ansazt proposed in Ref. [5] for the interpolative self-energy is:

$$\Sigma_\sigma(\omega) = \frac{\Sigma_\sigma^{(2)}(\omega)}{1 - \alpha\, \Sigma_\sigma^{(2)}(\omega)} \tag{4}$$

where $\alpha = (\epsilon - \tilde{\epsilon}_\sigma - U(1 - n_{\bar{\sigma}}))/(U^2 n_{\bar{\sigma}}(1 - n_{\bar{\sigma}}))$. This ansazt has the desired property $\Sigma \to \Sigma^{(2)}$ when $U \to 0$ and $\Sigma \to \Sigma^{(at)}$ when $\Gamma \to 0$.

The final step is to impose the proper self-consistent condition for determining the effective level $\tilde{\epsilon}$. At zero temperature, from the Luttinger-Ward relations [13] one can derive the Friedel sum rule for the Anderson model [14]

$$n_\sigma = -\frac{1}{\pi}\mathrm{Im}\ln G_\sigma^r(E_F)$$

which imposes an exact relation between the dot-level charge and the phase shift at the Fermi energy. The effective level can thus be determined in order to fulfill the Friedel sum rule. This condition is, however, not valid at finite temperature. In Ref. [4] we show that the condition $n_\sigma = \tilde{n}_\sigma$, i.e. imposing the same charge in the effective system as in the interacting system, is approximately equivalent to the Friedel sum rule at zero temperature but can be also used at finite temperature.

In order to check the accuracy of the interpolative method we have considered a simple two-sites problem that can be diagonalized exactly. One of the sites would describe the metallic leads and the other site corresponds to the dot. In order to analyze the more general situation we impose a finite splitting $\Delta = \epsilon_\sigma - \epsilon_{\bar{\sigma}}$ between the two spin orientations on the dot. Within this toy model the second order self-energy can be evaluated analytically.

In figure 1 we show the charge on the two dot levels as a function of gate voltage (the gate voltage is the distance between the lower dot level and the leads level). As can be observed, in the exact solution there is a blocking of the upper level charge until the gate voltage becomes larger than $\Delta + U$. The exact behavior is accurately reproduced by the interpolative method. It is instructive to consider another simple approximation widely used in the literature, which consist in just broadening the poles in the atomic Green function (2) by the non-interacting tunneling rates. This approximation corresponds to the so-called Hubbard I [12]. As can be observed in the lower panel of Fig. 1, this approximation fails to give the blocking of the upper level found in the exact solution.

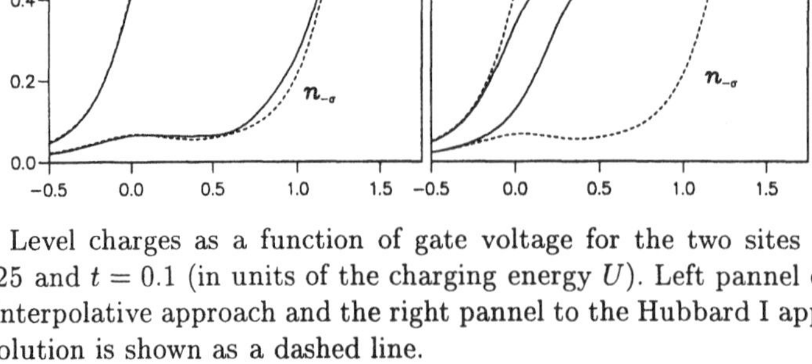

Fig. 1. Level charges as a function of gate voltage for the two sites model with $\Delta = 0.25$ and $t = 0.1$ (in units of the charging energy U). Left pannel corresponds to the interpolative approach and the right pannel to the Hubbard I approach. The exact solution is shown as a dashed line.

2.2 Multilevel case

The multilevel version of the interpolative method is somewhat more complex [10]. In the first place, the atomic limit Green functions do not contain just two poles but several poles corresponding to the various different charge states of the dot. The corresponding expression can be obtained using the equation of motion technique and is given by

$$G_m^{(at)}(\omega) = \frac{< \prod_{l\neq m}(1-\hat{n}_l) >}{\omega - \epsilon_m + i0^+} + \sum_{l\neq m} \frac{< \hat{n}_l \prod_{(s\neq l)\neq m}(1-\hat{n}_s) >}{\omega - \epsilon_m - U + i0^+} + \dots$$
$$+ \frac{< \prod_{l\neq m} \hat{n}_l >}{\omega - \epsilon_m - (M-1)U + i0^+}, \tag{5}$$

The evaluation of this expression requires the knowledge of up to $M-1$-body correlations functions $< \hat{n}_1\hat{n}_2 >$, $< \hat{n}_1\hat{n}_2\hat{n}_3 >$, ..., etc. However, for sufficiently large U fluctuactions in the dot charge by more of one electron with respect to the mean charge $\mathcal{N}$ are strongly inhibited. One can thus approximate Eq. (5) as follows

$$G_m^{(at)}(\omega) \simeq \frac{A^m_{N-1}}{\omega - \epsilon_m - U(N-1) + i0^+} + \frac{A^m_N}{\omega - \epsilon_m - UN + i0^+}$$
$$+ \frac{A^m_{N+1}}{\omega - \epsilon_m - U(N+1) + i0^+}, \tag{6}$$

where $N = Int[\mathcal{N}]$. In order to yield the exact first three momenta of the exact spectral density the weight factors A^m_N should satisfy the following sum rules

$$A^m_{N-1} + A^m_N + A^m_{N+1} = 1$$
$$(N-1)A^m_{N-1} + NA^m_N + (N+1)A^m_{N+1} = \sum_{l\neq m} n_l$$
$$(N-1)^2 A^m_{N-1} + N^2 A^m_N + (N+1)^2 A^m_{N+1} = \sum_{l\neq m} n_l + < \hat{n}\hat{n} >_m, \tag{7}$$

where $< \hat{n}\hat{n} >_m = \sum_{(l\neq k)\neq m} < \hat{n}_l\hat{n}_k >$. For the special case $N = 0$ ($N = M-1$) one has $A^m_{N-1} = 0$ ($A^m_{N+1} = 0$) and only the first two Eqs. in (7) have to be considered.

This approximated expression for $G_m^{(at)}(\omega)$ is now fully determined by the average charges n_l and the two-body correlation functions $< \hat{n}_l\hat{n}_k >$. As in the single-level case one can define an atomic self-energy, $\Sigma_m^{(at)} = \omega - \epsilon_m^{HF} - \left[G_m^{(at)}(\omega)\right]^{-1}$, which can be written as the ratio of two polynomials in ω

$$\Sigma_m^{(at)} = \frac{a_m U^2(\omega - \epsilon_m + i0^+) + b_m U^3}{(\omega - \epsilon_m + i0^+)^2 + c_m U(\omega - \epsilon_m + i0^+) + d_m U^2}, \tag{8}$$

where $a_m = (\mathcal{N} - n_m)\left[1 - (\mathcal{N} - n_m)\right] + < \hat{n}\hat{n} >_m$; $c_m = \mathcal{N} - n_m - 3N$; $d_m =< \hat{n}\hat{n} >_m + 3N^2 - 1 - (3N-1)(\mathcal{N} - n_m)$ and $b_m = N^2(1-N) - (\mathcal{N} - n_m)d_m$.

On the other hand, in the $U/\Gamma \to 0$ limit the self-energy is accurately given by second order perturbation theory as in the single level case. The second order self-energy $\Sigma_m^{(2)}$ now takes into account the interaction of an electron on the dot level m with electron-hole pairs on each one of the other channels.

For the interpolation one notices that both $\Sigma^{(2)}$ and $\Sigma^{(at)}$ have the same functional form when extrapolated to the corresponding opposite limit. The natural generalization of the ansazt in the single level case now has the form of a continued fraction

$$\Sigma_m(\omega) = \frac{\alpha_m \Sigma_m^{(2)}(\omega)}{1 - \beta_m \Sigma_m^{(2)}(\omega) - R_m(\omega)} \quad , \quad R_m(\omega) = \frac{\gamma_m (\Sigma_m^{(2)}(\omega))^2}{1 - \delta_m \Sigma_m^{(2)}(\omega)}$$

with coefficients $\alpha_m = U^2 a_m/\Delta_m$, $\beta_m = (\epsilon_m - \tilde{\epsilon}_m + (b_m/a_m - c_m)U)/\Delta_m$, $\gamma_m = ((c_m - b_m/a_m)b_m/a_m - d_m)U^2/\Delta_m^2$ and $\delta_m = (\epsilon_m - \tilde{\epsilon}_m + U b_m/a_m)/\Delta_m$, where $\Delta_m = U^2 \sum_{l \neq m} \tilde{n}_l(1 - \tilde{n}_l)$.

As in the single-level case the final step is to determine the effective levels self-consistently. In the multilevel case one has, in addition to self-consistently determine the two-body correlations $< \hat{n}\hat{n} >_m$ that appear in the atomic self-energy. This can be done by means of the relation

$$\sum_{l \neq m} < \hat{n}_l \hat{n}_m >= -\frac{1}{\pi U} \int_{-\infty}^{\infty} f(\omega) \mathrm{Im}\left[(\omega - \epsilon_m - \Gamma_m) G_m(\omega)\right] d\omega, \tag{9}$$

connecting the two-body correlations and the Green functions that can be derived from the equation of motion for $G_m(\omega)$. This step turns out to be essential in order to obtain the correct values of the charges in the large U limit.

Finally, the temperature dependent dot linear conductance can be obtained using the expression [15]

$$G = \frac{e^2}{h} \sum_m \frac{|\Gamma_{m,L}\Gamma_{m,R}|}{(|\Gamma_{m,L}| + |\Gamma_{m,R}|)} \int_{-\infty}^{\infty} \left(\frac{\partial f}{\partial \omega}\right) \mathrm{Im} G_m^r(\omega) d\omega$$

3 Results

The multilevel formalism allows to study the importance of the multilevel structure in the QD transport properties. For this purpose we have analyzed the $M = 4$ case which corresponds to two consecutive dot levels plus spin degeneracy. We have studied this case as a function of the level separation Δ.

Figure 2 shows the dot conductance as a function of Fermi energy and temperature for the cases $\Delta = 0$, 0.1 and 0.5 (in units of U).

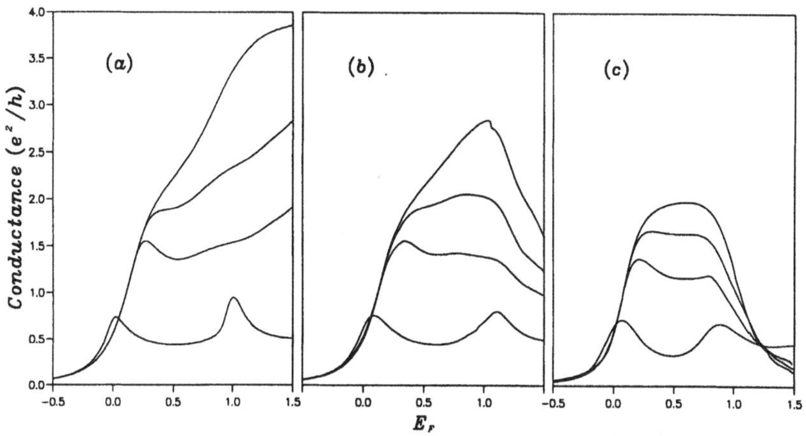

Fig. 2. Dot conductance as a function of Fermi energy for increasing temperatures with $\Gamma_L = \Gamma_R = 0.075$ and $\Delta = 0$ (a), 0.1 (b) and 0.5 (c). The temperature values are 0.0005, 0.0025, 0.005 and 0.03 in units of U

This figure illustrates the transition from a two-fold degenerate situation ($\Delta = 0$), where the conductance reaches a maximum value $4e^2/h$ for the half-filled case at zero temperature, to the case of well separated dot levels, where the maximum conductance $2e^2/h$ is reached for the quarter and three quarter filling case. The increase of conductance with decreasing temperature is due to the Kondo effect. While in the case of well separated levels one observes only the Kondo effect due to the spin-degeneracy of the individual levels, when Δ is small compared to Γ one can observe Kondo features involving the two nearby dot levels. When the temperature is raised above the Kondo temperature (which is around 0.005 for the parameters used in this figure) one recovers the sequence of dot resonances at the charge degeneracy points characteristic of the Coulomb blockade regime.

The Kondo effect should manifest also as a zero-bias anomaly in the dot non-linear conductance. This anomaly is directly related to the appearance of a narrow peak around the Fermi energy in the dot spectral density. In cases where the splitting between dot levels is of the order of Γ we expect to have a zero-bias anomaly not only between dot resonances corresponding to the same dot level but also in between resonances corresponding to different levels. This feature is illustrated in Fig. 3 where we plot the density of states around E_F for the same three cases of Fig. 2 with $E_F = 1.5$. The appearance of a zero-bias anomaly in between resonances corresponding to different levels is a clear manifestation of the multilevel structure of the QD which has already been observed in recent experiments on semiconducting quantum dots [16].

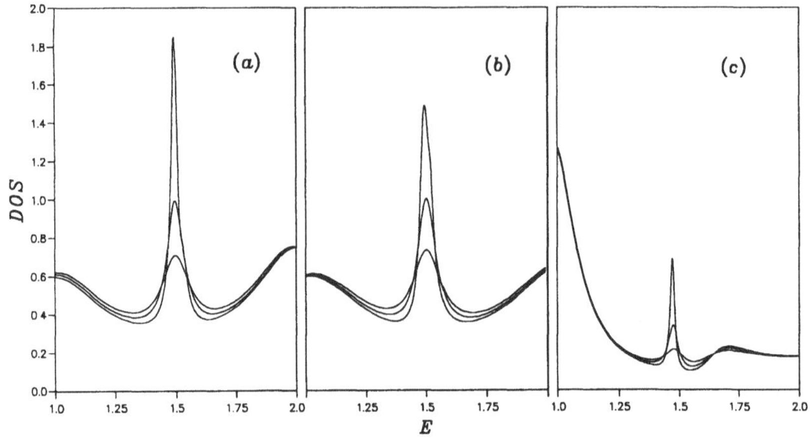

Fig. 3. Density of states around the Fermi energy for increasing temperatures values (0.01, 0.02 and 0.03) for the same three cases in Fig. 2 and $E_F = 1.5$

This work has been funded by the Spanish CICyT under contracts PB97-0028 and PB97-0044.

References

[1] D. Goldhaber-Gordon et al., Nature **391**, 156 (1998) and D. Goldhaber-Gordon et al. Phys. Rev. Lett. **81**, 5225 (1998).
[2] S.M. Cronenwett et al, Science **281**, 540 (1998).
[3] T.K. Ng and P.A. Lee, Phys. Rev. Lett. **61**, 1768 (1988); L.I. Glazman and M.E. Raikh, JETP Lett. **47**, 452 (1988); S. Hershfield, J.H. Davis and J.W. Wilkins, Phys. Rev. Lett. **67**, 3720 (1991); Y. Meir, N.S. Wingreen and P.A. Lee, Phys. Rev. Lett. **70**, 2601 (1993).
[4] A. Levy Yeyati, A. Martín-Rodero and F. Flores, Phys. Rev. Lett. **71**, 2991 (1993).
[5] A. Martín-Rodero et al., Solid State Commun. **44**, 911 (1982).
[6] A. Martín-Rodero et al., Phys. Rev. B **33**, 1814 (1986).
[7] H. Kajueter and G. Kotliar, Phys. Rev. Lett. **77**, 131 (1996); M. Potthoff, T. Wegner and V. Nolting, Phys. Rev. B **55**, 16132 (1997).
[8] A.I. Lichtenstein and M.I. Katsnelson, Phys. Rev. B **57**, 6884 (1998).
[9] R. Lopez et al. Phys. Rev. Lett. **81**, 4688 (1998).
[10] A. Levy Yeyati, F. Flores and A. Martín-Rodero, Phys. Rev. Lett. **83**, 600 (1999).
[11] B. Bell and A. Madhukar, Phys. rev. B **14**, 4281 (1976).
[12] J. Hubbard, Proc. Roy. Soc. A **276**, 238 (1963).
[13] J.M. Luttinger and J.C. Ward, Phys. Rev. **118**, 1417 (1960).
[14] D.C. Langreth, Phys. Rev. **150**, 516 (1966).
[15] Y. Meir and N.S. Wingreen, Phys. Rev. Lett. **68**, 2512 (1992).
[16] J. Schmid, J. Weis, K. Ederl and K. v. Klitzing, Physica B **256-258**, 182 (1998).

A New Tool for Studying Phase Coherent Phenomena in Quantum Dots

R.H. Blick[1], A.W. Holleitner[1], H. Qin[1], F. Simmel[1], A.V. Ustinov[2], K. Eberl[3], and J.P. Kotthaus[1]

[1] Center for NanoScience and Sektion Physik, Ludwig-Maximilians-Universität, Geschwister-Scholl-Platz 1, 80539 München, Germany.
[2] Physikalisches Institut III, Universität Erlangen-Nürnberg, Erwin-Rommel-Str. 1, 91058 Erlangen, Germany.
[3] Max–Planck–Institut für Festkörperforschung, Heisenbergstr. 1, 70569 Stuttgart, Germany.

Abstract. Microwave transport spectroscopy on quantum dots with near-field microwave sources is presented. As microwave sources we employ two different Josephson oscillators, being integrated with the AlGaAs/GaAs-heterostructure in which the dots are formed. We observe photon assisted tunneling induced by the Josephson oscillators and compare the results with those using an externally operated microwave source.

1 Introduction

Microwave spectroscopy on quantum dots promises to probe the dynamics of these few electron systems. Most of the experimental work conducted to date has been focused on rather simple spectroscopic tools: a microwave signal is coupled via an antenna or a stripline to the mesoscopic system under test. The photon-induced current through the dots is measured and allows to probe the discrete states of the quantum system directly (1), (2), (3). These results can be described by the Tien-Gordon theory (4) originally developed for a superconductor tunnel junction and more recent theoretical models (5), (6), (7). In order to reveal the dynamics of electrons confined in tunnel coupled dot systems, more intricate spectroscopic tools are required. In the work by Nakamura et al. (8) it was shown how to monitor a single tunneling Cooper pair in a superconducting tunnel junction transistor in the time domain. This spurs the interest in tracing electrons in coupled quantum dots, since in this case a similar tunnel splitting of the discrete states was found (9), (10).

Here we present a near-field microwave oscillator integrated in a single low-temperature setup with a coupled quantum dot structure. Integrating on-chip microwave sources has the advantage of combining advanced spectrometers easily with mesoscopic devices. Furthermore, the influence of blackbody radiation is minimized, since all the electrical connections to the outside world are essentially dc-lines and can thus be heavily filtered (11), (12). As a microwave source we employ a long Josephson tunnel junction (JTJ) with well-defined emission characteristics. These junctions radiate commonly in the range from few GHz up to 600 GHz (13). Their typical radiation linewidth can be as small as 10^{-6} relative to the emission frequency (14). Recently, this triggered rapid progress in using these devices in integrated sub-millimeter wave superconducting receivers (15). The quantum dots are defined in a two-dimensional electron gas (2DEG) by electron beam written lateral Schottky gates. The versatility of these devices is the possibility to tune the tunnel contacts in a wide range. This allows the straight forward realization of tunnel coupled quantum dots ('covalent artificial molecule') or decoupled dots ('ionic artificial molecule') (10).

In contrast to the excitation spectrum of real atoms or molecules, the spectrum of single or even coupled quantum dots reveals a striking difference in the discrete level structure. For quantum dots it has been shown in a whole variety of experiments that Kohn's theorem (16) prevails (17), (18). The theorem states that only the center-of-mass (cm) degree of freedom couples to a spatially homogenous electromagnetic field. In previous studies of excitations in quantum dots by coupling radiation via antennas only cm excitations were found. Here we 'move' the radiation source close to the quantum dots, in an attempt to observe additional near-field excited modes.

2 Experimental Setup and Sample Characterization

The setup we used is shown in Fig. 1: the Si-chip with the Nb/Al-AlO_x/Nb Josephson junction is glued on top of the quantum dot AlGaAs/GaAs-chip with photo resist. In the inset of Fig. 1 the quantum dot gate structure used in this work is depicted: application of an appropiate negative gate voltage defines two quantum dots in the 2DEG of the AlGaAs/GaAs heterostructure with an electron density of 1.7×10^{15} m^{-2} at 35 mK (marked by the upper white circle in Fig. 1). By variation of the voltages V_{gA} and V_{gB} applied to the plunger gates denoted in the inset of Fig. 1 the electron configuration of the double dot can be changed. Plotting the current through the system as a function of V_{gA} and V_{gB} results in the charging diagram of Fig. 2. Subsequently, we obtain charging energies of the individual dots of $E_C^A = e^2/2C_\Sigma = 220$ μeV and $E_C^B = 205$ μeV, respectively. Thus the 'electronic' radii are $r_A = 400$ nm and $r_B = 430$ nm. For the absolute number of electrons in each dot we find: $n_A = 850$ and $n_B = 980$. Hence, the dots are rather classical systems in which the resolution of discrete single particle energies due to the confinement potential is not expected. In the current measurements the interdot coupling is chosen to be weak ($C_{A-B} = 2$ aF), i.e. we see a hexagonal array of points for the charging diagram (19) which corresponds to the 'ionic' coupling limit. All results were obtained in the linear regime (drain/source bias $V_{ds} = 19$ μV).

The JTJ we use in this experiment is an overlap Nb/Al-AlO_x/Nb long Josephson junction with dimensions 20×400 μm^2 (width$\times$ length). In the

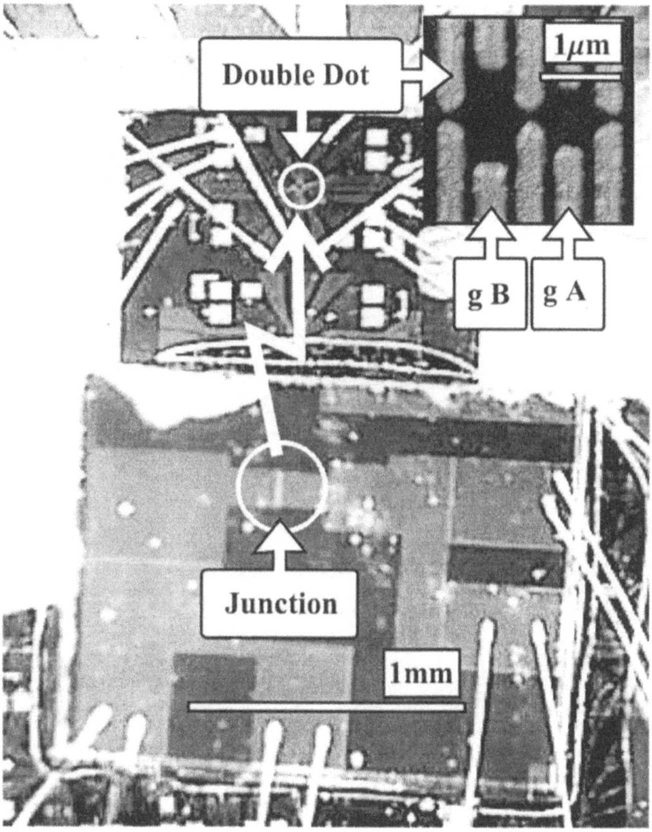

Fig. 1. Top view of the circuit: The quantum dots are located in the center of the Hall-bar (marked by the upper white circle). An AFM micrograph of the gates forming the double dots is seen in the inset on the upper right side (gA, gB denote the plunger gates operated in the measurements). The Josephson oscillator is placed on top of the quantum dot chip and glued to it with photo resist. The lower white circle indicates the position of the junction itself, while the white arrow represents the radiation coupling through the GaAs substrate to the quantum dots (see text for details).

measurements presented, tuning of the frequency was possible by varying the JTJ critical current with the applied magnetic field and selecting the appropriate bias point at a resonant state. When operated in the flux-flow regime, the JTJ radiation frequencies $f = 2eV/h$ are on the order of 100 – 500 GHz. This is above the Coulomb energy for this particular quantum dot. We employ a finite magnetic field to operate the junction at a Fiske step of the current/voltage characteristic, a self-resonant state. The fundamental cavity resonance frequency of the junction is $f = \bar{c}/(2\ell)$, where $\bar{c}$ is the Swihart velocity and ℓ is the junction length.

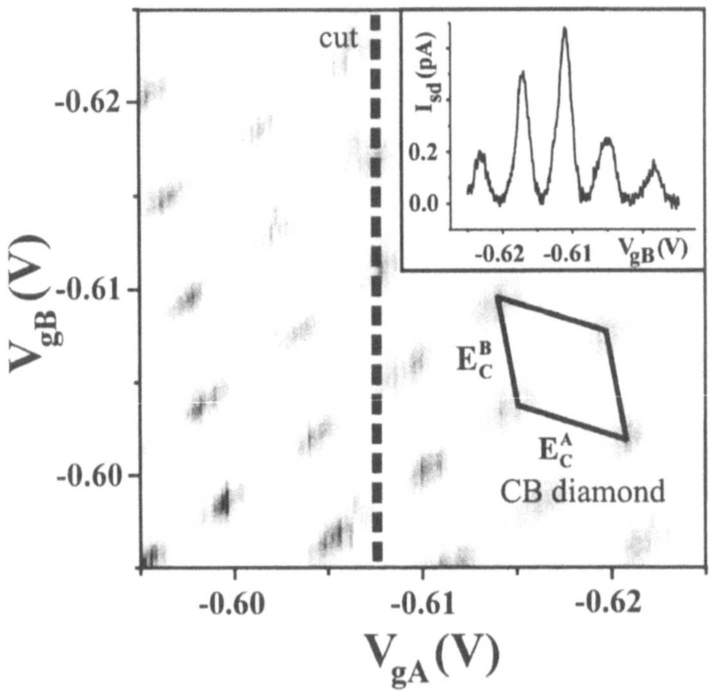

Fig. 2. Charging diagram of the double quantum dot as a grayscale plot without microwave radiation (white: $I < 0$ pA, black: $I > 0.5$ pA). A small forward bias of $V_{ds} = 19\ \mu$V is applied to monitor the current. The two dots are weakly coupled by the tunnel barrier and produce a periodic lattice with E_C^A and E_C^B denoting the charging energies, as marked by the diamond. The two gate voltages V_{gA} and V_{gB} span the charging diagram. Inset gives a line plot of the direct-current through the double dot.

3 Microwave Transport Spectroscopy

In order to characterize the microwave response of the coupled quantum dots we first studied them with a conventional microwave source. The radiation was transduced by an antenna upon the gate structure of the quantum dots. The charging diagram of the double dot with microwave excitation is shown as a grayscale plot in Fig. 3. The radiation of the far-field source is fixed at 10 GHz. The sidebands are found on only one side, depending on the tuning of the barrier transmission coefficients. In measurements on a similar two-dot device we obtained symmetrically as well as asymmetrically induced sidebands in the charging diagram (20). However, since we are interested in the alteration of the microwave coupling by the JTJ it was important to maintain the tuning with only one sideband. This finally ensures that we can directly compare the far- and near-field measurements. The inset shows one of the typical traces from the grayscale plot with the sidebands induced by the frequency dependent absorption. The induced sidebands are marked by arrows – the peak height modulation is due to the specific trace taken out of the charging diagram (marked by the dashed line). A cut along one of the two periodic resonance lines would yield peaks of similar amplitude. The net pumping of electrons leads to a reduction of the absolute current value down to only some 100 fA. Also the noise floor is slightly enhanced by coupling the radiation.

4 Integrated Sources

When the near-field Josephson oscillator is operated as a source with a typical emission frequency of $f \cong 10$ GHz, we observe charging diagrams as the one shown in Fig. 4. The JTJ emission frequency was determined by taking the IV-characteristics. Biasing the JTJ with a current of $I = 1\ \mu$A we are able to detect sidebands which resemble the ones induced by the far-field source (compare insets of Fig. 3 and 4). The power emitted is then on the order of $P_{\mathrm{dc}} \sim 20\ \mu\mathrm{V} \times 1\ \mu\mathrm{A} \approx 20\,\mathrm{pW}$, where $20\ \mu$V is the voltage drop over the JTJ at 10 GHz. Moreover, the peak broadening is almost identical to the one determined before.

As seen, the resonances of the current (marked by lines in the plot) and the induced sidebands (marked by arrows) possess a long term stability. Since the observed resonances for the near-field as well as the far-field source are almost identical, we conclude that the photon absorption process only depends on the shape of the local electrostatic potential. Varying the frequency of the radiation for such a tunnel coupled dot system results in the well-known linear relation between the position of the sideband relative to the ground state and the frequency for the case of weak coupling (10), (20).

In another setup we focused on defining a weak link as a Josephson oscillator with a modified atomic force microscope tip directly into the Al-contacts

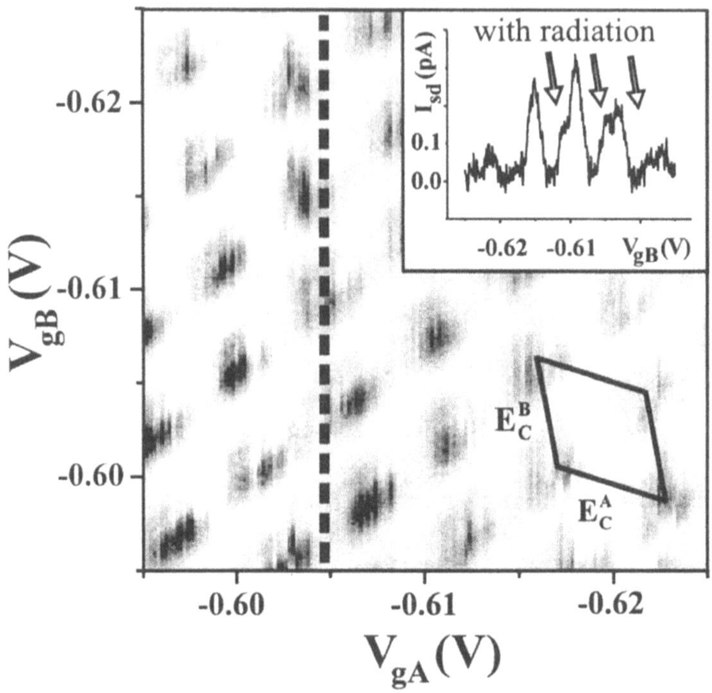

Fig. 3. Charging diagram with an identical scan range as in Fig. 2, but under microwave radiation by a far-field source at 10 GHz. The sidebands are found on only one side, depending on the tuning of the tunnel barriers. Inset: Photo current with the sidebands induced by the frequency dependent absorption. Pumping of the electrons leads to a reduction of the absolute current value.

forming the Schottky gates (21), (22). This is of great advantage for probing the microwave response of quantum dots in the absolute near-field limit, since the photon source is located only 100 nm apart from the tunnel barriers. Recent examples of the gate structures fabricated are depicted in Fig. 5: As seen we have chosen two different polarizations of how the radiation emitted from the junction is impinging on the quantum dots. In the first case (upper left) the junction is aligned with the direction of transport through the dots, while it is oriented perpendicular in the second case (upper right). These micrographs were taken with an atomic force microscope prior to the definition of the weak links. In the lower center part of Fig. 5 one of the junctions is depicted after scratching the Al layer.

The heterostructure applied in these first measurements is identical to the one characterized above. In one of the devices similar to the one in the upper

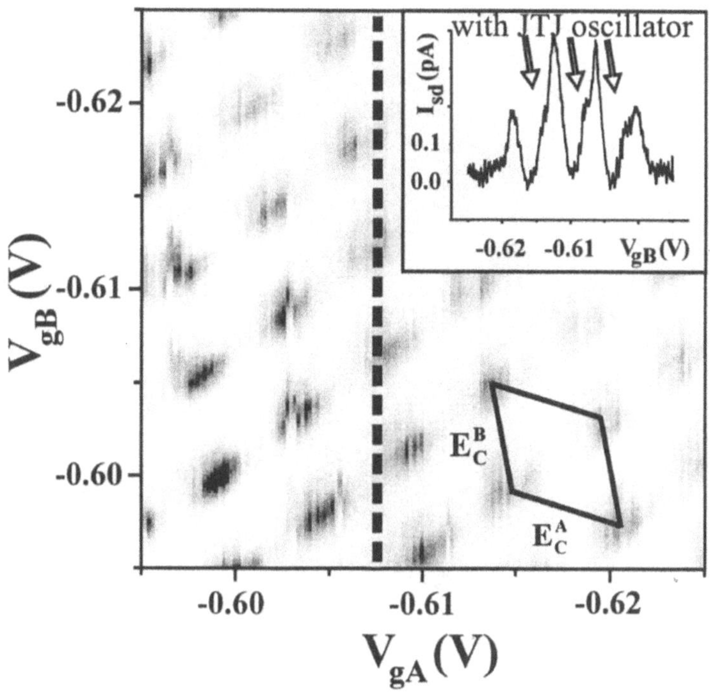

Fig. 4. Charging diagram: Josephson oscillator is operated as a microwave source at $f \cong 10$ GHz. The inset exemplifies a single trace from the gray scale plot with photon-induced sidebands. The peak heights and positions are identical to the plots obtained with the far-field source (compare Fig. 3).

right part of Fig. 5, an impurity dot below a single gate was formed with a typical conductance resonance. However, with the source turned on we find a clear response to the microwave signal. The single resonance trace is changed into broadened double peak, which is the signature of adiabatic response of the quantum dot. This is exemplified by the fit to the experimental data in inset (b) of Fig. 6. Due to the increase in background conductance the peak heights are not reproduced exactly. In the inset (a) of Fig. 6 a fit to the conductance resonance by the derivative of the Fermi/Dirac function is given. This result clearly demonstrates that the weak link is functioning as a Josephson oscillator. However, within the current resolution we have to assume that not only a single mode is excited, but a number of emission frequencies. This would explain the adiabatic response of the junction.

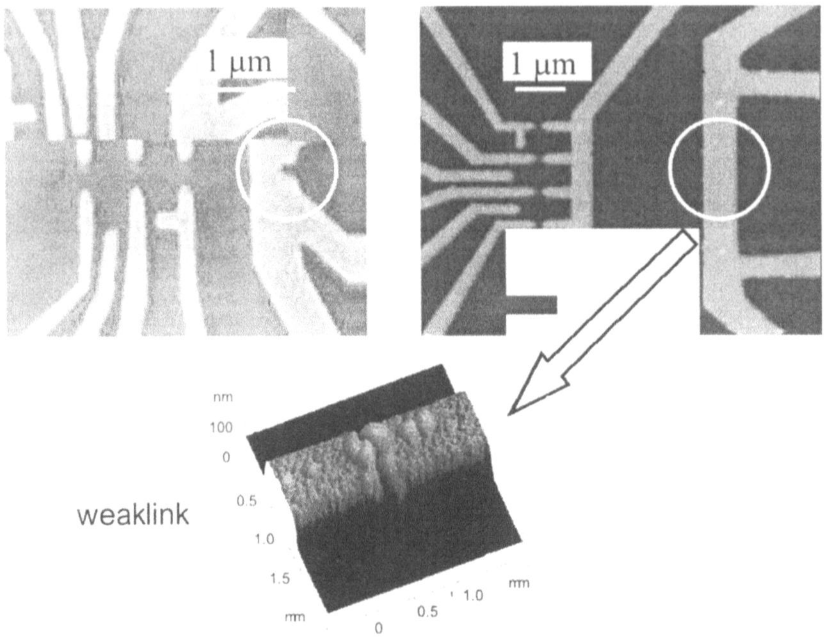

Fig. 5. Top view of the two types of junctions prior to the weak link definition, allowing different microwave polarizations to couple to the quantum dots. Shown below is a scratched Al weak link.

5 Summary

In summary, we find photon assisted tunneling in a weakly coupled double quantum dot, induced by a near-field source. This source is realized as a long Josephson junction placed on top of the chip carrying the double dot. We find nearly identical coupling of the radiation with the near-field source and with that of a far-field source. We conclude from this comparison that the photon absorption process depends only on the local electrostatic environment of the quantum dots. Furthermore, this result confirms that Kohn's theorem is valid in the near-field regime, as long as a spatially homogenous radiation field across the excited electronic system is provided. This we expect to change when the microwave radiation couples predominantely to one of the barriers, which can be achieved by a Josephson junction embedded in the dot's gate structure as shown.

We like to thank T. Klapwijk, S.J. Allen, and D. Grundler for extended discussions and Nano-Tools GmbH (http://www.nano-tools.com) for tech-

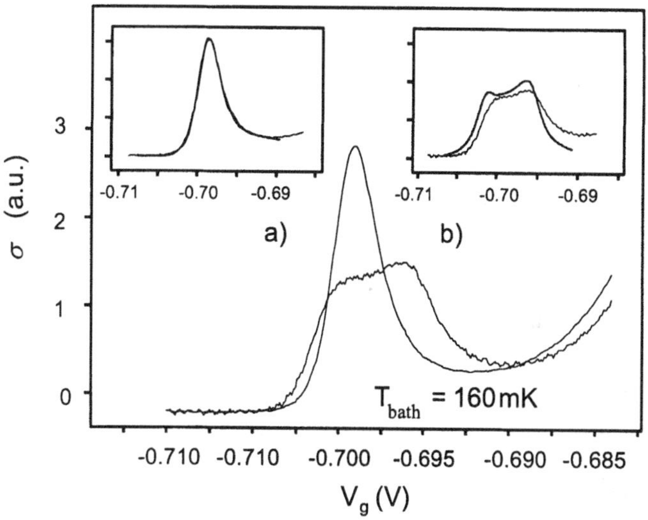

Fig. 6. Conductance resonance versus gate voltage of a single dot formed in the gate structure of Fig. 5. The resonance is modified to a broadened double peak by the microwave radiation being emitted from the weak link. The two insets give theoretical fits to the conductance resonance with (a) and without (b) radiation.

nical support. This work was funded in part by the Deutsche Forschungsgemeinschaft (DFG), the Volkswagen-Stiftung, and the Defense Advanced Research Projects Agency (DARPA).

References

Kouwenhoven, L.P., Jauhar, S., Orenstein, J., McEuen, P.L., Nagamune, Y., Motohisa, J., Sakaki, H., (1994), Phys. Rev. Lett. **73**, 3443

Blick, R.H., Haug, R.J., van der Weide, D.W., von Klitzing, K., Eberl, K., (1995), Appl. Phys. Lett. **67**, 3924

Oosterkamp, T.H., Kouwenhoven, L.P., Koolen, A.E.A., van der Vaart, N.C., Harmans, C.J.P.M., (1997), Phys. Rev.Lett. **78**, 1536

Tien, P.K., Gordon, J.P., (1963), Phys. Rev. **129**, 647

Stafford, C.A., Wingreen, N.S., (1996), Phys. Rev. Lett. **76**, 1916

Ivanov, T., (1997), Phys. Rev. B **56**, 12339; Stoof, T.H., Nazarov, Yu.V., (1996), Phys. Rev. B **53**, 1050

Inarrea, J., Platero, G., Tejedor, C., (1994), Phys. Rev. B **50**, 4581

Nakamura, Y., Pashkin, Yu.A., Tsai, J.S., (1999), Nature **398**, 786

Blick, R.H., van der Weide, D.W., Haug, R.J., Eberl, K., (1998), Phys. Rev. Lett. **81**, 689
Oosterkamp, T.H., Fujisawa, T., van der Wiel, W.G. Ishibashi, K., Hijman, R.V., Tarucha, S., Kouwenhoven, L.P., (1998), Nature **395**, 873
Adourian, A.S., Yang, S., Westervelt, R.M., Chapman, K.L., Gossard, A.C., (1998), J. Appl. Phys. **84**, 5808
Visscher, E.H., Schraven, D.M., Hadley, P., Mooij, J.E., cond-mat/9904382.
Ustinov, A.V., (1998), Physica D **123**, 315-329
Koshelets, V.P., Shitov, S.V., Shukin, A.V., Filippenko, L.V., Mygind, J., Ustinov, A.V., (1997), Phys. Rev. B **56**, 5572
Koshelets, V.P., Shitov, S.V., Filippenko, L.V., Baryshev, A.M., Golstein, H., de Graauw, T., Luinge, W., Schaeffer, H., van de Stadt, H., (1996), Appl. Phys. Lett. **68**, 1273
Kohn, W., (1961), Phys. Rev. **123**, 1242
Sikorski, C.H., Merkt, U., (1989), Phys. Rev. Lett. **62**, 2164; Demel, T., Heitmann, D., Grambow, P., Ploog, K., (1990), Phys. Rev. Lett. **64**, 788; Merkt, U., (1996), Phys. Rev. Lett. **76**, 1134; Bakshi, P., Broido, D.A., Kempa, K., (1990), Phys. Rev. B **42**, 7416
Pfannkuche, D., Ulloa, S.E., (1995), Phys. Rev. Lett. **74**, 1194; Weis, J., Haug, R.J., von Klitzing, K., Ploog, K., (1993), Phys. Rev. Lett. **71**, 4019
Livermore, C., Crouch, C.H., Westervelt, R.M., Campman, K.L., Gossard, A.C., (1996) Science **274**, 1332
Qin, H., Blick, R.H., Simmel, F., Holleitner, A.W., Kotthaus, J.P., Eberl, K., to be submitted (1999)
Irmer, B., Blick, R.H., Simmel, F., Gödel, W., Lorenz, H., Kotthaus, J.P., (1998) Appl. Phys. Lett. **73**, 2061
Irmer, B., Simmel, F., Blick, R.H., Lorenz, H., Kotthaus, J.P., Bichler, M., Wegscheider, W., (1999), Superlattices and Microstructures **25**, 785

Part II

Quantum Chaos

Quantum Chaos and Spectral Transitions in the Kicked Harper Model

Karsten Kruse, Roland Ketzmerick, and Theo Geisel

Max-Planck-Institut für Strömungsforschung und Fakultät Physik der Universität Göttingen, Bunsenstraße 10, D-37073 Göttingen, Germany

Abstract. In contrast to bounded systems, quantum chaos in extended systems may be associated with fractal spectra, metal-insulator transitions due to avoided band crossings, and spreading wave packets. In this lecture we point out the role of avoided band crossings for spectral transitions in the example of the kicked Harper model. We explain the coexistence of localized and extended eigenfunctions off the critical line as well as changes of the fractal dimension on the critical line. Avoided band crossings thus provide a common cause for various phenomena observed numerically in the spectrum of the kicked Harper model.

1 Introduction

In the realm of quantum chaos [1,2] the kicked Harper model [3–14] has become a paradigm, because its spectrum may be of any type, i.e., pure point, absolutely continuous or singular continuous, and because it is easily treated numerically. In particular, it exhibits phenomena that can only occur in *extended* systems like (multi-)fractal spectra, transitions between pure point and absolutely continuous spectra, and spreading wave packets. As for many kicked system, the time evolution of wave packets [15] and the spectrum [14] can be calculated very efficiently by using the Fast Fourier Transform algorithm. Furthermore, as a generalization of the Harper model, the kicked Harper model possesses a well-studied basis to start from. Here we will study its spectrum under the aspect of avoided band crossings. In this case, avoided band crossings are a signature of a non-integrable classical limit analogous to avoided level crossings in bounded quantum systems which have a discrete spectrum. As we will see, they provide a unifying explanation for a whole set of numerical observations made on the kicked Harper model in the past.

Before introducing the kicked Harper model formally, let us recall some facts about the Harper model. It was first derived to describe electrons moving in a two-dimensional periodic potential, so-called Bloch electrons, in a magnetic field directed perpendicularly to the potential plane [16,17]. Its Hamiltonian is discrete and reads

$$H = \sum_n \left\{ V \cos\left(2\pi\sigma n + \nu\right) a_n^\dagger a_n + a_{n+1}^\dagger a_n + a_{n-1}^\dagger a_n \right\}, \tag{1}$$

where V is the potential strength, σ a measure of the magnetic flux, ν a phase, and $a_n^\dagger, a_n$ are the creation and annihilation operators at site n. In the course of the time the Harper model has attracted a lot of attention in the mathematical community, too, because of its interesting spectral structure. This structure was first explored in a numerical study by Hofstadter [18] and because of the form of the gaps the graph in the (σ, E)-plane for $V = 2$ is now called Hofstadter's butterfly. The spectrum strongly depends on the parameter σ: If σ is rational, the potential is periodic and the spectrum consists of Bloch bands. If σ is irrational, the spectrum is independent of ν, but one has to distinguish three different regimes depending on the parameter V: If $V < 2$ the spectrum is absolutely continuous [19] (loosely speaking it consists of intervals), for $V > 2$ it is singular [20,21] (discrete), and for $V = 2$ singular continuous (fractal) [22–27]. The last property is strongly linked to the so-called Aubry duality of the Harper model [28]. It roughly states that the Fourier transform of an eigenfunction of (1) with eigenvalue E is a solution of the eigenvalue equation of (1) for the eigenvalue $2E/V$ where V has to be replaced by $4/V$. Note that $V = 2$ is a self dual point.

One can associate a classical Hamiltonian to the system:

$$H = 2\cos p + V \cos x, \tag{2}$$

that is clearly integrable, because it has one degree of freedom. Among other reasons, the kicked Harper model (KHM) was introduced in order to study the influence of a classical chaotic limit on a quantum system with a fractal spectrum [6]. Its Hamiltonian differs from (2) essentially by a time-dependence of the potential:

$$H = K \cos p + L \cos x \sum_{n=-\infty}^{\infty} \delta(t-n). \tag{3}$$

As can be seen in Fig. 1 it indeed generates chaotic dynamics. Its quantized version is most easily described by the one-step time evolution operator

$$U = \exp\left\{-i\frac{K}{\hbar}\cos p\right\} \exp\left\{-i\frac{L}{\hbar}\cos x\right\}, \tag{4}$$

that in contrast to the Hamiltonian (3) is independent of time. Since U is unitary, the elements of its spectrum may be written in the form $\exp(-i\omega)$, where ω is called a quasienergy.

In order to establish the link with the Harper model, we rewrite the time evolution operator (4) using the Baker Campbell Hausdorff-formula:

$$U = \exp\left\{-i\frac{K}{2\hbar}\left[2\cos p + \frac{2L}{K}\cos x\right] + O\left(\frac{KL}{\hbar^2}\right)\right\}. \tag{5}$$

One thus finds that when $K, L \to 0$ with $K/L = const.$ for each ω in the quasienergy spectrum of the KHM there is a value E in the spectrum of the Harper model with $V = 2L/K$ and $\sigma = \hbar/2\pi$ such that $\omega = \frac{K}{2\hbar}E$, see Fig. 2.

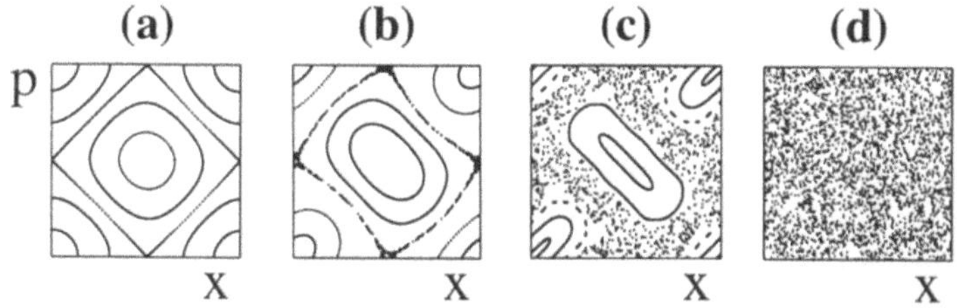

Fig. 1. A cell of the phase space of the classical kicked Harper model for (a) $K = L = 0.1$, (b) $K = L = 1$, (c) $K = L = 2$, and (d) $K = L = 5$. For increasing K and L the stochastic layer grows around the former separatrix until finally the whole phase space is chaotic. For $K = L = 0.1$ the stochastic layer is too small to be seen.

Consequently, in this limit the spectral properties of the KHM are the same as for the Harper model. For larger values of K the following observations have been made numerically: i) For $K = L$ the spectrum remains fractal, but its dimension changes [7]. ii) For some $K \neq L$ continuous and discrete spectral regions, i.e., extended and localized eigenfunctions coexist [5,8,13]. iii) For some $K \neq L$ even a singular continuous component was inferred [13]. All these observations find a common explanation in the presence of avoided band crossings in the quasienergy spectrum of the KHM. What avoided band crossings are and how they change the spectral structure will be explained in the next section.

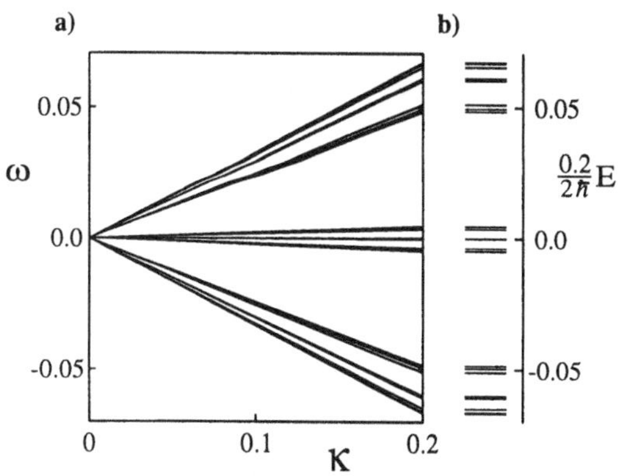

Fig. 2. a) Quasienergy spectrum of the KHM with $K = L = \kappa$ and $\hbar = 2\pi\frac{34}{55}$. b) Energy spectrum of the Harper model with $V = 2$ and $\sigma = \frac{34}{55}$. In the given resolution the eigenvalues $e^{-i\omega}$ of the KHM and the eigenvalues E of the Harper model are linked by the relation $\omega = \frac{K}{2\hbar}E$.

2 Avoided Band Crossings

One prominent feature of quantum systems with a chaotic classical limit and a discrete spectrum are avoided level crossings [2]. That is, when changing an external parameter two approaching levels will repel each other. While a single avoided level crossing is not very interesting, they are in their totality responsible for the universal statistical properties of such spectra for small level spacings [2]. If a spectrum contains other spectral components then one has to consider avoided band crossings, i.e., avoided crossings of entire spectral regions. Contrary to avoided level crossings even a single avoided band crossing may have interesting consequences [29], as we will see in the following.

In the case of a periodic system the spectrum consists of Bloch bands, each of which may be parameterized by a Bloch phase k. An isolated avoided crossing of Bloch bands is most easily modeled by the Hamiltonian

$$H = \begin{pmatrix} E'(k) - \lambda & \varepsilon \\ \varepsilon & E''(k) + \lambda \end{pmatrix}, \tag{6}$$

where the crossing bands are given by E' and E'', the external parameter λ linearly shifts the individual Bloch bands and the coupling between them is given by the constant ε. Note, for simplicity we have neglected a k-dependence of ε. In Fig. 3 we present the spectrum of (6), where the dispersion relations have been chosen to be (a) $E'(k) = E''(k) = -\cos k$ and (b) $E'(k) = -E''(k) = -\cos k$. These two cases represent the 'extreme' cases of an overall same dependence of E' andE'' on k, i.e., $\frac{dE'}{dk}\frac{dE''}{dk} \geq 0$ for $k \in [-\pi, \pi]$ and an overall opposite dependence of E' and E'' on k, i.e., $\frac{dE'}{dk}\frac{dE''}{dk} \leq 0$ for $k \in [-\pi, \pi]$. The spectrum in Fig. 3a simply resembles that of an avoided level crossing, but in Fig. 3b a new feature appears: There the two bands are twisted, i.e., for each band the derivative of energy with respect to the Bloch phase changes its sign in course of the avoided crossing.

Two subsequent avoided crossings of Bloch bands may be described by

$$H = \begin{pmatrix} E^+(k) + A - \lambda & \varepsilon & 0 \\ \varepsilon & E^0(k) + \lambda & \varepsilon \\ 0 & \varepsilon & E^-(k) - A - \lambda \end{pmatrix}. \tag{7}$$

Here the rising band E^0 tends to cross subsequently the bands E^+ and E^-, that are separated by an offset $2A$. The spectrum of (7) is shown in Fig. 4 for $E^0 = -\cos(k)$, $E^\pm(k) = \cos(k)$, and four different values of the coupling parameter ε. For $\varepsilon < 1$ the two avoided crossings are well seperated (a). If the coupling is increased (b,c) the two twists of the intermediate band approach each other, such that its width diminishes. Increasing the coupling further the twists eventually merge and annihilate (d).

The previous model of two subsequent avoided crossings of Bloch bands describes a periodic system. Thus, the energies can still be described as depending on a Bloch phase and the spectrum remains absolutely continuous.

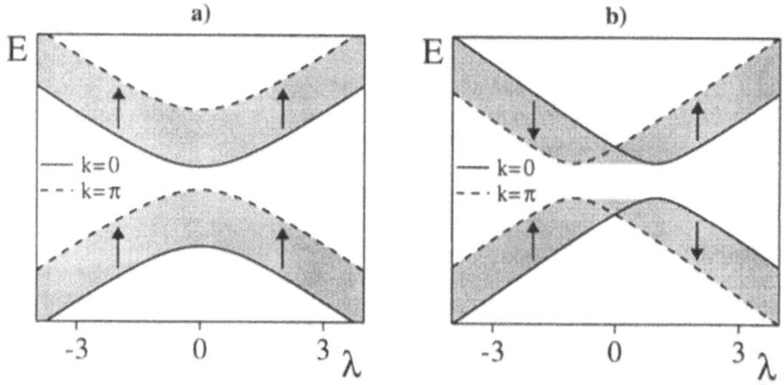

Fig. 3. a) Avoided crossing of two Bloch bands without twist modeled by Hamiltonian (6) with $E' = E'' = -\cos k$ and $\varepsilon = 1.4$. b) Avoided crossing of two Bloch bands with twist modeled by Hamiltonian (6) with $E' = -\cos k$, $E'' = \cos k$, and $\varepsilon = 0.6$. Arrows indicate the change of energies as k is varied from 0 to π. Solid and dashed lines represent the eigenvalues for $k = 0$ and $k = \pi$, respectively.

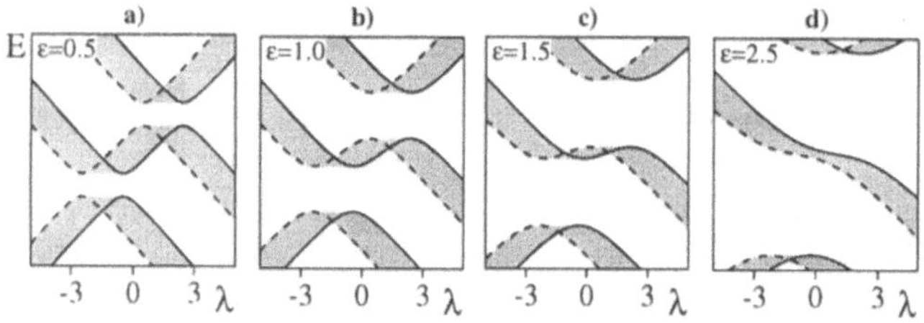

Fig. 4. Two subsequent twists modeled by Hamiltonian (7) with $E^0 = -\cos k$, $E^{\pm} = \cos k$ and $A = 3$ for increasing coupling. a) $\varepsilon = 0.5$: The twists are well separated. b), c) $\varepsilon = 1, 1.5$: The two twists of the intermediate band approach each other and reduce its width. d) $\varepsilon = 2.5$: The twists have annihilated.

For non-periodic systems the situation is different. There the *bands*, i.e., spectral regions, that are analogous to *Bloch* bands in periodic systems, have to be described by Hamiltonians. Straightforward generalization of operator (7) leads to the following model describing two subsequent avoided band crossings

$$H = \begin{pmatrix} H^+ + A - \lambda & \varepsilon & 0 \\ \varepsilon & H^0 + \lambda & \varepsilon \\ 0 & \varepsilon & H^- - A - \lambda \end{pmatrix} . \tag{8}$$

Here, the significance of λ, ε and A is the same as in (7). In what follows, the operators $H^{\pm,0}$ will be taken to be of the form of tight-binding models, i.e.

$$H^{\pm,0} = \sum \left\{ V_n^{\pm,0} a_n^\dagger a_n + t^{\pm,0} \left(a_{n+1}^\dagger a_n + a_{n-1}^\dagger a_n \right) \right\} , \tag{9}$$

where $a_n^\dagger$ and a_n are creation and annihilation operators, $V_n^{\pm,0}$ are the potential strengths at site n, and $t^{\pm,0}$ are the hopping amplitudes which are assumed to be constant along the chain. The Harper model introduced above is one example of a one-band Hamiltonian. Other examples are provided by the Anderson model for disordered systems [30] or the Fibonacci chain model for quasicrystals [31,32]. In the first case the potential terms are uniformly distributed random variables, while in the second they are determined by the Fibonacci sequence $ABAAB\ldots$ to be $+V$ or $-V$. Note that when setting $V_n^{\pm,0} \equiv 0$ operator (8) equals operator (7) with $E^{\pm,0} = 2t^{\pm,0}\cos k$. In particular, twists are generated by choosing $t^{\pm} = -t^0$.

In Fig. 5a the spectrum is shown for the case where H^0 is given by the Harper model and $H^{\pm}$ describe free particles, i.e., $V_n^0 = V^0\cos(2\pi\sigma n)$ with $\sigma = (\sqrt{5}-1)/2$ and $V_n^{\pm} \equiv 0$. We have chosen $V^0/t^0 = 1.5$ such that the spectra of the uncoupled Hamiltonians are absolutely continuous and the corresponding 'eigenfunctions' are extended; typical examples of eigenfunctions are presented. Surprisingly, around $\lambda = 0$ the eigenfunctions of the middle band are localized, indicating a pure point spectrum. That is, a metal-insulator transition has occurred due to the avoided band crossing! For the spectrum presented in Fig. 5b we have chosen the Anderson model for H^0, where the V_n^0 are uniformly disributed random variables in the interval $[-V^0, V^0]$. It can be clearly read off from the figure, that avoided band crossings with free particle bands lead in this case to a decrease in the localization lengths of the eigenfunctions. Figure 6 demonstrates that the same effect can be found if we choose for $H^{\pm}$ the Harper or the Anderson model, too. Indeed, this situation is more realistic, because the anti-crossing bands are now of the same kind.

Until now the hopping terms were chosen such that $t^{\pm} = -t^0$ in order to generate avoided band crossings with twist. What happens in the case $t^{\pm} = t^0$? Two examples are shown in Fig. 7, where it can be seen that avoided band crossings may also lead to a delocalization of eigenfunctions, i.e., a change from a pure point to an absolutely continuous spectrum, or to an increase of the localization lengths of the eigenfunctions. The same effects exist if we choose the Harper or the Anderson model for $H^{\pm}$. Furthermore, the fractal dimensions of a spectrum may change due to an avoided band crossing, see Fig. 8b.

An intuitive explanation of the localization of eigenfunctions due to avoided band crossings is provided by reconsidering avoided crossings of Bloch bands as described by the Hamiltonian (8) with $H^{\pm,0} = \sum_n t^{\pm,0}\left(a_{n+1}^\dagger a_n + a_{n-1}^\dagger a_n\right)$ and $t^{\pm} = -t^0$. Then, around $\lambda = 0$, the intermediate Bloch band has narrowed and is thus described by H^0 with a smaller hopping parameter $\bar{t}^0$. In the general case, around $\lambda = 0$ the middle band is also approximately described by H^0 – again, by analogy to the case considered before, with reduced hopping terms. This leads to an increase of the ratio V_n^0/t^0 and, if large enough, explains the metal-insulator transition found for the Harper band and the decrease of the localization lengths of the

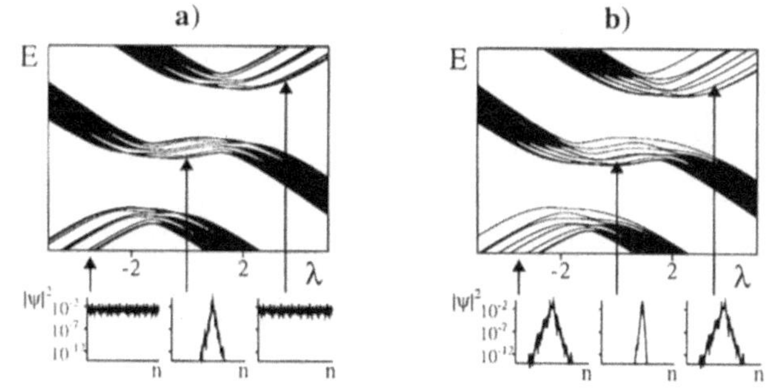

Fig. 5. The spectrum of operator (8) and typical eigenfunctions corresponding to the spectral regions indicated by the arrows. a) H^0: the Harper model with $V^0/t^0 = 1.5$, $H^{\pm}$: free particle with $t^{\pm} = -t^0$, $\varepsilon = 1.5$, $A = 3$. b) H^0: the Anderson model with $V^0/t^0 = 2.5$, $H^{\pm}$: free particle with $t^{\pm} = -t^0$, $\varepsilon = 1.5$, $A = 3$.

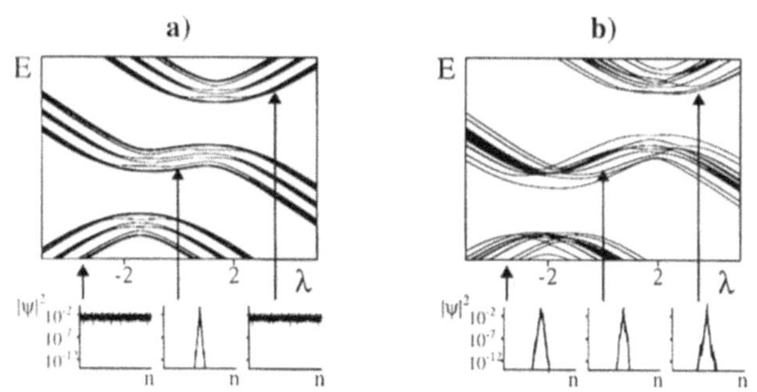

Fig. 6. As Fig. 5. a) H^0: the Harper model with $V^0/t^0 = 1.5$, $H^{\pm}$: the Harper model with $V^0/t^0 = 1.2$ and $t^{\pm} = -t^0$, $\varepsilon = 1.5$, $A = 3$. b) H^0: the Anderson model with $V^0/t^0 = 2.5$, $H^{\pm}$: the Anderson model with $V^0/t^0 = 2.5$ and $t^{\pm} = -t^0$, $\varepsilon = 1.5$, $A = 3$.

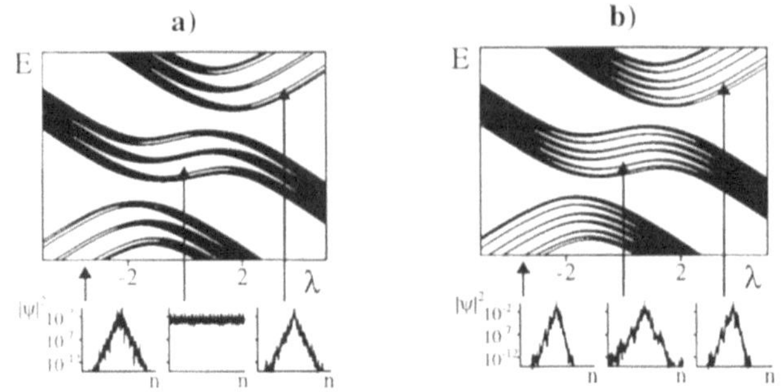

Fig. 7. As Fig. 5. a) H^0: the Harper model with $V^0/t^0 = 2.5$, $H^{\pm}$: free particle with $t^{\pm} = t^0$, $\varepsilon = 1.5$, $A = 3$. b) H^0: the Anderson model with $V^0/t^0 = 2.5$, $H^{\pm}$: free particle with $t^{\pm} = t^0$, $\varepsilon = 1.5$, $A = 3$.

eigenfunctions in the Anderson band, see Fig. 5. In the case of avoided band crossings without twist the hopping terms should not alter [see Fig. 3] and thus the observed changes cannot be understood within this simple picture. For this reason we will now calculate the effective Hamiltonian describing the middle band for $\lambda = 0$ by a perturbation calculation.

To this end we choose the basis such that the operator (8) takes the form

$$H' = \begin{pmatrix} \ddots & \ddots & & & \\ \ddots & \mathbf{W}_{n-1} & \mathbf{T} & & \\ & \mathbf{T} & \mathbf{W}_n & \mathbf{T} & \\ & & \mathbf{T} & \mathbf{W}_{n+1} & \ddots \\ & & & \ddots & \ddots \end{pmatrix} \tag{10}$$

where

$$\mathbf{W}_n = \begin{pmatrix} V_n^+ + A & 0 & 0 \\ 0 & V_n^0 & 0 \\ 0 & 0 & V_n^- - A \end{pmatrix} + \varepsilon \begin{pmatrix} 0 & 1 & 0 \\ 1 & 0 & 1 \\ 0 & 1 & 0 \end{pmatrix} \tag{11}$$

and

$$\mathbf{T} = \begin{pmatrix} t^+ & 0 & 0 \\ 0 & t^0 & 0 \\ 0 & 0 & t^- \end{pmatrix} . \tag{12}$$

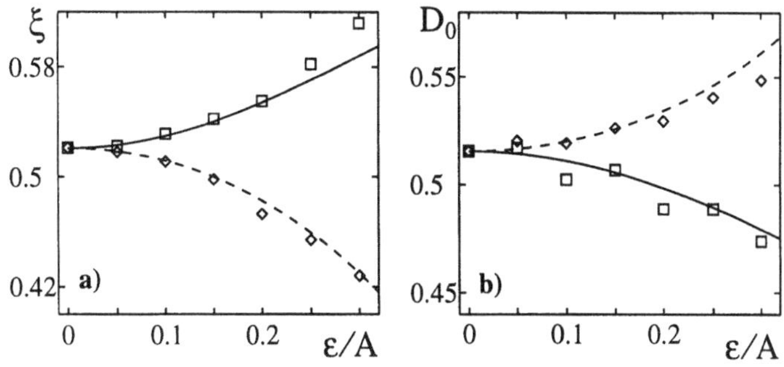

Fig. 8. a) Average inverse participation ratio $\xi = \sum_n |\phi_n|^4$ of the eigenfunctions ϕ of the intermediate band of operator (8) at $\lambda = 0$ for $A = 10$ as a function of ε. H^0 was chosen to be the Anderson model with $V^0 = 5$ and $t^0 = 1$ while $H^{\pm}$ represents a free particle with $t^{\pm} = -1$ (squares) and $t^{\pm} = 1$ (diamonds), respectively. Solid and dashed lines are the results for an isolated Anderson model with the respective parameters obtained from Eqs. (15) and (18a). **b)** Same as (a) but for the fractal (box-counting) dimension of the intermediate band, where H^0 was chosen to be the on-site Fibonacci chain model [31,32] ($V^0 = 1.5$, $t^0 = 1$, $A = 100$).

By the similarity transformation $\mathbf{S}_n^{-1}\mathbf{W}_n\mathbf{S}_n$ with

$$\mathbf{S}_n = \begin{pmatrix} 1 - \frac{\varepsilon^2}{2A^2} & \frac{\varepsilon}{A} & -\frac{\varepsilon^2}{2A^2} \\ -\frac{\varepsilon}{A} & 1 - \frac{\varepsilon^2}{A^2} & \frac{\varepsilon}{A} \\ \frac{\varepsilon^2}{2A^2} & -\frac{\varepsilon}{A} & 1 - \frac{\varepsilon^2}{2A^2} \end{pmatrix} \tag{13}$$

we obtain in second order in ε/A

$$\mathbf{S}_n^{-1}\mathbf{W}_n\mathbf{S}_n = \begin{pmatrix} \bar{V}_n^+ & 0 & 0 \\ 0 & \bar{V}_n^0 & 0 \\ 0 & 0 & \bar{V}_n^- \end{pmatrix} + O\left(\left(\frac{\varepsilon}{A}\right)^4\right) \tag{14}$$

where

$$\bar{V}_n^0 = V_n^0\left(1 - \frac{2\varepsilon^2}{A^2}\right) + \left(V_n^+ + V_n^-\right)\frac{\varepsilon^2}{A^2}. \tag{15}$$

The explicit expressions for $\bar{V}_n^{\pm}$ will not be used and are therefore not given here. In order to calculate $\mathbf{S}^{-1}\mathbf{H}\mathbf{S}$ with

$$\mathbf{S} = \begin{pmatrix} \ddots & \ddots & & & \\ \ddots & \mathbf{S}_{n-1} & 0 & & \\ & 0 & \mathbf{S}_n & 0 & \\ & & 0 & \mathbf{S}_{n+1} & \ddots \\ & & & \ddots & \ddots \end{pmatrix} \tag{16}$$

we now have to evaluate $\mathbf{S}_{n\pm1}^{-1}\mathbf{T}\mathbf{S}_n$. We obtain

$$\bar{\mathbf{T}} = \begin{pmatrix} \bar{t}^+ & t_1^+ & t_2 \\ t_1^+ & \bar{t}^0 & t_1^- \\ t_2 & t_1^- & \bar{t}^- \end{pmatrix} \tag{17}$$

with

$$\bar{t}^0 = t^0\left(1 - 2\frac{\varepsilon^2}{A^2}\right) + \left(t^+ + t^-\right)\frac{\varepsilon^2}{A^2} + O\left(\left(\frac{\varepsilon}{A}\right)^4\right), \tag{18a}$$

$$t_1^{\pm} = \left(t^{\pm} - t^0\right)\frac{\varepsilon}{A} + O\left(\left(\frac{\varepsilon}{A}\right)^3\right), \tag{18b}$$

$$t_2 = \left[t^0 + \frac{1}{2}\left(t^+ - t^-\right)\right]\frac{\varepsilon^2}{A^2} + O\left(\left(\frac{\varepsilon}{A}\right)^4\right). \tag{18c}$$

The non-diagonal terms t_2 give a correction of the order of ε^4/A^4 to the potential terms coupled by them, because the eigenvalues of

$$\begin{pmatrix} \bar{V}_n^+ & t_2 \\ t_2 & \bar{V}_{n\pm1}^- \end{pmatrix} \tag{19}$$

are

$$\bar{V}_n^+ + \frac{\left[t^0 + \frac{1}{2}\left(t^+ - t^-\right)\right]^2}{V_n^+ - V_{n\pm1}^-}\frac{\varepsilon^4}{A^4}, \tag{20a}$$

$$\bar{V}_{n\pm1}^- - \frac{\left[t^0 + \frac{1}{2}\left(t^+ - t^-\right)\right]^2}{V_n^+ - V_{n\pm1}^-}\frac{\varepsilon^4}{A^4}. \tag{20b}$$

The effects of the terms $t_1^{\pm}$ are determined by considering the coupling of $\bar{V}_n^0$ and $\bar{V}_{n-1}^+ + A$. In lowest non-vanishing order of ε^2/A^2 the correction is given by

$$\left(t^+ - t^-\right)^2 \frac{\varepsilon^2}{\left(\bar{V}_{n-1}^+ + \bar{V}_n^0 + A\right)A^2}. \tag{21}$$

Being of the order of ε^2/A^2 it cannot be neglected, *a priori*. However, the prefactor is of the order of $1/A$. Since A had to be chosen not too small as to seperate well the two falling bands, this term can thus also be neglected. Figure 8 shows that the results of the perturbation calculation are in good agreement with numerical results.

The effects found in the three-band model can also be found in single avoided band crossings. In that case a change from absolutely continuous to pure point or vice versa will perhaps occur only in a part of a band, but in principle all of the other effects described above may show up. How can this be described? In fact, avoided crossings of two bands fit in the three-band model by choosing a small value for the offset parameter A. Then, however, the perturbation approach is no longer useful. It should also be noted, that *isolated* avoided band crossings are rather the exception than the rule and further studies are needed to investigate the case of non-isolated crossings.

3 The Spectrum of the Kicked Harper Model

Now we come back to the spectrum of the kicked Harper model. Let us first note that its time evolution operator U (4) is periodic in x. Therefore, the eigenfunctions are of the form $\phi(x) = e^{i\theta_x x} u_{\theta_x}(x)$ with $u_{\theta_x}(x + 2\pi) = u_{\theta_x}(x)$ and U can be represented as an infinite matrix depending on the Bloch phase θ_x. This phase corresponds to the phase ν in the Harper Hamiltonian (1) and will be fixed to be zero in the following. We have stated already in section 1 that for $K, L \to 0$ with $K/L = const.$ the spectrum of the KHM is given by the Harper spectrum. Concerning the properties (i)-(iii) it is already clear from the analysis in the preceding section that they could all be due to avoided band crossings. But why should they appear in the quasienergy spectrum of the KHM? For small values of K and L the quasienergies are proportional to the parameter K. Quasienergies are defined modulo 2π only, such that some bands will necessarily approach each other when K is increased (and K/L kept fixed). Since the classical limit of the KHM is non-integrable, the expected crossings of bands will be avoided. In Fig. 9 it can be clearly seen that there are indeed avoided band crossings – with and without twist.

3.1 The case $K = L$

In Ref. [6] the metamorphosis of Hofstadter's butterfly due to classical chaos, i.e., when $K = L$ is increased, was investigated. It was found that the big gaps are closing and that the hierarchical band structure of the spectrum vanishes on large scales. It remains fractal on small scales presumably because of a duality property analogous to the Aubry duality in the Harper model. Its fractal dimension D_0, however, changes. In Ref. [7] an increase of D_0 with $K = L$ has indeed been reported. In the light of avoided band crossings the reason is obvious: due to avoided band crossings the fractal dimension will *locally* change as it was the case for the three-band model for a 'Fibonacci-band', see Fig. 8. In some regions it will increase, in others it will decrease.

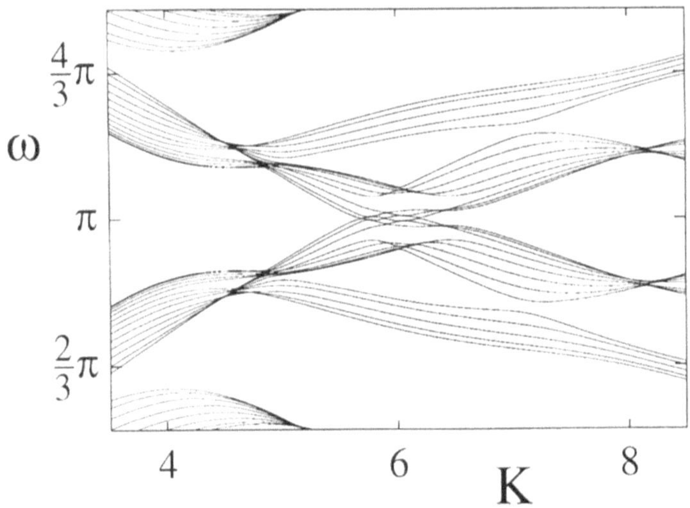

Fig. 9. Detail of the quasienergy spectrum of the KHM for $K/L = \frac{1}{3}$ and $\hbar = 2\pi\frac{34}{55}$. Several avoided band crossings can be seen – most of them have twists.

Since the spectrum as a whole has a fractal dimension equal to the largest 'local' D_0, its value will increase.

3.2 The case $K \neq L$

For a given value of the potential strength the eigenfunctions of the Harper model are either all localized ($V > 2$) or all extended ($V < 2$). Therefore the same holds for the eigenfunctions of the KHM in the cases $L > K$ and $L < K$, resp., as long as the values of K and L are small. By investigating the dynamics of wave packets the coexistence of localized and extended eigenfunctions for larger $K \neq L$ was revealed [5,8]. Later this was confirmed by Borgonovi and Shepelyansky [13] who studied the scaling of the inverse participation ratio of the eigenfunctions with system size when approaching an irrational value of $\hbar/2\pi$ by rationals. Again this effect is due to the now familiar avoided band crossings, see Fig. 10. The bands of the quasienergy spectrum undergoing avoided crossings are approximately given by the Harper model [33]. Some of these crossings change the effective potential strength such that metal-insulator transitions are induced.

The data of Ref. [13] also convincingly led to the conclusion of the existence of critical eigenfunctions and therefore of a singular continuous spectral component. However, by using the Lanczos algorithm, that permits to treat much larger systems sizes than those considered in Ref. [13], it was concluded later that spectral regions are either pure point or absolutely continuous [14]. How can one reconcile these results? When increasing or decreasing the poten-

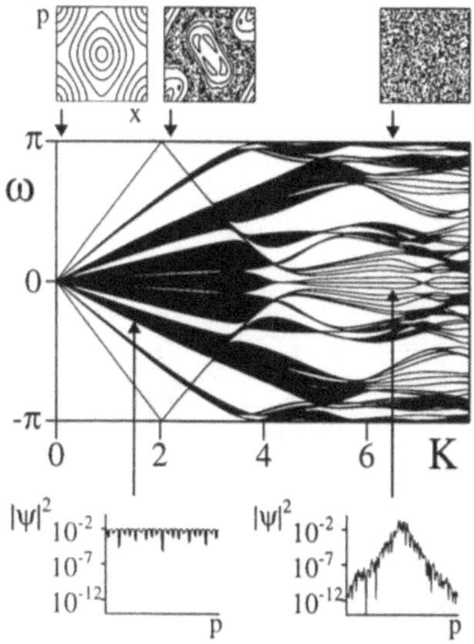

Fig. 10. Metal-insulator transitions in the kicked Harper model upon increasing the parameter K with $K/L = 7/4$ and $\hbar/2\pi = 8/61$, a rational approximant of $1/(7+\sigma_G)$, where $\sigma_G = (\sqrt{5}-1)/2$ is the Golden Mean. For small K the spectrum is, up to a scaling, given by the Harper model and all eigenfunctions are extended as expected, for larger K extended and localized eigenfunctions coexist. Typical examples are shown. As can be seen on top, the classical phase space shows chaotic dynamics, which leads to the avoided band crossings.

tial strength to approach $V = 2$ in the Harper model, its spectrum, although it is known to be either pure point or absolutely continuous, looks fractal on some scales. The closer one is to the transition point, the deeper one has to dwell into the spectrum in order to determine its type. The same holds for the eigenfunctions: the closer they are to the transition point the larger the system size has got to be to reveal their localization properties. For the KHM the same is true. Due to avoided band crossings there are many transition points where eigenfunctions change from localized to extended or vice versa. Close to them eigenfunctions seem to be critical on many scales and the spectrum, too, has a hierarchical structure on many scales, see Fig. 11. However, unlike in the Harper model, transitions occur for many different values of K ($K/L = const.$) and for one specific value of K and L the transition points are expected to form a set of measure zero, because a transition line parallel to the energy axis is very unlikely. Nevertheless note, that even for the resolution obtained by using the approximant of Fig. 11 some regions show a hierarchical structure on all scales. But for a still larger approximant they should lose this feature. Therefore, we expect that the spectrum has no singular continuous component for $K \neq L$.

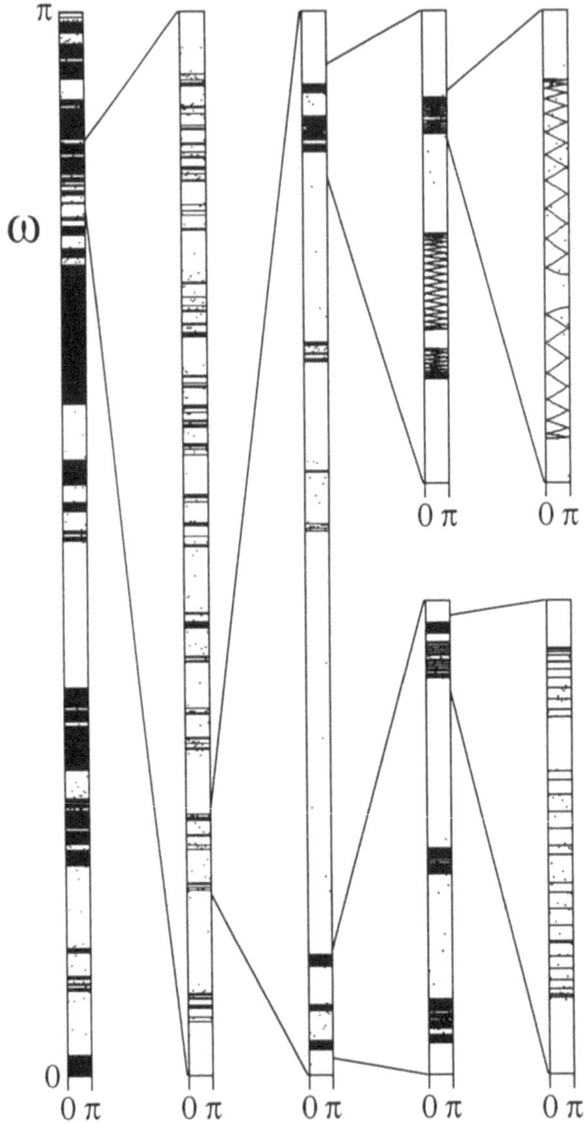

Fig. 11. Quasienergies ω of the KHM for $K = 4$, $L = 7$ and $\hbar = 2\pi 6765/51536$ vs. the Bloch phase θ_y. The latter appears, because we have used a rational approximant of $\hbar/2\pi$, resulting in a periodic system. Several successive zooms show at first a self-similar structure, that vanishes later. The isolated points are 'ghosts' produced by the Lanczos algorithm.

In conclusion, we have analyzed the kicked Harper model as a paradigm for extended classically chaotic quantum systems. In particaluar, we have shown that avoided band crossings provide a unified explanation of various phenomena that had been observed numerically in the past.

References

1. M. C. Gutzwiller (1991) Chaos in classical and quantum mechanics. Springer-Verlag, New York.
2. F. Haake (1991) Quantum signatures of chaos, Springer Series in Synergetics Vol. 54. Springer-Verlag, Berlin Heidelberg.
3. For a review see R. Artuso, G. Casati, F. Borgonovi, L. Rebuzzini, and I. Guarneri (1994). Int. J. Mod. Phys. B **8**, 207.
4. P. Lebœuf, J. Kurchan, M. Feingold, and D. P. Arovas (1990). Phys. Rev. Lett. **65**, 3076.
5. R. Lima and D. Shepelyansky (1991). Phys. Rev. Lett. **67**, 1377.
6. T. Geisel, R. Ketzmerick, and G. Petschel (1991). Phys. Rev. Lett. **67**, 3635.
7. R. Artuso, G. Casati, and D. Shepelyansky (1992). Phys. Rev. Lett. **68**, 3826.
8. R. Artuso, F. Borgonovi, I. Guarneri, L. Rebuzzini, and G. Casati (1992). Phys. Rev. Lett. **69**, 3302.
9. I. Guarneri and F. Borgonovi (1993). J. Phys. A: Math. Gen. **26**, 119.
10. R. Roncaglia, L. Bonci, F. M. Izrailev, B. J. West, and P. Grigolini (1994). Phys. Rev. Lett. **73**, 802.
11. P. Lebœuf and A. Mouchet (1994). Phys. Rev. Lett. **73**, 1360.
12. I. Dana (1994). Phys. Rev. Lett. **73**, 1609.
13. F. Borgonovi and D. Shepelyansky (1995). Europhys. Lett. **29**, 117.
14. R. Ketzmerick, K. Kruse, and T. Geisel (1999). Physica D **131**, 247.
15. F. M. Izraelev and D. L. Shepelyanskii (1979). Sov. Phys. Dokl. **24**, 996.
16. P. G. Harper (1955). Proc. Roy. Soc. London **A68**, 874.
17. M. Ya. Azbel (1964). Sov. Phys. JETP **19**, 634.
18. D. R. Hofstadter (1976). Phys. Rev. B**14**, 2239.
19. F. Delyon (1987). J. Phys. A **20**, L21.
20. B. Simon (1982). Adv. Appl. Math. **3**, 463.
21. J. Avron and B. Simon (1983). Duke Math. J. **50**, 369.
22. J. Bellissard and B. Simon (1982). J. Funct. Anal. **48**, 408.
23. S. C. Bell and R. B. Stinchcombe (1989). J. Phys. A: Math. Gen. **22**, 717.
24. C. Tang and M. Kohmoto (1986). Phys. Rev. B **34**, 2041.
25. M. Wilkinson and E. J. Austin (1994). Phys. Rev. B 50, 1420.
26. A. Rüdinger and F. Piéchon (1997). J. Phys. A: Math. Gen. **30**, 117.
27. R. Ketzmerick, K. Kruse, F. Steinbach, and T. Geisel (1998). Phys. Rev. B **58**, 9881.
28. S. Aubry and G. André (1980). Ann. Isr. Phys. Soc. **3**, 133.
29. R. Ketzmerick, K. Kruse, and T. Geisel (1998). Phys. Rev. Lett. **80**, 137.
30. P. W. Anderson (1958). Phys. Rev. **109**, 1492.
31. M. Kohmoto, L. P. Kadanoff, and C. Tang (1983). Phys. Rev. Lett. **50**, 1870.
32. S. Ostlund, R. Pandit, D. Rand, H. J. Schellnhuber, and E. D. Siggia (1983). Phys. Rev. Lett. **50**, 1873.
33. M. Wilkinson and R. J. Kay (1996). Phys. Rev. Lett. **76**, 1896.

Quantum Chaos Effects in Mechanical Wave Systems

Stephen W. Teitsworth

Department of Physics, Duke University, Durham, NC 27708-0305, USA

Abstract. Quantum chaos effects, such as wavefunction scarring and Wigner-Dyson level-spacings distributions, have been observed in a number of analogue experimental systems, including microwave cavities, water surface waves, acoustic cavities, and mechanical elastic wave systems. We describe experiments on two such mechanical wave systems, 1) elastic membranes - i.e., drumheads - which obey an underlying Helmholtz equation, and 2) thin metal plates undergoing transverse vibrations which are described by a biharmonic wave equation. For the elastic membrane system we have extensively studied the spectral statistics of both circular and stadium-shaped geometries with clamped boundary conditions; limited information about the spatial structure of high order eigenmodes has also been obtained. In plate experiments, a thin stadium-shaped plate is vibrated mechanically subject to fully clamped boundaries. High order eigenmodes are imaged using time-averaged holographic interferometry and show clear evidence of scarring. Adopting an eikonal form of the solution along classical trajectories of the stadium billiard and treating the problem exactly at the boundaries, we have obtained and experimentally verified novel quantization conditions for the observed modes.

1 Introduction

Qauntum chaos effects, which are manifested in spatial properties of wavefunctions and the statistics of energy spectra, have been reported over the last ten years for a number of experiments on *classical* wave systems that obey wave equations with identical or similar form to the Schrödinger equation that describes a point particle moving in two or three dimensions. These are called analogue experiments and most often they model quantum properties of chaotic billiards, i.e., point particles moving in two- or three-dimensional boxes that have appropriately irregular shape to guarantee that the classical motion is chaotic. Two examples are the Bunimovich stadium and the Sinai billiard in two dimensions (see, e.g., Berry 1987). The basis for this analogy is the Helmholtz equation which, with correct boundary conditions, serves as the time-independent wave equation for many classical wave problems, and also serves as the time-independent Schrödinger equation for a particle in a box. Numerical studies of the Helmholtz equation have confirmed the validity of random matrix theory - i.e., the Gaussian Orthogonal Ensemble (GOE) - in describing the statistics of the energy spectra (Berry 1987), as well as the presence of amplitude build-up in high order eigenfunctions along the unstable periodic orbits of the corresponding classical billiard (i.e., scars) (Heller 1984). Several analogue experiments that are exactly modelled by the Helmholtz equation have reported quantum chaotic spectral and spatial properties including the electromagnetic field modes of quasi-two-dimensional microwave cavities (Stöckmann and Stein 1990; Sridhar 1991; Gräf et al. 1992; Stein and Stöckmann 1992), stationary capillary waves on water (Blümel et al. 1992), and three-dimensional acoustic resonances in water-filled cavities (Chinnery and Humphrey 1996).

More recently, researchers have explored possible extension of quantum chaos methods to the high frequency limit of wave problems with modes that are not exactly described by the Helmholtz equation. The statistical properties of spectra as predicted by random matrix theory have since been confirmed in experiments on elastic waves in three-dimensional metal blocks (Delande et al. 1994; Ellegard et al. 1995), and in the electromagnetic modes of three-dimensional microwave cavities (Deus et al. 1995; Alt et al. 1996). In both of these cases, the vectorial nature of the classical fields implies that the eigenmodes cannot generally be described with a scalar Helmholtz problem. Very recently, scar-like phenomena have been reported for the electromagnetic modes of three-dimensional microwave cavities; however, the role of underlying classical periodic orbits on the observed amplitude build-up is not fully understood (Dörr et al. 1998).

In this paper, I describe the methods and results of analogue experiments on two types of mechanical wave system - drumheads and plates - that I have carried out with my students at Duke University over the past several years.

2 Drumhead Experiments

The small-amplitude transverse vibrational modes of an ideal elastic membrane are described by the following equation:

$$(\Delta + \frac{\omega^2}{v^2})u(x,y) = 0, \tag{1}$$

where $u(x,y)$ denotes the transverse displacement and ω is the frequency. The quantity v denotes the wave velocity and is defined by $v = \sqrt{T/\rho}$, where T is the membrane tension and ρ is the mass per unit area. This wave velocity appears in the dispersion relation for waves on the membrane, $\omega = vk$ where k denotes wavevector. Note that because the dispersion relation is linear, one must square the frequencies to make a comparison with energy spectra of an equivalent quantum system. Furthermore, we assume that $u(x,y)$ is zero on the boundary.

The membrane experiment was constructed as follows. We started with a commercially available circular drumhead (made for a bass drum) of diameter 72.4 cm, and made of mylar sheet with thickness 0.023 cm and areal mass density 0.0316 gm/cm^2. The drumhead assembly included a circular boundary attachment for mounting and tensioning. The approximate tension in the membrane used in our experiments was 94.1 N/cm. This assembly was then stretched over a metal frame molded into the shape under study. These frames were made of Al plates (typically 0.3 cm thickness) that were thin enough to bend into the desired shape, and thick enough to retain their rigidity as the membrane was stretched over them. We studied two frames - one in the shape of quarter circle with radius 35 cm, and the other in the shape of a quarter stadium with radius 27 cm and straight section length 10 cm (Thurstone 1992). These dimensions were selected so that the areas of the two shapes were identical which implies similar asymptotic density of modes per unit frequency. The drumhead was driven with a speaker modified so that the voicecoil was attached to hollow steel rod that was pressed against the membrane itself. A small piece of felt was used at the rod tip to prevent scratching the membrane surface. The speaker was driven by a variable frequency sine-wave that was amplified. Signals were detected using a phonograph cartridge (inductive pickup) which was sensitive to velocity. The output of the cartridge was amplified and filtered before being fed to an HP3561 signal analyzer which performed a fast-Fourier transform of the response. Using this method, resonances were measured in the range 50 to 1800 Hz.

Due to rather low quality factors inherent in this experiment - mostly due to air resistance, we were able to measure approximately 50 resonant frequencies for each shape. These frequencies were then squared to get a set of effective energies - the spectrum of our drumhead. From this set we computed level-spacings histograms, spectral rigidity, and Fourier transform of the spectrum (Berry 1987; Stöckmann and Stein 1990) for both quarter circle and quarter stadium. For the quarter stadium we found that the level-spacings distribution could be fit to a Wigner-Dyson distribution - which follows from the GOE - with a chi-square statistic of 0.8, while attempting to fit to the Poisson distribution (expected for an integrable system) gave a chi-square of 1.4. The measured spectral rigidity was also quite close to GOE prediction which grows logarithmically with energy spacing.

For the quarter circle, the level-spacings histogram was fit to the Wigner-Dyson distribution with a chi-square of 0.8, and to the Poisson distribution with a chi-square of 0.4. Spectral rigidity results were more ambiguous in this case: for small energy spacing they were quite close to the linear dependence expected for integrable systems, but as the energy spacing increased there were significant deviations towards sublinear behavior. This reflects the experimental difficulty of making a system that is exactly integrable in the classical limit. For example, a small bump or other imperfection in the circle boundary could be expected to yield a billiard with classically chaotic behavior.

We also studied the Fourier transform of the spectra for both drumhead shapes. In the case of the quarter stadium, strong peaks were identified corresponding to bouncing ball, rectangular, bow-tie, and hexagonal orbits from the underlying classical billiard. In the case of the quarter circle, the spectrum did not show isolated strong peaks, presumably reflecting the absence of isolated periodic orbits.

The vibrational modes were imaged using two methods. The first method was inspired by the technique attributed to Chladni in his early study of plate vibrations: finely chopped pine needles were sprinkled over the drumhead surface as it vibrated and these moved quickly to the nodal lines. (Pine needles were used because they are relatively insensitive to the build-up of static electricity on the insulating membrane.) This technique gave clear nodal line images only for low-order modes. The results showed a clear tendency for nodal lines of the quarter stadium to follow apparently random patterns and to repel one another. The second imaging method involved manually scanning the phonograph needle detector in a raster pattern across the entire membrane and recording the resonant response at each point. This technique provided a clear contour plot of the amplitude response. A particular mode imaged with this technique at 746 Hz showed strong amplitude build-up along the outer edges of quarter stadium, as one would expect for a "whispering gallery" mode (Thurstone 1992). For collecting many images in a short time, optical imaging is much better than the rastering scheme above. Unfortunately, optical imaging based on interference of light waves (i.e., holography) is not feasible for real drumheads because the vibrations have much too large an amplitude. However, in metal plates and blocks, the transverse vibrational amplitudes are often on the order of optical wavelengths.

3 Plate Experiments

The transverse vibrational waves of a thin metal plate provide an important example of a wave system for which the stationary modes are not exactly described by a scalar Helmholtz equation: the plate equation contains fourth-order spatial derivatives. In a recent numerical study, the computed asymptotic spectrum for a fully clamped thin plate was shown to possess statistical properties predicted by random matrix theory; furthermore, strong evidence was found for scarring in some of the high frequency modes (Legrand et al. 1992). More recently, a general theory of scarring and spectral properties has been developed by Bogomolny and Hughes (1998) for plate vibrations.

The equation governing the driven plate system that we study experimentally can be written in the form (Starobin and Teitsworth 1999):

$$\rho \frac{\partial^2 \psi}{\partial t^2} = -\frac{D}{h} \Delta^2 \psi - \frac{\rho}{\tau} \frac{\partial \psi}{\partial t} + f_\omega \delta(\mathbf{r} - \mathbf{r}_s) \cos(\omega t), \tag{2}$$

where ψ is the vibrational amplitude, ρ is the mass density of the plate, h is the plate thickness, $\mathbf{r}_s$ is the position where the driving force (described below) is applied, f_ω is the amplitude of the driving term, and D denotes the flexural rigidity. The effects of damping are described by lumping all losses into a velocity-dependent damping term of the form $-\frac{\rho}{\tau}\frac{\partial \psi}{\partial t}$, where τ ($\tau \gg T$) is the characteristic damping time, T is the oscillation period. In the limit of very short wavelengths and neglecting the damping we obtain the dispersion relation for plate waves (Landau and Lifshitz 1970):

$$f = \frac{1}{2\pi}\sqrt{\frac{Eh^2}{12\rho(1-\sigma^2)}}\,k^2 = \frac{D}{2\pi\rho h}k^2 \equiv Ck^2, \tag{3}$$

where E is the Young's modulus and σ is Poisson's ratio.

Also, in the limit of short wavelengths we expect that an eikonal approach will provide a good description of solutions. On this basis, we have derived a simple and useful quantization condition for the eigenmodes of a fully clamped plate that can be associated with periodic orbits of the underlying classical system (Starobin and Teitsworth 1999). In particular, for orbit Γ with N boundary reflections we have:

$$k_{\|}L - 2\sum_{i=1}^{N} \arccos\left(\sqrt{\frac{1+\sin^2\alpha_i}{2}}\right) = 2\pi n, \tag{4}$$

where $k_{\|}$ is a wavevector component along the periodic trajectory, L is the total orbit length, α is the reflection angle measured relative to the boundary normal vector, and n is an integer "quantum" number. It is interesting to note that (4) contains angle-dependent phase shifts associated with each boundary reflection. The specific form of the phase shift depends on details of the boundary conditions. This expression for the phase shift is obtained by matching incoming and outgoing plane wave solutions (Bogomolny and Hughes 1998).

For our experiment, we have studied a stainless steel plate of thickness $h = 0.305$ mm. Using (3) we obtain the value of $C = 767$ cm^{2}sec^{-1}. The plate is in the shape of a full stadium and consists of a square central section of side 8.00 cm and two semi-disks on each side of radius $R = 4.00$ cm. Contour plots of vibrational amplitude for individual modes are obtained using the technique of time-averaged holographic interferometry (Vest 1979; Moyar 1994). To achieve a fully clamped boundary condition (i.e., both amplitude and its normal derivative tend to zero at the boundary), the plate edges were carefully epoxied to a massive aluminum support. To drive the plate vibrations harmonically, we use a modified audio speaker (frequency range from 50 Hz to 12 kHz) which is coupled to the plate via a thin steel rod which acts as the driving needle. We have modelled this driving force as harmonic in time and δ-like in space in (2) above. A scannable phonograph needle is used to monitor the amplitude response of the plate at various positions. Measurements are carried out in the frequency range 100 Hz to 12 kHz. The fundamental frequency of the plate is approximately 200 Hz.

Within the frequency range 7 – 8 kHz, five vibrational resonances are detected at 7272 Hz, 7443 Hz, 7552 Hz, 7622 Hz, and 7874 Hz. Two of these modes - at 7443 Hz and 7622 Hz - reveal prominent scarring in their holographic interferograms (Starobin and Teitsworth, 1999). The 7443 Hz mode shows prominent scarring by a "bouncing ball" trajectory across the width of the stadium plate; inspection of the image shows that there are 7 antinode pairs as one traverses the complete periodic trajectory. The 7443 Hz mode also shows some scarring by a V shaped orbit that has its vertex at the center of the top straight section of the plate boundary and which passes through both semi-disk centers. A total of 17 anti-node pairs are crossed as the complete V shaped orbit is traced. The 7622 Hz mode shows prominent scarring effects due to a rectangular orbit that follows along the edges of the plate rather closely. In this case, we find 16 antinode pairs along the orbit. The 7622 Hz also appears to be scarred by the same V shaped orbit that appears in the 7443 Hz mode. For this reason, we have tentatively associated the V orbit with the 7552 Hz mode (Starobin and Teitsworth 1999).

We can use (4) to estimate the frequency of the scarred modes. The quantum number n is associated with the number of antinode pairs as determined above. For the bouncing ball trajectory with $n = 7$ and appropriate α values (i.e., there are two bounces such that $\alpha_1 = \alpha_2 = 0$), we obtain $k_{\|} = 295\ \mathrm{m}^{-1}$. Then, we use (3)to estimate a frequency $f = 6675$ Hz, somewhat less than the measured value of 7443 Hz. This discrepancy results from the fact that in using (3) we have neglected the contribution of the wavevector perpendicular to the orbit, and this is non-negligble for the relatively low-order modes we are studying. Including this "lateral confinement" contribution gives a much better agreement between the predicted and measured frequencies (Starobin and Teitsworth 1999). For the V shaped orbit we have $n = 17$ and α values are given by $\alpha_1 = \alpha_3 = \pi/4$ and $\alpha_2 = \alpha_4 = 0$. Following the same procedure as before we estimate a frequency of $f = 6463$ Hz, again somewhat less than the measured value. Finally, for the rectangular orbit, there are four bounces each with $\alpha_i = \pi/4$. Using a quantum number of $n = 16$, we estimate a frequency of $f = 5644$ Hz, considerably less than the measured frequency of 7662 Hz. For the rectangular orbit, the lateral confinement effects are expected to be strongest, because the eigenfunction is confined to a rather narrow strip that hugs the plate perimeter (Starobin and Teitsworth 1999).

We have observed scarring effects in both aluminum and steel plates that have non-stadium boundary shapes, but which are still predicted to be chaotic, e.g., one-quarter of a "bow-tie" shape. In contrast, we studied the spatial properties of vibrational modes in clamped plates with a boundary in the shape of a quarter circle expected to correspond to an integrable billiard; in this case no significant scarring of *isolated* periodic orbits was found (Chen

1996). We measured spectral statistics for all of the plates, but the present experimental set-up makes it difficult to obtain complete and accurate results due to missed modes and non-negligible effects of the driver position on the measured spectrum.

4 Acknowledgements

John Straw and Chris Thurstone played important roles in setting up the drumhead experiments and collecting data for drumhead studies. Lester Chen, Dean Moyar, and Andre Starobin played key roles in the set-up, measurement, and analysis of plate experiments.

References

Alt, H. *et al.* (1996): Phys. Rev. E **54**, 2303
Berry M. V. (1987): Proc. R. Soc. London, Ser. A **413**, 183
Blümel R., Davidson I. H., Reinhardt W. P., Lin H., Sharnoff M. (1992): Phys. Rev. A **45**, 2641
Bogomolny E., Hughes E. (1998): Phys. Rev. E **57**, 5404
Chen L. H. (1996): Undergraduate thesis, Duke University
Chinnery P. A., Humphrey V. F. (1996): Phys. Rev. E **53**, 272
Deland D., Sornette D., Weaver R. (1994): J. Acoust. Soc. Am. **96**, 1873
Deus S., Koch P., Sirko L. (1995): Phys. Rev. E **52**, 1146
Dörr U., Stöckmann H.-J., Barth M., Kuhl U. (1998): Phys. Rev. Lett. **75**, 1030
Ellegard C., Guhr T., Lindemann K., Lorensen H. Q., Nygaard J., Oxborrow M. (1995): Phys. Rev. Lett. **75**, 1546
Gräf H.-D. et al. (1992): Phys. Rev. Lett. **69**, 1296
Heller E. J. (1984): Phys. Rev. Lett. **53**, 1515
Landau L. D., Lifshitz E. M. (1970): *Theory of Elasticity, 2nd ed.* (Pergammon, London)
Legrand O., Schmitt C., Sornette D. (1992): Europhys. Lett. **18**, 101
Moyar Dean (1994): Undergraduate thesis, Duke University
Sridhar S. (1991): Phys. Rev. Lett. **67**, 785
Starobin A. J. (1997): Undergraduate thesis, Duke University
Starobin A. J., Teitsworth S. W. (1999): *submitted for publication in Phys. Rev. E*
Stein J., Stöckmann H. J. (1992): Phys. Rev. Lett. **68**, 2867
Stöckmann H. J., Stein J. (1990): Phys. Rev. Lett. **64**, 2215
Thurstone C. (1992): Undergraduate thesis, Duke University
Vest C. M. (1979): *Holographic interferometry* (Wiley, New York)

Magnetoconductance in Chaotic Quantum Billiards

E. Louis[1] and J.A. Vergés[2]

[1] Departamento de Física Aplicada, Universidad de Alicante, Apartado 99, E-03080 Alicante, Spain
[2] Instituto de Ciencia de Materiales de Madrid, Consejo Superior de Investigaciones Científicas, Cantoblanco, E-28049 Madrid, Spain

Abstract. Magnetotransport through quantum chaotic billiards is investigated by means of new model which incorporates chaoticity by introducing Anderson disorder on a number of sites (either at the surface or within the bulk) proportional to the linear size of the system L. In particular weak localization effects and the selfsimilarity of magnetoconductance fluctuations are studied. The results indicate that the controlling parameter in both phenomena is W/L, where W is the leads width.

1 Introduction

Transport properties of mesoscopic systems is a subject of much current interest (Datta (1995)). One of the issues that attracts a great deal of attention is the identification of signatures of chaos in systems whose classical counterparts show chaotic behavior (Guhr et al. (1998)). Present technology permits the fabrication microcavities with reproducible shapes, thus allowing to obtain either regular or chaotic quantum dots. Some outstanding differences between regular and chaotic cavities have already been found, the most remarkable of which is the shape of the weak localization peak observed at magnetic fluxes below one flux quanta (Chang et al. (1994)). In this work we discuss several aspects of transport through quantum cavities by means of a new model which incorporates chaoticity by introducing Anderson impurities at a number of sites proportional to the linear size of the system. The latter condition guarantees the ballistic character of the system no matter how big it is.

2 Theoretical Methods

2.1 Lattice Model Hamiltonian

The quantum dot is described by means of a tight-binding Hamiltonian with a single atomic level per lattice site on $L \times L$ clusters of the square lattice:

$$\widehat{H} = \sum_{m,n\,\in\,IS} \epsilon_{m,n}|m,n><m,n| - \sum_{<m,n;m',n'>} t_{m,n;m',n'}|m,n><m',n'|, \tag{1}$$

where $|m,n>$ represents an atomic orbital on site (m,n). Indexes run from 1 to L, and the symbol $<>$ denotes that the sum is restricted to nearest-neighbors. Using Landau's gauge the hopping integral is given by,

$$\begin{aligned} t_{m,n;m',n'} &= t\exp\left(2\pi i\frac{m}{(L-1)^2}\frac{\Phi}{\Phi_0}\right), \quad m=m' \\ &= t, \quad \text{otherwise} \end{aligned} \tag{2}$$

where $(L-1)^2$ is the area of the cluster and $\Phi_0 = h/e$ the quantum of magnetic flux. The hopping integral t will be used as the unit of energy, whereas lengths will be measured in terms of the lattice constant a (t=1 and a=1). The energy $\epsilon_{m,n}$ of atomic levels at impurity sites (IS) is randomly chosen between $-\Delta/2$ and $\Delta/2$, whereas at other sites $\epsilon_{m,n} = 0$. Impurities were taken on a number of sites proportional to L. They were distributed either at the surface as in Cuevas et al. (1996) or within the bulk (either randomly or forming a cross). It should be noted that the ballistic character of the cavity and its chaoticity are not affected by the spatial distribution of the impurities as long as their amount remains proportional to L (Vergés et al. (1999)). Some differences between surface and bulk impurities may, however, appear in the magnetoconductance at medium/high magnetic fields (see Section 4).

2.2 Kubo Formula for the Conductance

The conductance (measured in units of the quantum of conductance $G_0 = e^2/h$) was computed by using an efficient implementation of Kubo formula. A detailed description of the method can be found in (Vergés (1999)), whereas applications to mesoscopic systems are described in (Vergés et al. (1998)). For a current propagating in the x–direction, the static electrical conductance is given by:

$$G = -2\left(\frac{e^2}{h}\right)\mathrm{Tr}\left[(\hbar\hat{v}_x)\mathrm{Im}\,\widehat{\mathcal{G}}(E)(\hbar\hat{v}_x)\mathrm{Im}\,\widehat{\mathcal{G}}(E)\right], \tag{3}$$

where $\mathrm{Im}\,\widehat{\mathcal{G}}(E)$ is obtained from the advanced and retarded Green functions:

$$\mathrm{Im}\,\widehat{\mathcal{G}}(E) = \frac{1}{2i}\left[\widehat{\mathcal{G}}^R(E) - \widehat{\mathcal{G}}^A(E)\right], \tag{4}$$

and the velocity (current) operator $\hat{v}_x$ is related to the position operator $\hat{x}$ through the equation of motion.

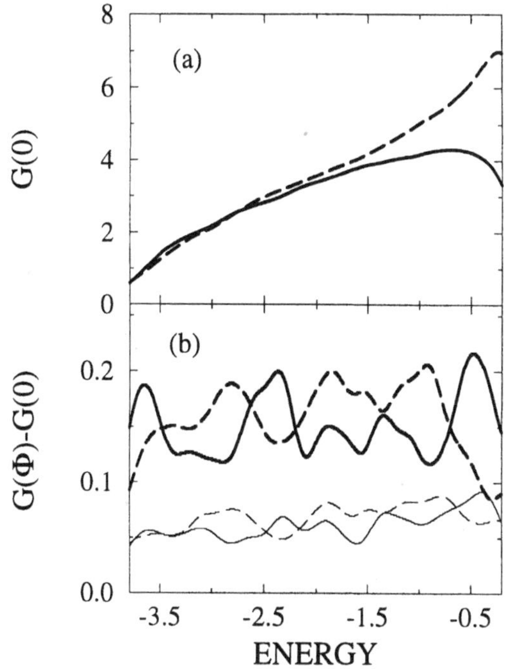

Fig. 1. Magnetoconductance as a function of energy in 137 × 137 chaotic cavities with leads of width 17 connected at opposite corners (broken lines) or contiguous corners (continuous lines) of the dot. Chaoticity was induced by introducing disorder ($\Delta = 6$, see text) at all surface sites but those at the leads entrance sites. a) Magnetoconductance at zero magnetic field, and b) Increment in the magnetoconductance at magnetic fluxes $\Phi = 0.2\Phi_0$ (thin lines) and $\Phi = 0.6\Phi_0$ (thick lines). The results were averaged by convolving with the derivative of the Fermi function.

Numerical calculations were carried out connecting quantum dots to semi-infinite leads of width W in the range 1–L. The hopping integral inside the leads and between leads and dot at the contact sites is taken equal to that in the quantum dot. Assuming the validity of both the one-electron approximation and linear response, the exact form of the electric field does not change the value of G. An abrupt potential drop at one of the two junctions provides the simplest numerical implementation of the Kubo formula Vergés (1999) since, in this case, the velocity operator has finite matrix elements on only two adjacent layers and, consequently, Green functions are just needed for this restricted subset of sites. In particular, the velocity operator is given by:

$$i\hbar v_x = -\sum_{j=1}^{W} (|lc, j >< 1, j| - |1, j >< lc, j|) \tag{5}$$

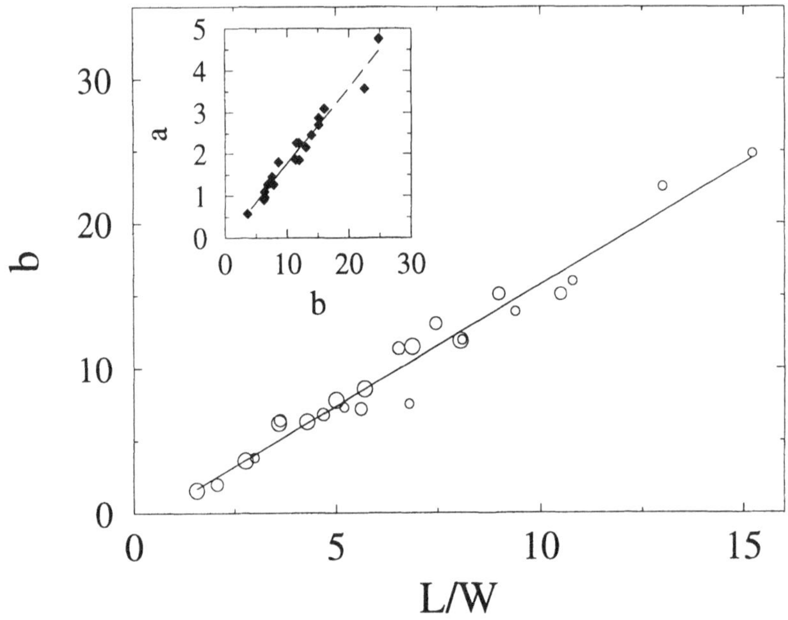

Fig. 2. Constant b in Eq. (6) as a function of L/W, where L is the linear size of the cavity and W the leads width, for chaotic cavities with $\Delta = 6$ and leads connected at contiguous corners of the dot. The results correspond to L in the range 27–137, and W=9, 17 and 23 (the size of the symbols is proportional to W). The fitted straight line is: $b = -1 + 1.67L/W$. The inset shows a vs. b in Eq. (6) for all combinations of L and W. The straight line is $a = -0.05 + 0.18b$. The results were obtained from fittings of the magnetoconductance averaged over the whole band.

where ($|lc, j >$ are the atomic orbitals at the left contact sites nearest neighbors to the dot. Green functions are calculated using standard methods.

3 Weak Localization Effects

Several experimental studies of magnetoconductance in quantum dots indicate that, at low magnetic fields (typically below one flux quantum), the conductance increases monotonically with the field (Marcus et al. (1992), Chang et al. (1994), Taylor et al. (1997), Sachrajda et al. (1998)). The effect has been investigated theoretically (Chang et al. (1994), Guhr et al. (1998)) and related to a similar behavior observed in disordered metallic conductors in the diffusive regime that is commonly referred to as weak localization (Datta (1995), Efetov (1997)). In chaotic cavities the increment in the magnetoconductance as a function of magnetic flux Φ is given by,

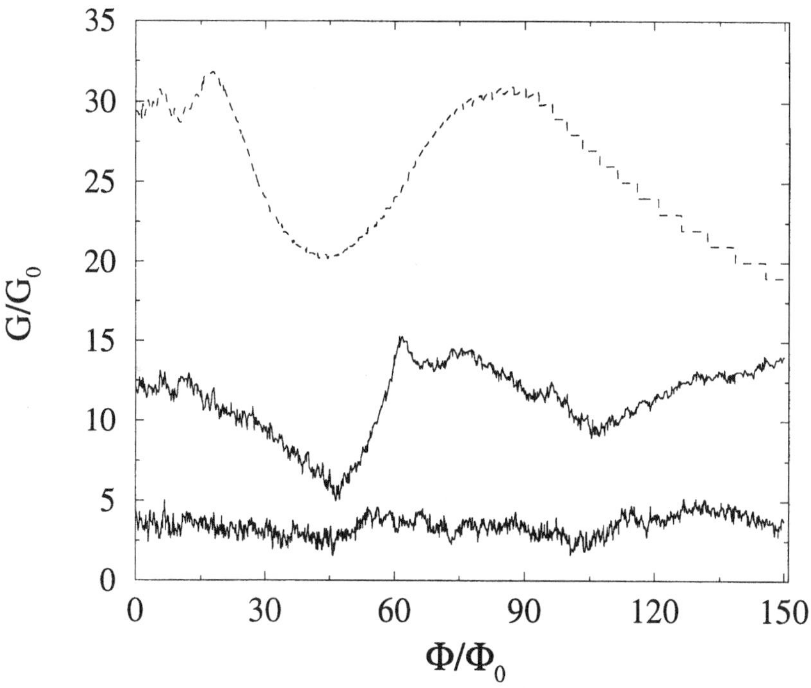

Fig. 3. Magnetoconductance versus magnetic flux in 97 × 97 regular dots with leads of width 9 (lower curve), 25 (middle curve) and 57 (upper curve) connected at opposite corners of the dot. The results correspond to the energy $E = -\pi/3$.

$$\delta G = G(\Phi) - G(0) = \frac{a\Phi^2}{1 + b\Phi^2} \tag{6}$$

where the conductance and the magnetic flux are given in units of their respective quanta, $G_0 = e^2/h$ and $\Phi_0 = h/e$. The constant b gives the critical flux at which the time–reversal symmetry is effectively destroyed, $\Phi_c = 1/\sqrt{b}$. The supersymmetric σ–model predicts that constants a and b should be inversely proportional to the number of channels N_{ch} that contribute to the current (Efetov (1997)),

$$b \propto \frac{D_0}{N_{ch}} \propto \frac{L}{N_{ch}} \tag{7}$$

where L is the linear size of the cavity. The size dependence arises from the expression for the diffusion coefficient $D_0 = v_F l/2$, where v_F is the Fermi velocity and l the elastic mean free path, and from the fact that in a two–dimensional ballistic system, $l \propto L$ (Louis et al. (1997)). The two constants a and b are proportional to each other. In particular random matrix theory gives,

$$\frac{a}{b} = \frac{N_{ch}}{4N_{ch} + 2} \tag{8}$$

Random matrix theory cannot reproduce, however, the size dependence predicted by the supersymmetric σ–model.

Results for the magnetoconductance increment (referred to its value at zero field) at Φ= 0.2 and 0.6 as a function of energy are shown in Fig. 1. In order to smooth out fluctuations, the results were convoluted with the derivative of the Fermi energy to simulate finite temperature, with $kT = 200\delta E$, δE being the average level spacing ($8/L^2$). The upper panel of the Figure shows the conductance at zero field averaged in the same way. The results clearly indicate the presence of the weak localization effect for all energies. On the other hand, no monotonic relationship between the magnetoconductance increment and the conductance at zero field (or the number of open channels) can be noted. This suggest that the number of channels may not be a relevant parameter as far as the weak localization effect is concerned. Similar results were obtained for leads connected either at contiguous or at opposite corners of the dot.

The Lorentzian shape of the weak localization peak in chaotic quantum dots has been investigated in more detail. Calculations were carried out for linear sizes of the cavity in the range L=27– 137, and leads of width W=9, 13 and 17, connected at contiguous corners of the dot. The results for the constants a and b of Eq. (6) are shown in Fig. 2. As no monotonic dependence on the Fermi energy was found, the fittings were done on averages of G over the whole energy band and in the flux range where G follows a Lorentzian (approximately from 0 to 1 flux quanta). Although the numerical results show an appreciable dispersion, they indicate that both constants are proportional to L/W. While the dependence on L agrees with Efetov (1997), the leads width W replaces the number of channels in Eq. (7). This indicates that what matters as far as the shape of the weak localization peak is concerned is the geometrical ratio W/L rather than the actual conductance (number of open channels) of the system. On the other hand a is proportional to b with a slope ≈ 0.2, not far from the random matrix result (in the range 1/6–1/4).

4 Selfsimilar Magnetoconductance Fluctuations

Conductance fluctuations in quantum dots is the subject of much current interest (See, for example, Marcus et al. (1992), Taylor et al. (1997), Sachrajda et al. (1998), Ketzmerick (1996), Guhr et al. (1998)). Experimental evidence indicates that the graph of conductance — versus the applied magnetic field have a selfsimilar character (Taylor et al. (1997), Sachrajda et al. (1998)). More recently, it has been observed that magnetoconductance fluctuations have a fractal character with a fractal dimension which decreases with the gate voltage applied to the cavity (Sachrajda et al. (1998)). The effect of the latter is to decrease the potential well experienced by the carriers (soft–wall potential).

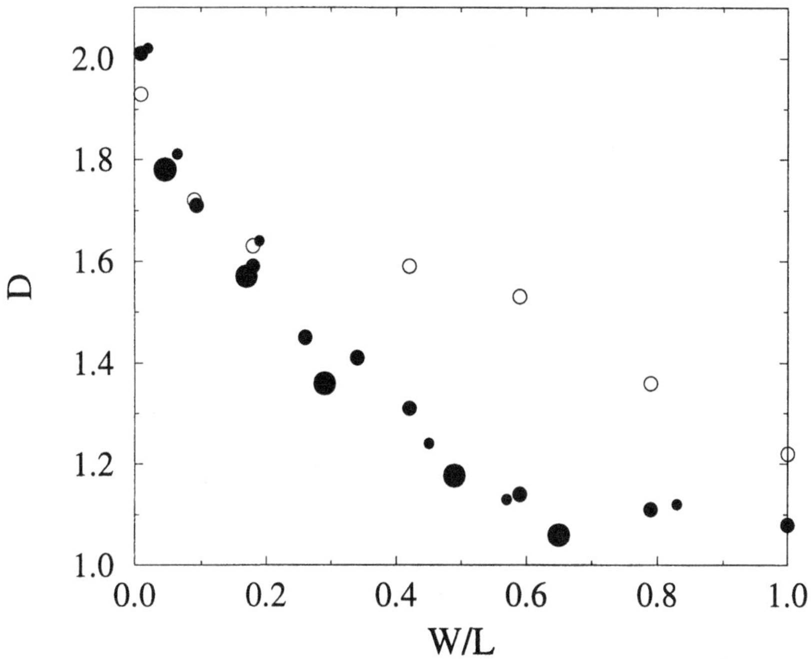

Fig. 4. Fractal dimension of magnetoconductance fluctuations, versus the ratio W/L, where W is the width of the leads, and L the linear dimension of the system. The size of the circles is proportional to the system size. The results correspond regular (filled circles) and chaotic (empty circles) cavities of linear sizes $L = 47, 97, 197$. For chaotic cavities Δ=6 was used and the impurities were distributed on a cross centered in the cavity.

These experimental findings seem in agreement with theoretical studies which predict a selfsimilar character of conductance fluctuations in mixed phase space systems (Ketzmerick (1996)). In mixed systems the distribution of enclosed areas larger than A follows a power law $P(A) \propto A^{-\gamma}$. A semiclassical argument (Ketzmerick (1996)) allows to conclude that the change in the conductance due to a small change in the magnetic field $\Delta G = G(E, B + \Delta B) - G(E, B)$ is a random variable with zero mean and a variance given by:

$$\langle (\Delta G)^2 \rangle = (\Delta B)^\gamma \qquad (9)$$

Random processes having such a variance and zero mean are known as fractional Brownian motion its fractal dimension being $D = 2 - \gamma/2$. We have investigated how D varies with the degree of opening of the system which we have tuned by changing the leads width W. The results indicate that D decreases as the opening increases (increasing W) in line with Sachrajda et al. (1998) measurements.

The effect of the leads width W on magnetoconductance fluctuations is illustrated in Fig. 3. As W decreases the graph becomes more noisy mainly due to the reduction of the momentum constraint. This applies to cavities with or without disorder and independently of other parameters of the model (energy, leads configuration, ...). For large W and above a given flux, the conductance coincides with that obtained for $W = L$ (a case without interference effects). In the latter case G decreases stepwise, each step (of one conductance quantum) being a consequence of transversal modes successively crossing the Fermi energy.

In Fig. 4 we plot the fractal dimension for three cluster sizes versus the ratio W/L. The collapse of the numerical results indicate that the relevant parameter is in fact the ratio W/L. D monotonically decreases from 2 to 1 as W/L increases from $1/L$ to 1. In the latter case we do not obtain $D = 1$ for regular cavities due to numerical inaccuracies. Our results clearly show that results for D greater than 3/2 are possible, in open contradiction with the remarks made in Sachrajda et al. (1998), and with the fractal dimension derived from Eq. (9), namely, $D = 2 - \gamma/2$, if the bounds for γ suggested by those authors, $1 < \gamma < 2$, do in fact hold. Introducing disorder increases noise and thus D, although no qualitative change in the behavior of D is observed. We have checked that the spatial distribution of the impurities does not qualitatively change these results, although actual values of D may of course be affected. This indicates that the nature of the closed cavity at $B = 0$ is not a determining factor as far as the selfsimilarity of fluctuations is concerned. For very closed cavities (small leads width, low wall softening, etc..) the fractal dimension is near 2 (no fractal character) no matter the type of the cavity, and D decreases down to ≈ 1 with the degree of opening, also for all cavities, although the limit $D = 1$ is probably never reached in chaotic cavities. It is also interesting to note that the fact that D does not depend on the cavity size indicates that it is neither affected by the strength of the magnetic field. In fact the cyclotron radius varies in a factor of 4 from the smallest to the largest cavity investigated in this work, without a significant change in D.

An interesting feature of the results reported by Sachrajda et al. (1998) is that for sufficiently soft walls the magnetoconductance increases monotonically in approximately $8G_0$ over $50\Phi_0$, with a low degree of noise. In Fig. 5 we plot G versus Φ for a cavity with leads of approximate width $0.6L$. Results for a regular cavity and chaotic cavities with $\Delta = 6$ and two types of disorder, surface or bulk, are shown (no effects due to the spatial distribution of bulk

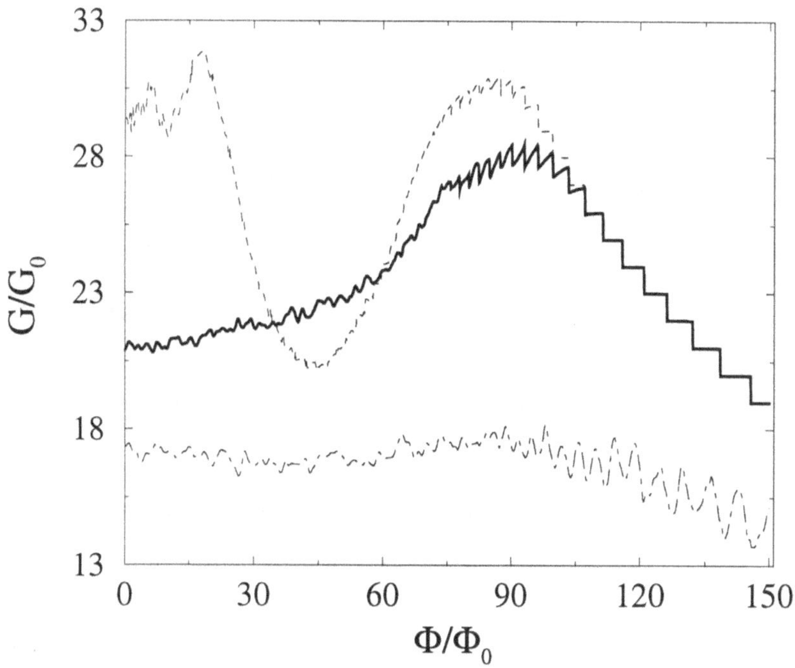

Fig. 5. Magnetoconductance versus magnetic flux in 97 × 97 dots with leads of width 57 connected at opposite corners of the dot. The results correspond to dots with no disorder (broken line), and chaotic cavities with Δ=6 and impurities either on all surface sites but those directly connected with the leads (continuous line) or $2L$ bulk impurities distributed on a cross centered at the cavity (chain line).

impurities were observed). The results indicate that the cavity having surface disorder is the only one that reproduces the experimental results, in line with the fact that the latter is the adequate Cuevas et al. (1996) to simulate cavities in which chaotic behavior is induced by the boundaries (as in Sachrajda et al. (1998)). Note that at large fields the result for the cavity with surface disorder coincides with that for the regular cavity. The difference between bulk and surface disorder is a consequence of the fact that surface disorder cannot scatter carriers between opposite sides of the cavity. Semiclassically one can view carriers motion as short orbits bouncing off the same boundary. The associated quantum states have chirality and are thus commonly refer to as chiral states.

5 Concluding Remarks

We have investigated magnetotransport through chaotic quantum dots by means of a model which accounts for chaoticity by introducing disorder on a number of sites proportional to the linear size of the system. Our numerical results indicate that the critical magnetic flux which characterizes the weak localization peak is proportional to $\sqrt{W/L}$. While this size dependence is in agreement with results obtained by means of the supersymmetric σ–model, the leads width replaces the number of channels in the result of the latter theory. Preliminar results indicate that a correlation between the shape of the weak localization peak and the number of open modes may exist in almost closed systems (small leads width) with homogeneous disorder (bulk, instead of surface, disorder). A detailed analysis of the selfsimilarity of magnetoconductance fluctuations indicates that the fractal dimension of the conductance vs. flux curve is also controlled by the geometric parameter W/L. Increasing W/L or, alternatively, the degree of opening of the system, decreases the fractal dimension, in agreement with experimental observations.

References

Chang, A.M., Baranger, H.U., Pfeiffer, L.N., & West, K.W. (1994): Phys. Rev. Lett. 73, 2111

Cuevas, E., Louis, E., & Vergés, J.A. (1996): Phys. Rev. Lett. 77, 1970

Datta, S. (1995): *Electronic Transport in Mesoscopic Systems* (Cambridge University Press, Cambridge)

Efetov, K. (1997): *Supersymmetry in disorder and chaos* (Cambridge University Press, Cambridge)

Guhr, T.,. Müller-Groeling, A., & Weidenmüller, H.A. (1998): Phys. Rept. 299, 189

Ketzmerick, R. (1996): Phys. Rev. B 54, 10841

Louis, E., Cuevas, E., Vergés, J.A. & Ortuño, M. (1997): Phys. Rev. B 56, 2120

Marcus, C.M., Rimberg, A.J., Westervelt, R.M., Hopkins, P.F., & Gossard, A.C. (1992): Phys. Rev. Lett. 69, 506

Taylor, R.P. et al. (1997): Phys. Rev. Lett. 78, 1952

Sachrajda, A.S., Ketzmerick, R., Gould, C., Feng, Y., Kelly, P.J., Delage, A. & Wasilewski, Z. (1998): Phys. Rev. Lett. 80, 1948

Vergés, J.A. (1999): Comput. Phys. Commun. 118, 71

Vergés, J.A., Cuevas, E., Ortuño, M. Louis, E. (1998): Phys. Rev. B 58, R10143

Vergés, J.A. & Louis, E. (1999): Phys. Rev. B 59, R3803

Part III

Time-Dependent Phenomena

Shot Noise Induced Charge and Potential Fluctuations of Edge States in Proximity of a Gate

Markus Büttiker

Département de Physique Théorique, Université de Genève,
CH-1211 Genève 4, Switzerland

Abstract. We evaluate the RC-time of edge states capacitively coupled to a gate located away from a QPC which allows for partial transmission of an edge channel. At long times or low frequencies the RC-time governs the relaxation of charge and current and governs the fluctuations of the equilibrium electrostatic potential. The RC-time in mesoscopic structures is determined by an electrochemical capacitance which depends on the density of states of the edge states and a charge relaxation resistance. In the non-equilibrium case, in the presence of transport, the shot noise leads to charge fluctuations in proximity of the gate which are again determined by the equilibrium electrochemical capacitance but with a novel resistance. The case of multiple edge states is discussed and the effect of a dephasing voltage probe on these resistances is investigated. The potential fluctuations characterized by these capacitances and resistances are of interest since they determine the dephasing rate in Coulomb coupled mesoscopic conductors.

1 Introduction

Dynamic fluctuations in mesoscopic conductors have attracted considerable attention. Most of the work has focused on the low frequency white noise limit of the current fluctuations that can be measured at the terminals of a conductor [1]. Much less is known, if we ask about fluctuations at higher frequencies. To be sure, there are a number of questions which can be asked in a frequency range for which the scattering matrix of the conductor can still be taken energy independent. All that matters in this regime is the frequency dependence of Fermi functions which govern the occupation of the states incident form a reservoir. Much more interesting problems arise if we ask questions which directly probe the energy dependence of the scattering matrix.

In this work we are concerned with charge and potential fluctuations in Coulomb coupled systems. Such systems are of increasing interest because one of the systems can serve as a measurement probe of the other system [2, 3]. Coulomb coupled mesoscopic systems are also of interest in the investigation of dephasing: through the long range Coulomb interactions the proximity of a mesoscopic conductor affects the dephasing rate in the other conductor [4, 5]. The dephasing rate is essentially determined by the fluctuations of the electrostatic potential which leads directly to the fluctuation of the phase of a carrier. Thus a theoretical description and experimental characterization of potential fluctuations is essential for an understanding of such problems. Perhaps the simplest Coulomb coupled system consists of a mesoscopic capacitor: two small plates, separated by a barrier which is too high to permit carrier exchange, are each separately coupled to a reservoir. Such a system permits no dc-transport, but exhibits an ac-conductance and exhibits frequency dependent charge, potential and current fluctuations [6, 7]. ¿From the point of view of the scattering theory of electrical transport, it is a simple example, in which the energy dependence of the scattering matrix is crucial. We are not merely testing the transmission probability of a conductor, nor the frequency dependence of the Fermi functions, but are now asking a question that is sensitive to the charge distribution and its dynamics. The questions we whish to address and illustrate with a simple example in this article are of this nature.

The dynamic behavior of a capacitor is determined by its RC-time. At long times, the relaxation of charge and current and the electrostatic potential is determined by this time. Thus it is intersting to ask: What is the RC-time of a phase-coherent conductor? Ref. [6, 7] considered two small conductors each of which is connected only via a single lead to an electron reservoir. The two conductors interact only via the long range Coulomb force. Assuming that the main effect of the Coulomb interaction is the energy cost to charge the system, Ref. [6] presents an answer in terms of the geometrical capacitance and the energy derivatives of the scattering matrix. The resulting capacitance is called an *electrochemical capacitance* C_μ, and the resistance of the structure is called a charge relaxation resistance R_q, to distinguish them from their geometrical and classical counterparts. Note that such a system has an infinite dc-resistance and thus the expression for the resistance R_q which gov-

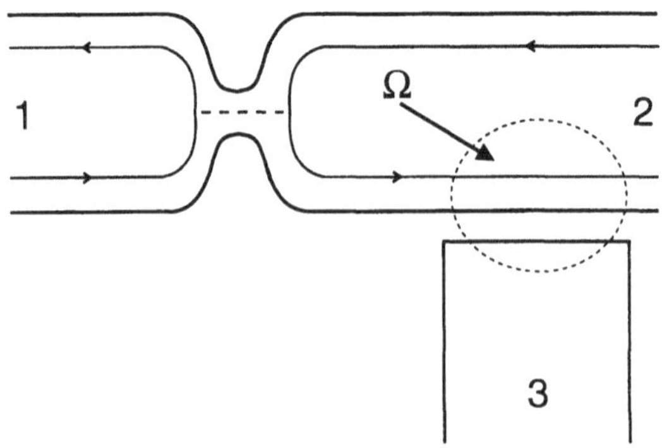

Fig. 1. Hall bar with a quantum point contact and a gate overlapping the edge of the conductor.

erns the relaxation of charge looks very different from the scattering matrix expressions for dc-resistances of conductors with non-vanishing transmission probability.

The RC-time plays a central role also for mutually Coulomb coupled multiprobe conductors. In multiterminal structures, especially if they are ballistic, additional inductive-like time-scales appear [8, 10]. However, as soon as we consider such a conductor not in isolation, but coupled to another gate or conductor, the RC-time remains a fundamental quantity: if we keep all external potentials of the conductor at the same value, we are again faced with a purely capacitive question: To what extend can we charge this conductor against the other nearby conductor or gate?

A closely related phenomenon occurs if we drive the conductor out of equilibrium by applying a dc voltage to it. Now at zero temperature the conductor exhibits shot noise [9, 1] which in addition to the usually investigated current fluctuations at the contacts of the conductor, generates charge fluctuations. These charge fluctuations depend again crucially on the capacitance of the mesoscopic conductor vis-a-vis other nearby conductors or gates. For small driving voltages, we find in fact that the capacitance is C_μ as in the equilibrium system. But a novel resistance appears [11], which we call R_v to indicate that it is connected to a non-equilibrium state obtained by applying a voltage V to one of the conductors.

The example which we treat in this work is shown in Fig. 1. A conductor subject to a high magnetic field with a quantum point contact (QPC) is capacitively coupled to a gate. The contacts of the conductor are labeled 1 and 2 and the gate contact is labeled 3. We assume that the magnetic field is in a range at which the only states at the Fermi energy which connect contacts 1 and 2 are edge states [12]. A similar geometry without the QPC was investigated by Chen et al. [13]. In this work it was shown that an oscillating voltage applied to the gate (contact 3) generates a current only at contact 2 but if the magnetic field polarity is reversed the induced current is found only at contact 1. Since coupling between the gate and the mesoscopic sample is purely capacitive, this experiment verifies a prediction [14] that capacitance coefficients are in general not even functions of magnetic field. The geometry with the QPC is inspired by a recent experiment of Sprinzak et al. [5] which investigates the dephasing of a double quantum dot due to the charge fluctuations generated by a current through the QPC. Here we will consider the geometry with the gate, instead of the double quantum dot. The conductor of Fig. 1 permits an investigation of the electrochemical capacitance C_μ and the resistances R_q and R_v of this structure. The relationship of these transport coefficients to the dephasing time is the subject of Ref. [15]. We will not review this part of Ref. [15] but only mention that related work [16] addresses this question invoking only the fluctuations of non-interacting electrons. Here we treat the fluctuations within a charge and current conserving self-consistent random-phase approximation (RPA) which represents a dynamical extension [7, 11, 15] of Ref. [17].

2 The Scattering Matrix

To be specific we consider the conductor shown in Fig. 1. Of interest is the current $dI_\alpha(\omega)$ at contact α of this conductor if an oscillating voltage $dV_\beta(\omega)$ is applied at contact β. Here α and β label the contacts of the conductor and the gate and take the values $1, 2, 3$. Furthermore, we are interested in the current noise spectrum $S_{I_\alpha I_\beta}(\omega)$ defined as $2\pi S_{I_\alpha I_\beta}(\omega)\delta(\omega+\omega') = 1/2\langle \hat{I}_\alpha(\omega)\hat{I}_\beta(\omega') + \hat{I}_\beta(\omega')\hat{I}_\alpha(\omega)\rangle$ and the fluctuation spectrum of the electrostatic potential. We assume that the charge dynamics is relevant only in the region underneath the gate. Everywhere else we assume the charge to be screened completely. This is a strong assumption: In reality the QPC is made with the help of gates (capacitors) and also exhibits its own capacitance [8]. Edge states might generate long range fields, etc. Thus the results presented below can only be expected to capture the main effects but can certainly be refined. We assume that the gate is a macroscopic conductor and screens perfectly.

The scattering matrix of the QPC alone can be described by $r \equiv s_{11} = s_{22} = -i\mathcal{R}^{1/2}$ and $t \equiv s_{21} = s_{12} = \mathcal{T}^{1/2}$ where $\mathcal{T} = 1 - \mathcal{R}$ is the transmission probability through the QPC. Here the indices 1 and 2 label the reservoirs (see Fig. 1). A carrier traversing the region underneath the gate acquires a phase $\phi(U)$ which depends on the electrostatic potential U in this region. Since we consider only the charge pile up in this region all additional phases in the scattering problem are here without relevance. The total scattering matrix of the QPC and the traversal of the region Ω is then simply

$$\mathrm{s} = \begin{pmatrix} r & t \\ te^{i\phi} & re^{i\phi} \end{pmatrix}. \tag{1}$$

If the polarity of the magnetic field is reversed the scattering matrix is given by $s_{\alpha\beta}(B) = s_{\beta\alpha}(-B)$, i. e. in the reversed magnetic field it is only the second column of the scattering matrix which contains the phase $\phi(U)$. In what follows, the dependence of the scattering matrix on the phase ϕ is crucial. We emphasize that the approach presented here can be generalized by considering all the phases of the problem and by considering these phases and the amplitudes to depend on the entire electrostatic potential landscape [7].

3 Density of States Matrix Elements

To describe the charge distribution due to carriers in an energy interval dE in our conductor, we consider the Fermi-field [1]

$$\hat{\Psi}(\mathbf{r},t)=\sum_{\alpha m}\int dE\psi_{\alpha m}(\mathbf{r},E)\hat{a}_{\alpha m}(E)exp(-iEt/\hbar) \tag{2}$$

which annihilates an electron at point $\mathbf{r}$ and time t. The Fermi operator Eq. (2) is built up from all scattering states $\psi_{\alpha m}(\mathbf{r},E)$ which have unit incident amplitude in contact α in channel m. The operator $\hat{a}_{\alpha m}(E)$ annihilates an incident carrier in reservoir α in channel m. The local carrier density at at point $\mathbf{r}$ and time t is determined by $\hat{n}(\mathbf{r},t) = \hat{\Psi}^{\dagger}(\mathbf{r},t)\hat{\Psi}(\mathbf{r},t)$. We will investigate the density operator in the frequency domain, $\hat{n}(\mathbf{r},\omega)$. It is now very convienient and instructive to consider an expression for the density operator not in terms of wave functions but more directly in terms of the scattering matrix. It can be shown [7], that the density operator $\hat{n}(\mathbf{r},\omega)$, in the zero frequency limit, can be written in the form

$$\hat{n}(\mathbf{r})=\sum_{\alpha\gamma\delta}\int dE\hat{a}^{\dagger}_{\gamma m}(E)n_{\gamma m\delta n}(\alpha,\mathbf{r})\hat{a}_{\delta n}(E) \tag{3}$$

where the elements $n_{\gamma m\delta n}$ form a matrix of dimensions $M_{\gamma} * M_{\delta}$. Here M_{γ} is the number of channels at the Fermi energy in contact γ. This matrix is given by [7]

$$n_{\beta\gamma}(\alpha,\mathbf{r})=-(1/4\pi i)[s^{\dagger}_{\alpha\beta}(\partial s_{\alpha\gamma}/\partial eU(\mathbf{r}))-(\partial s^{\dagger}_{\alpha\beta}/\partial eU(\mathbf{r}))s_{\alpha\gamma}]. \tag{4}$$

The low frequency charge dynamics can be found if these density of states matrix elements are known. Eq. (4) tells us that in order to find the carrier distribution and its fluctuations, we should introduce a small potential perturbation into the sample and find the scattering matrix which belongs to this perturbation. Clearly, such a detailed information requires a considerable effort and even more so, if we subsequently should solve the Poisson equation to find the electrostatic potential landscape which belongs to this density distribution. To proceed we introduce the simplifying assumption that it is only the charge pile-up near the gate which counts and moreover that the potential in this region Ω can be described with a single potential parameter U. All we need then is the density elements integrated over the region Ω. Instead of Eq. (3) we want to find

$$\begin{aligned}\hat{N}(\mathbf{r})&=\sum_{\alpha\gamma\delta}\int_{\Omega}d^3\mathbf{r}\int dE\hat{a}^{\dagger}_{\gamma m}(E)n_{\gamma m\delta n}(\alpha,\mathbf{r})\hat{a}_{\delta n}(E)\\&\equiv\sum_{\alpha\gamma\delta}\int dE\hat{a}^{\dagger}_{\gamma m}(E)N_{\gamma m\delta n}(\alpha)\hat{a}_{\delta n}(E)\end{aligned} \tag{5}$$

with

$$N_{\beta\gamma}(\alpha)=-(1/4\pi i)[s^{\dagger}_{\alpha\beta}(ds_{\alpha\gamma}/edU)-(ds^{\dagger}_{\alpha\beta}/edU)s_{\alpha\gamma}]. \tag{6}$$

Thus it is sufficient to find the variation in the scattering matrix for a potential that is uniform over the region of interest. In our example it is only the phase ϕ in Eq. (1) which depends on U. Thus we can evaluate the density of states elements if we know $d\phi/edU$. But in the WKB-limit, which is sufficient for our purpose, $d\phi/edU = -d\phi/dE$. However, $d\phi/dE = 2\pi N$ where N is just the density of states of the edge state underneath the gate.

We are now ready to evaluate the density of states elements Eq. (4). For the specific example given by Eq. (1) we find that all elements with $\alpha = 1$ vanish: $N_{11}(1) = N_{21}(1) = N_{12}(1) = N_{22}(1) = 0$. There are no carriers incident from contact 1 or 2 which pass through region Ω and leave the conductor through contact 1. The situation is different if we demand that the current leaves the sample through contact 2. Now we find

$$N_{\beta\gamma}(2) = \begin{pmatrix} \mathcal{T}N & t^*rN \\ r^*tN & \mathcal{R}N \end{pmatrix}, \tag{7}$$

where, as already mentioned, N is the density of states of carriers in the edge state underneath the gate. For the reverse magnetic field polarity all components of the matrix vanish except the elements $N_{22}(1) = \mathcal{T}N$ and $N_{22}(2) = \mathcal{R}N$.

For the charge and its fluctuations underneath the gate it is not relevant through which contact carriers leave. The charge pile up and its fluctuations are thus governed by a matrix

$$N_{\beta\gamma} = \sum_{\alpha} N_{\beta\gamma}(\alpha) \tag{8}$$

which is obtained by summing over the contact index α from the elements given by Eq. (4). For our example the density matrix elements for the charge are thus evidently given by $N_{\beta\gamma} = N_{\beta\gamma}(2)$ whereas for the reversed magnetic field polarity we have $N_{11} = \mathcal{T}N$, $N_{22} = \mathcal{R}N$ and $N_{21} = N_{21} = 0$.

Furthermore, we will make use of the *injectivity* of a contact into the region Ω and will make use of the emissivity of the region Ω into a contact. The injectivity of contact α is the charge injected into a region in response to a voltage variation at this contact, independently through which contact the carriers leave the sample [14]. The injectivities of contact 1 and 2 are

$$N_1 = N_{11}(1) + N_{11}(2) = \mathcal{T}N \tag{9}$$

$$N_2 = N_{22}(1) + N_{22}(2) = \mathcal{R}N \tag{10}$$

Note that the sum of the injectivities of both contacts is just the density of states N underneath the gate. The *emissivity* of region Ω is the portion of the density of states of carriers which leave the conductor through contact α irrespectively from which contact they entered the conductor [14]. We find emissivities

$$N^{(1)} = N_{11}(1) + N_{22}(1) = 0, \tag{11}$$

$$N^{(2)} = N_{11}(2) + N_{22}(2) = N. \tag{12}$$

Any charge accumulation or depletion is only felt in contact 2. The injectivities and emissivities in the magnetic field B are related by reciprocity to the emissivities and injectivities in the reversed magnetic field, $N_\alpha(B) = N^{(\alpha)}(-B)$ and $N^{(\alpha)}(B) = N_\alpha(-B)$. In contrast, the density of states N is an even function of magnetic field.

4 The Poisson Equation: The Effective Interaction

Thus far we have only considered bare charges. The true charge, however, is determined by the long range Coulomb interaction. First we consider the screening of the average charges and in a second step we consider the screening of charge fluctuations. We describe the long range Coulomb interaction between the charge on the edge state and on the gate with the help of a geometrical capacitance C. The charge on the edge state beneath the gate is determined by the voltage difference between the edge state and the gate $dQ = C(dU - dV_g)$, where dU and dV_g are deviations from an equilibrium reference state. On the other hand the charge beneath the gate can also be expressed in terms of the injected charges $e^2N_1dV_1$ in response to a voltage variation at contact 1 and $e^2N_2dV_2$ in response to a voltage variation at contact 2. Furthermore, the injected charge leads to a response in the internal potential dU which in turn generates a screening charge $-e^2NdU$ proportional to the density of states. Thus the Poisson equation for the charge underneath the gate is

$$dQ = C(dU - dV_g) = e^2N_1dV_1 + e^2N_2dV_2 - e^2NdU \tag{13}$$

and the charge on the gate is given by $-dQ = C(dV_g - dU)$. Solving Eq. (13) for dU gives

$$dU = G_{eff}(CdV_g + e^2N_1dV_1 + e^2N_2dV_2), \tag{14}$$

where $G_{eff} = (C + e^2N)^{-1}$ is an effective (RPA) interaction which gives the potential underneath the gate in response to an increment in the charge.

5 Admittance

Consider now the low-frequency conductance: To leading order in the frequency ω we write

$$G_{\alpha\beta}(\omega) = G_{\alpha\beta}(0) - i\omega E_{\alpha\beta} + \omega^2 K_{\alpha\beta} + O(\omega^3). \tag{15}$$

Here $G_{\alpha\beta}(0)$ is the dc-conductance matrix, $E_{\alpha\beta}$ is the *emittance* matrix, and $K_{\alpha\beta}$ is a second order dissipative contribution to the frequency dependent admittance. The zero-frequency dc-conductance matrix has only four non-vanishing elements which are given by $G \equiv G_{11} = G_{22} = -G_{12} = -G_{21} = (e^2/h)\mathcal{T}$. Ref. [14] showed that the emittance matrix $\mathbf{E}$ is given by

$$E_{\alpha\beta} = e^2 N_{\beta\beta}(\alpha) - e^2 N^{(\alpha)} G_{eff} N_{\beta} \tag{16}$$

As it is written, Eq. (16) applies only to the elements where α and β take the values 1 or 2. The remaining elements can be obtained from current conservation (which demands that the elements of each row and column of this matrix add up to zero) or can be obtained directly by using a more general formula [8]. For our example we find an emittance matrix,

$$\mathbf{E} = C_\mu \begin{pmatrix} 0 & 0 & 0 \\ \mathcal{T} & \mathcal{R} & -1 \\ -\mathcal{T} & -\mathcal{R} & 1 \end{pmatrix}, \tag{17}$$

with an electrochemical capacitance of the conductor vis-a-vis the gate given by $C_\mu = Ce^2N/(C + e^2N)$. Eq. (17) determines the displacement currents in response to an oscillating voltage at one of the contacts. There is no displacement current at contact 1 (the elements of the first row vanish) which is consequence of our assumption that charge pile up occurs only underneath the gate. The emittance matrix in the magnetic field B and in the magnetic field $-B$ are related by reciprocity, $E_{\alpha\beta}(B) = E_{\beta\alpha}(-B)$. For the reverse polarity, a voltage oscillation at contact 1 generates no displacement currents (the elements of the first column vanish).

The emittance matrix element E_{21} is positive and thus has the sign not of a capacitive but of an inductive response. The elements of row 3 and column 3 are a consequence of purely capacitive coupling and have the sign associated with the elements of a capacitance matrix. Thus these elements represent the capacitance matrix elements which can be measured in an ac-experiment. Note that the capacitances $E_{31} \equiv C_{31}$ and $E_{32} \equiv C_{32}$ depend not only on the density of states and geometrical capacitances but also on transmission and reflection probabilities. Measurement of these capacitances provides thus a direct confirmation of the concept of *partial* density of states [14, 8]. Furthermore, we see that for instance $C_{31}(B) \equiv E_{31} = \mathcal{T} C_\mu$ but $C_{31}(-B) = 0$. A similarly striking variation of the capacitance coefficients was observed in the experiment of Chen et al. [13] in the integer quantum Hall effect and in Refs. [19] in the fractional quantum Hall effect.

6 Bare Charge Fluctuations

Let us now turn to the charge fluctuations. With the help of the charge density matrix the low frequency limit of the bare charge fluctuations can be obtained [6, 11, 15]. It is given by

$$S_{NN}(\omega) = h \sum_{\delta\gamma} \int dE\, F_{\gamma\delta}(E,\omega) Tr[N_{\gamma\delta}(E, E+\hbar\omega) N^{\dagger}_{\gamma\delta}(E, E+\hbar\omega)] \tag{18}$$

where the elements of $N_{\gamma\delta}$ are in the zero-frequency limit of interest here given by Eq. (8) and $F_{\gamma\delta} = f_\gamma(E)(1 - f_\delta(E+\hbar\omega)) + f_\delta(E+\hbar\omega)(1 - f_\gamma(E))$

is a combination of Fermi functions. Using only the zero-frequency limit of the elements of the charge operator determined above gives,

$$S_{NN}(\omega) = hN^2 \left[\mathcal{T}^2 \int dE\; F_{11}(E,\omega) + \mathcal{TR} \int dE\; F_{12}(E,\omega) + \mathcal{TR} \int dE\; F_{21}(E,\omega) + \mathcal{R}^2 \int dE\; F_{22}(E,\omega)\right]. \quad (19)$$

At equilibrium all the Fermi functions are identical and we obtain $S_{NN}(\omega) = hN^2 \int dE\; F(E,\omega)$ which in the zero-frequency limit is

$$S_{NN}(\omega) = hN^2 kT \quad (20)$$

and at zero-temperature to leading order in frequency is,

$$S_{NN}(\omega) = hN^2 \hbar\omega. \quad (21)$$

In the zero-temperature, zero-frequency limit, in the presence of a current through the sample, we find for the charge fluctuations associated with shot noise

$$S_{NN}(\omega) = hN^2 \mathcal{TR} e|V|. \quad (22)$$

However, the bare charge fluctuations are not by themselves physically relevant.

7 Fluctuations of the True Charge

To find the fluctuations of the true charge we now write the Poisson equation for the fluctuating charges. All contact potentials are at their equilibrium value, $dV_1 = dV_2 = dV_g = 0$. The fluctuations of the bare charge now generate fluctuations in the electrostatic potential. Thus the electrostatic potential has also to be represented by an operator $\hat{U}$. Furthermore, the potential fluctuations are also screened. As in the case of the average charges we take the screening to be proportional to the density of states N but replace the c-number U by its operator expression $\hat{U}$. The equation for the fluctuations of the true charge is thus

$$d\hat{Q} = Cd\hat{U} = e\hat{\mathcal{N}} - e^2 N\hat{U} \quad (23)$$

whereas the fluctuation of the charge on the gate is simply $-d\hat{Q} = -Cd\hat{U}$.

Thus dQ is the charge operator which determines the *dipole* which forms between the charge on the edge state and the charge on the gate. Solving Eq. (23) for the potential operator $\hat{U}$ and using this result to find the fluctuations of the charge dQ gives

$$S_{QQ}(\omega) = e^2 C'^2 G_{eff}^2 S_{NN}(\omega) = 2C_\mu^2 (1/2e^2)(S_{NN}(\omega)/N^2). \quad (24)$$

We now discuss three limits of this result.

8 Equilibrium and Non-Equilibrium Charge Relaxation Resistance

At equilibrium, in the zero-frequency limit, the charge fluctuation spectrum can be written with the help of the equilibrium charge relaxation resistance [6, 7, 11] R_q,

$$S_{QQ}(\omega) = 2C_\mu^2 R_q kT. \tag{25}$$

For our specific example[15], we find using Eqs. (20) and (24),

$$R_q = h/2e^2. \tag{26}$$

The charge relaxation resistance is universal and equal to *half* a resistance quantum as expected for a single edge state [8]. At equilibrium the fluctuation spectrum is via the fluctuation dissipation theorem directly related to the dissipative part of the admittance. We could also have directly evaluated the element K_{33} of Eq. (15) to find $K_{33} = C_\mu^2 R_q$. Second at equilibrium, but for frequencies which are large compared to the thermal energy, but small compared to any intrinsic excitation frequencies, we find that zero-point fluctuations give rise to a noise power spectral density

$$S_{QQ}(\omega) = 2C_\mu^2 R_q \hbar\omega \tag{27}$$

which is determined by the charge relaxation resistance Eq. (26). Third, in the presence of transport, we find in the zero-frequency, zero-temperature limit, a charge fluctuation spectrum [11],

$$S_{QQ}(\omega) = 2C_\mu^2 R_v e|V|, \tag{28}$$

where $|V|$ is the voltage applied between the two contacts of the sample and a non-equilibrium charge relaxation resistance which for our example is given by [15]

$$R_v = (h/e^2)\mathcal{T}\mathcal{R}. \tag{29}$$

It is maximal for a semi-transparent QPC, $\mathcal{T} = \mathcal{R} = 1/2$.

The current at the gate due to the charge fluctuations is $dI_g = -i\omega dQ(\omega)$ and thus its fluctuation spectrum is given by $S_{I_g I_g}(\omega) = \omega^2 S_{QQ}$. The potential fluctuations are related to the charge fluctuations by $d\hat{U} = d\hat{Q}/C$ and thus the spectral density of the potential fluctuations is $S_{UU}(\omega) = C^{-2} S_{QQ}$. Thus the charge relaxation resistance determines, together with the electrochemical and geometrical capacitance, the fluctuations of the charge, the potential and the current induced into the gate. Since dephasing rates can be linked to the low frequency limit of the potential fluctuations [18] the resistances R_q and R_v also determine the dephasing rate in Coulomb coupled mesoscopic conductors [15].

9 Several Edge States

Let us next consider the case, where there are several edge states. A QPC in a high magnetic field permits perfect transmission of the outer edge states (belonging to the lower Landau levels) and it is only the innermost edge state which is partially transmitted or reflected at the QPC. Let us just consider two edge states: the outer edge state labeled 1 is perfectly transmitted $T_1 = 1$, whereas the inner edge state labeled 2 has a transmission probability $T_2 \equiv T$ which might take any value between zero and one. The outer edge state, with transmission probability 1 is entirely noiseless as far as the shot noise in the total current is concerned [1]. One might thus be tempted to think that such a perfectly transmitted edge state plays no role at all. That however is not the case. Our result involves screening in an essential manner and the charge fluctuations in one of the edge states can now be screened by charge accumulation or depletion in the other edge state. The screening properties depend on the electrostatic interaction between the two edge states. Thus the answer we obtain depends on the detailed electrostatic assumptions which we invoke to treat this problem. Here, to provide a simple discussion, we assume that the two edge states are so close, that they can be described with a common electrostatic potential U. If we denote the density of states of the edge states 1 and 2 in the region Ω of interest by N_1 and N_2 a detailed consideration, repeating the procedure given above for one edge state only, leads to an equilibrium charge relaxation resistance [15]

$$R_q = \frac{h}{2e^2} \frac{N_1^2 + N_2^2}{(N_1 + N_2)^2} \tag{30}$$

Note that in contrast to the single edge state, now R_q depends explicitly on the densities of states. We can expect that the density of states N_2 of the inner edge state 2 is typically larger than the density of states of the outer edge state since the potential for the inner edge state is much shallower. In this case $N_2 >> N_1$ and R_q for the two edge states will in fact be the same as for one edge state only. In contrast, for samples with a sharp edge, we can expect that both density of states are comparable, and thus R_q for two edge states will be nearly a factor 2 smaller than the R_q of a single edge state only.

Similarly, if we investigate R_v for two edge states, we find [15]

$$R_v = \frac{h}{2e^2} \frac{N_2^2}{(N_1 + N_2)^2} TR. \tag{31}$$

Again the density of states of the two edge states appear now explicitly. The density of states of the outer edge state appears only in the denominator since it plays a role only in screening but it is not a primary source of charge fluctuations. In the limit $N_2 >> N_1$ of a shallow edge the outermost edge state is unimportant, whereas for a steep edge if both density of states are comparable, R_v is reduced by a factor 4 compared to the case of a single edge state only.

Form the above results it is obvious how the formulas must be written if there is one edge state which is partially reflected or transmitted and many edge states which are perfectly transmitted.

10 Phase Randomization

Is the result given above sensitive to phase? Experimentally this question is investigated by Sprinzak et al. [5]. Our result for one channel, Eq. (29), contains only transmission probabilities. To investigate this question, we consider, like the experiment, an additional contact between the QPC and the region Ω as shown in Fig. 2. The contact will be considered as a voltage probe. An ideal voltage probe exhibits infinite impedance at all frequencies. Consequently, the net current at the voltage probe vanishes at every instant of time. Thus the voltage of the probe becomes a fluctuating quantity. Despite the fact that the total current vanishes, carriers leave the sample through this contact, and are replaced by carriers which enter from the reservoir. Carriers leaving into the reservoir and carriers rentering the conductor from the reservoir have no phase relationship and consequently a voltage probe acts as a dephasor [12].

The voltage probe changes the conductor: if we include the gate we now deal with a four probe conductor. We keep for the gate the label 3 and designate the voltage probe as contact 4. Since the potential is a function of time, we must also know the dynamic conductance of the system. To begin we consider the general relation between currents and voltages of our four-

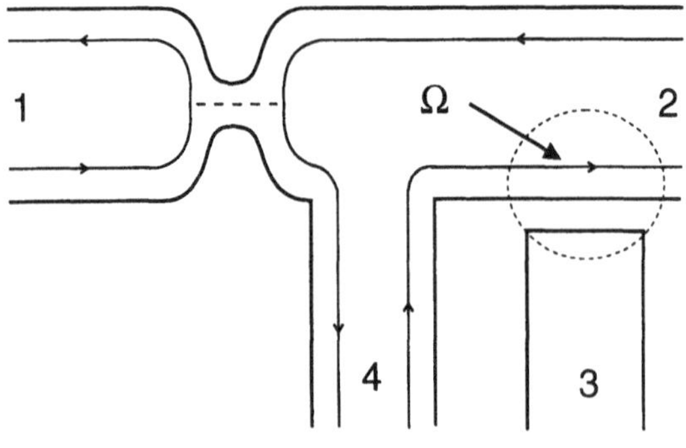

Fig. 2. Hall bar with a quantum point contact and a gate overlapping the edge of the conductor. A voltage contact between the QPC and the gate serves to destroy quantum coherent electron motion along the edge states.

terminal conductor. This relation takes the form of a Langevin equation which includes the fluctuating currents at the terminals as noise sources [1]

$$dI_\alpha(\omega) = \sum_\beta G_{\alpha\beta}(\omega) dV_\beta(\omega) + \delta I_\alpha(\omega). \tag{32}$$

Here $G_{\alpha\beta}(\omega)$ is the self-consistent dynamic conductance and $\delta I_\alpha(\omega)$ are the (self-consistent) frequency-dependent current fluctuations at the contacts of the conductor. Since the current spectrum at the gate is second order in frequency, it is sufficient to calculate the current amplitudes to first order in frequency. We thus need $G_{\alpha\beta}(\omega)$ only to first order in frequency and write $G_{\alpha\beta}(\omega) = G_{\alpha\beta}(0) - i\omega E_{\alpha\beta} + O(\omega^2)$. Here $G_{\alpha\beta}(0)$ is the dc-conductance which for $\nu - 1$ perfectly transmitted channels and one partially transmitted channel at the QPC is given by $G_{11} = -G_{12} = -G_{41} = (e^2/h)(\nu - 1 + T)$, $G_{22} = G_{44} = -G_{24} = -(e^2/h)\nu$ and $G_{42} = -(e^2/h)R$. All other elements vanish. Repeating the calculation which led to Eq. (16) for the conductor of Fig. (2), we find $E_{23} = -E_{24} = E_{33} = -E_{34} = -i\omega C_\mu$ with C_μ as given in Eq. (16). Inserting these results into Eq. (32) and holding all potentials, except dV_4 at their equilibrium value gives for I_3 and I_4,

$$\begin{aligned} I_3 &= -i\omega C_\mu dV_4 \\ I_4 &= \frac{e^2}{h}\nu dV_4 + \delta I_4 \end{aligned} \tag{33}$$

The noise spectrum at the voltage probe at low frequencies is just the spectrum of the noise of a QPC $S^0_{I_4I_4}(\omega) = 2\frac{e^2}{h}TRe|V|$ where we have added an upper index 0 to indicate that it is the spectrum for zero external impedance. Note that there is no noise source to order ω in the total current for I_3. (The lowest order in frequency which is dissipative is proportional to ω^2). For an ideal (infinite impedance) voltage probe we have $I_4 = 0$ and consequently

$$dV_4(\omega) = -\frac{h}{e^2\nu}\delta I_4(\omega) \tag{34}$$

Inserting this result in the equation for I_3 we find $S_{I_3I_3}(\omega) = \omega^2 C_\mu^2 S^0_{I_4I_4}(\omega)/\nu^2$. Using the shot noise power spectrum for $S^0_{I_4I_4}(\omega)$ gives for the spectrum at the gate $S_{I_3I_3}(\omega) = 2\omega^2 C_\mu^2 R_v e|V|$ with [15]

$$R_v = \frac{h}{e^2}\frac{1}{\nu^2}TR \tag{35}$$

Eq. (35) makes now an interesting prediction. For one edge state only, the dephasing voltage probe has no effect. The fluctuations observed at the gate remain unchanged. If there are several edge states, the voltage probe does have an effect since the voltage probe re-injects an equal current into all edge states. The difference between Eq. (35) and Eq.(30) is, however quite subtle. R_v as given above is simply inversely proportional to the square of the number of edge states. Without the voltage probe we have seen that R_v varies between $\frac{1}{e^2}TR$ for a steep edge and $\frac{1}{4e^2}TR$ for a shallow edge. Thus for a steep edge introducing a voltage probe has a considerable effect, whereas for a shallow edge introducing a voltage probe has no effect at all.

Apparently, in the experiment [5] the voltage probe is not ideal. Instead of an infinite impedance it might, at the relevant frequency, exhibit a finite impedance $Z_{ext}(\omega)$. We assume that the external impedance arises from a macroscopic circuit and its noise is voltage independent. In the presence of a finite impedance the current Eq. (33) can also be expressed as $I_4 = -Z_{ext}^{-1}(\omega)\delta V_4$. Consequently, instead of Eq. (34) we find

$$\delta V_4 = -\frac{Z_{ext}}{1 + G_0 Z_{ext}}\delta I_4 \tag{36}$$

where we have introduced the abbreviation $G_0 = \nu e^2/h$. Repeating the considerations given above, we find for the resistance R_v

$$R_v = \frac{e^2}{h}\frac{|Z_{ext}|^2}{|1 + G_0 Z_{ext}|^2}TR \tag{37}$$

This consideration shows that a finite external impedance reduces the current fluctuations induced into the gate. Clearly this is simply a consequence of the fact that for a finite external impedance part of the current is "lost" at the voltage probe. This effect becomes significant when $Z_{ext}(\omega)$ at the frequency of interest becomes smaller than G_0^{-1}.

11 Discussion

In this work, we have illustrated the calculation of charge and potential fluctuations for a simple problem: A Hall conductor with a QPC has on its side a gate which couples capacitively to the edge states. We have asked: What is the current induced into this gate due to the shot noise generated a the QPC. The simplifying assumption we have made is that the conductor remains charge neutral everywhere except near the gate where a charge pile-up limited by the Coulomb interaction between gate and edge is permitted. This allows a solution in terms of one fluctuating potential only.

Independent of the detailed discussion it is clear that the non-equilibrium resistance R_v reflects the shot noise. The theoretical question concerns only the factor of proportionality. If we measure R_v in units of $R_0 = h/2e^2\mathcal{TR}$, we find that for one edge state R_v/R_0 is universal, wheras in the presence of a number of edge states it is not-universal, except if an ideal voltage probe completely equilibrates different channels, in which case we find $R_v/R_0 = 1/\nu^2$, where ν is the number of edge states. In Ref. [5] it is argued that the dephasing rate (which is proportional to R_v) should be *periodic* in a phase with period π even for a single edge state. In contrast, in our our result [15], Eq. (29), such a periodic factor does not appear. We conclude by mentioning that the approach out-lined here can be generalized to hybrid normal-superconducting systems [20].

12 Acknowledgments

This work was supported by the Swiss National Science Foundation and by the TMR network Dynamics of Nanostructures.

References

1. M. Büttiker, Phys. Rev. B**46**, 12485 (1992).
2. M. Field, C. G. Smith, M. Pepper, D. A. Ritchie, J. E. F. Frost, G. A. C. Jones and D. G. Hasko, Phys. Rev. Lett. **70**, 1311 (1993).
3. M. Kataoka, C. J. B. Ford, G. Faini, D. Mailly, M. Y. Simmons, D. R. Mace, C.-T. Liang and D. A. Ritchie, Phys. Rev. Lett. **83**, 160 (1999).
4. E. Buks, R. Schuster, M. Heiblum, D. Mahalu and V. Umansky, Nature **391**, 871 (1998).
5. D. Sprinzak, E. Buks, M. Heiblum, H. Shtrikman, (unpublished). cond-mat/9907162
6. M. Büttiker, H. Thomas, and A. Prêtre, Phys. Lett. A **180**, 364 (1993).
7. M. Büttiker, J. Math. Phys., **37**, 4793 (1996).
8. T. Christen and M. Büttiker, Phys. Rev. Lett. **77**, 143 (1996): M. Büttiker and T. Christen, in *Mesoscopic Electron Transport*, NATO ASI, Series E: Applied Science, edited by L. L. Sohn, L. P. Kouwenhoven and G. Schoen, (Kluwer Academic Publishers, Dordrecht, 1997) Vol. 345. p. 259. cond-mat/9610025.
9. V. A. Khlus, Sov. Phys. JETP **66**, 1243 (1987); G. B. Lesovik, JETP Lett. **49**, 592 (1989).
10. Ya. M. Blanter, F.W.J. Hekking, and M. Büttiker, Phys. Rev. Lett. **81**, 1925 (1998).
11. M. H. Pedersen, S. A. van Langen and M. Büttiker, Phys. Rev. B **57**, 1838 (1998).
12. M. Büttiker, Phys. Rev. B**38**, 9375, (1988).
13. W. Chen, T. P. Smith, M. Büttiker, and M. Shayegan, Phys. Rev. Lett. **73**, 146 (1994).
14. M. Büttiker, J. Phys. Condensed Matter **5**, 9361 (1993).
15. M. Büttiker and A. M. Martin, (unpublished). cond-mat/9902320.
16. Y. B. Levinson, (unpublished). cond-mat/9907055; Europhys. Lett. **39**, 299 (1997).
17. Y. B. Levinson, Sov. Phys. JETP **68**, 1257 (1989).
18. B. L. Altshuler, A. G. Aronov and D. Khmelnitskii, J. Phys. C**15**, 7367 (1982).
19. J. S. Moon, J. A. Simmons, J. L. Reno, B. L. Johnson, Phys. Rev. Lett. **79**, 4457 (1997).
20. A. M. Martin, T. Gramespacher, M. Buttiker, (unpublished). cond-mat/9907240

Shot-Noise in Non-Degenerate Semiconductors with Energy-Dependent Elastic Scattering

H. Schomerus[1], E.G. Mishchenko[1,2], and C.W.J. Beenakker[1]

[1] Instituut-Lorentz, Universiteit Leiden, P. O. Box 9506, 2300 RA Leiden, The Netherlands
[2] L. D. Landau Institute for Theoretical Physics, Kosygin 2, Moscow 117334, Russia

Abstract. We investigate current fluctuations in non-degenerate semiconductors, on length scales intermediate between the elastic and inelastic mean free paths. The shot-noise power P is suppressed below the Poisson value $P_{\text{Poisson}} = 2e\bar{I}$ (at mean current $\bar{I}$) by the Coulomb repulsion of the carriers. We consider a power-law dependence of the elastic scattering time $\tau \propto \varepsilon^{\alpha}$ on kinetic energy ε and present an exact solution of the non-linear kinetic equations in the regime of space-charge limited conduction. The ratio P/P_{Poisson} decreases from 0.38 to 0 in the range $-\frac{1}{2} < \alpha < 1$.

1 Introduction

The noise power P of current fluctuations in an electron gas in thermal equilibrium (at temperature T) is related by the Johnson-Nyquist formula $P = 4kTG$ (with k Boltzmann's constant) to the linear-response conductance $G = \lim_{V\to 0} d\bar{I}/dV$ (with $\bar{I}$ the mean current in response to an applied voltage V). This formula can be generalized to a large applied voltage, $P = 4kT(V/\bar{I})(d\bar{I}/dV)^2$, provided the electron gas remains in local equilibrium with the lattice. Local equilibrium requires inelastic scattering. When the conductor is shorter than the inelastic mean free path l_{in} and the potential drop V is large enough, the Johnson-Nyquist formula no longer applies and a measurement of current noise (then also called shot noise) reveals more detailed information about the transport of charge carriers—in particular about their correlations. The maximal noise level $P_{\text{Poisson}} = 2e\bar{I}$ is attained in absence of all correlations (both in the injection process as well as in the subsequent transport). Examples are vacuum diodes at large bias in absence of space-charge effects and tunneling diodes with low transmissivity.

Here we consider the transport through a disordered semiconductor of length L terminated by two metal contacts, under the conditions of elastic scattering ($l \ll L \ll l_{\text{in}}$, with l the elastic mean free path). In a degenerate conductor correlations are induced by the Pauli exclusion principle (for a review of the theory of shot noise in this situation see Ref. [1]) and the shot noise has the universal value $P = \frac{1}{3}P_{\text{Poisson}}$ [2], [3].

At low carrier concentration the electron gas is non-degenerate, and the Pauli principle is ineffective. Because carriers can now accumulate, giving rise to space-charge effects, they become correlated through Coulomb repulsion. This is the situation which we want to study presently. In a recent Monte-Carlo simulation [4] a shot-noise suppression factor of about $P/P_{\text{Poisson}} = 1/3$ was found in the regime of space-charge limited transport; an energy-independent elastic scattering rate was assumed. The coincidence with the noise level obtained in the degenerate situation attracted a lot of attention [5]. The degree of universality is less pronounced here since the number actually depends on the geometry and dimensionality—as well as the scattering mechanism [6], [7], [8].

In Ref. [6] the problem was investigated for an energy-independent elastic scattering time τ, using the kinetic theory of non-equilibrium fluctuations (reviewed in Ref. [9]). The non-linear kinetic equations were solved in a certain approximation (the drift approximation), with the result $P/P_{\text{Poisson}} = 0.3410$. In Ref. [7] we obtained an exact solution, giving $P/P_{\text{Poisson}} = 0.3097$, and also considered a power-law dependence $\tau \sim \varepsilon^{\alpha}$ on the kinetic energy ε. For $\alpha = -\frac{1}{2}$ (corresponding to short-range impurity scattering or quasi-elastic acoustic phonon scattering [10]) we found the exact result $P/P_{\text{Poisson}} = 0.3777$. For other values of α we only presented results within the drift approximation. In this work we derive the exact solution in the range $-\frac{1}{2} < \alpha < 1$. As we will discuss, α should be in this range for space-charge limited conduction to be realized.

2 The Drift-Diffusion Equation

We consider a three-dimensional conductor of length L and cross-sectional area A terminated by two contacts. The equilibrium density ρ_{eq} of charge carriers (charge e, effective mass m) in the decoupled conductor is assumed to be much lower than the density ρ_c of those carriers that are energetically allowed (at a given voltage V) to enter the conductor from the contacts. (A possible realization would be an intrinsic or barely doped semiconductor between two metal contacts or two heavily doped semiconducting regions.) The dielectric constant of the conductor is κ. The temperature T is assumed to be so high that the electron gas is degenerate, and a large voltage drop $V \gg kT/e$ is maintained between the contacts. Transport is assumed to be diffusive and elastic, $l < L < l_{\text{in}}$. We assume a power-law energy dependence

$$\tau(\varepsilon) = \tau_0 \varepsilon^{\alpha} \tag{1}$$

of the elastic scattering time on the kinetic energy ε. We want to calculate the zero-frequency component

$$P = 2 \int_{-\infty}^{\infty} \mathrm{d}t' \, \overline{\delta I(t) \delta I(t+t')} \tag{2}$$

of the noise power of the fluctuations $\delta I(t)$ of the electric current $I(t) = \bar{I} + \delta I(t)$ around its mean $\bar{I}$.

We use Cartesian coordinates x, y, z with x parallel to the conductor (the current source is at $x = 0$, the drain at $x = L$). To linear order in the fluctuations, the transverse coordinates can be ignored. In the zero-frequency limit the current is independent on x because of the continuity equation and is given by the drift-diffusion equation [6], [7]

$$I(t) = -\frac{\partial}{\partial x} \int d\varepsilon\, D(\varepsilon)\rho(x,\varepsilon,t) + E(x,t) \int d\varepsilon\, \mathcal{F}(x,\varepsilon,t)\frac{d\sigma(\varepsilon)}{d\varepsilon} + \delta J(x,t). \quad (3)$$

The electric field $E(x,t)$ is related to the laterally integrated charge density $\rho(x,t)$ by the Poisson equation

$$\kappa \frac{\partial}{\partial x} E(x,t) = \frac{1}{A}\rho(x,t), \quad (4)$$

where we omitted the low background charge density $-\rho_{\mathrm{eq}}$. The fluctuating source $\delta J(x,\varepsilon,t)$ accounts for the stochasticity of individual scattering events and has the correlator

$$\overline{\delta J(x,t)\delta J(x',t')} = 2A\delta(t-t')\delta(x-x') \int d\varepsilon\, \sigma(\varepsilon)\bar{\mathcal{F}}(x,\varepsilon). \quad (5)$$

Here and in Eq. (3), $\mathcal{F}(x,\varepsilon,t) = \rho(x,\varepsilon,t)/e\nu(\varepsilon)$ with the density of states $\nu(\varepsilon) = 4\pi m(2m\varepsilon)^{1/2} = \nu_0\varepsilon^{1/2}$ (we set Planck's constant $h \equiv 1$). The conductivity $\sigma(\varepsilon) = e^2\nu(\varepsilon)D(\varepsilon) = \sigma_0\varepsilon^{\alpha+3/2}$ is the product of the density of states and the diffusion constant $D(\varepsilon) = v^2\tau/3 = D_0\varepsilon^{\alpha+1}$.

3 Space-Charge Limited Conduction

For a large voltage drop V between the two metal contacts and a high carrier density ρ_c in the contacts, the charge injected into the semiconductor is much higher than the equilibrium charge ρ_{eq}, which can then be neglected. For sufficiently high V and ρ_c the system enters the regime of space-charge limited conduction [11], defined by the boundary condition

$$E(x,t) = 0 \quad \text{at} \quad x = 0. \quad (6)$$

Eq. (6) states that the space charge $Q = \int_0^L \rho(x)\, dx$ in the semiconductor is precisely balanced by the surface charge at the current drain. At the drain we have the absorbing boundary condition

$$\rho(x,t) = 0 \quad \text{at} \quad x = L. \quad (7)$$

With this boundary condition we again neglect ρ_{eq}.

To determine the electric field inside the semiconductor we proceed as follows. Since scattering is elastic, the total energy $u = \varepsilon - e\phi(x,t)$ of each carrier is preserved. The potential gain $-e\phi(x,t)$ (with $E = -\partial\phi/\partial x$) dominates over the initial thermal excitation energy of order kT almost throughout the whole semiconductor; only close to the current source (in a thin boundary layer) this is not the case. We can therefore approximate the kinetic energy $\varepsilon \approx -e\phi$ and introduce this into $D(\varepsilon)$ and $d\sigma/d\varepsilon$. Substituting into Eq. (3) one obtains

$$\mathcal{F}(x,t) \approx e \int_x^L dx' \, \frac{I(t) - \delta J(x,t)}{\sigma_0[-e\phi(x',t)]^{\alpha+3/2}}, \tag{8}$$

$$\rho(x,t) \approx \frac{[-e\phi(x,t)]^{1/2}}{D_0} \int_x^L dx' \, \frac{I(t) - \delta J(x,t)}{[-e\phi(x',t)]^{\alpha+3/2}}, \tag{9}$$

where the absorbing boundary conditions have been used. From the Poisson equation (4) we find the third-order, non-linear, inhomogeneous differential equation

$$2(-\phi)^{\alpha}\phi'\phi'' + 4(-\phi)^{\alpha+1}\phi''' = B\bar{I}[1 + \delta i(x,t)], \tag{10}$$

$$\delta i(x,t) = \frac{I(t) - \delta J(x,t)}{\bar{I}}, \tag{11}$$

for the potential profile $\phi(x,t)$. Primes denote differentiation with respect to x, and $B = 6/e^{\alpha}\mu_0\kappa A$ with $\mu_0 = e\tau_0/m$.

Since the potential difference V between source and drain does not fluctuate, we have the two boundary conditions $\phi(0,t) = 0$, $\phi(L,t) = -V$. Eqs. (6) and (7) imply two additional boundary conditions, $\phi'(0,t) = 0$, $\phi''(L,t) = 0$.

The differential equation (10) and the accompanying boundary conditions possess two remarkable scaling properties: The product $B\bar{I}$ of material parameters and mean current $\bar{I}$ and the length L can be eliminated by introduction of the scaled potential

$$\chi(x,t) = -\left(L^3 B\bar{I}\right)^{-1/(\alpha+2)} \phi(xL,t). \tag{12}$$

The rescaled differential equation reads

$$2\chi^{\alpha}\chi'\chi'' - 4\chi^{\alpha+1}\chi''' = 1 + \delta i, \tag{13}$$

which has to be solved with the boundary conditions $\chi(0,t) = 0$, $\chi(1,t) = \left(L^3 B\bar{I}\right)^{-1/(\alpha+2)} V$, $\chi'(0,t) = 0$, $\chi''(1,t) = 0$. The scaling properties entail that the shot-noise suppression factor depends only on the exponent α, but no longer on the parameters L, A, V, τ_0, and κ.

We will solve this boundary value problem for $\chi = \bar{\chi} + \delta\chi$, first for the mean (Section 4) and then for the fluctuations (Section 5), in both cases neglecting terms quadratic in $\delta\chi$.

4 Average Profiles and the Current-Voltage Characteristic

The averaged equation (13) for the rescaled mean potential $\bar{\chi}(x)$ reads

$$2\bar{\chi}^{\alpha}\bar{\chi}'\bar{\chi}'' - 4\bar{\chi}^{\alpha+1}\bar{\chi}''' = 1. \tag{14}$$

We seek a solution which fulfills the three boundary conditions $\bar{\chi}(0) = 0$, $\bar{\chi}'(0) = 0$, $\bar{\chi}''(1) = 0$. The value of $\bar{\chi}$ at the current drain determines the current-voltage characteristic

$$\bar{I}(V) = \frac{1}{L^3 B}\left(\frac{V}{\bar{\chi}(1)}\right)^{\alpha+2}. \tag{15}$$

We now construct $\bar{\chi}(x)$. The function $\bar{\chi}_0(x) = a_0 x^{\beta}$ with $\beta = 3/(2+\alpha)$ and $a_0 = [2\beta(\beta-1)(4-\beta)]^{-\beta/3}$ solves the differential equation and satisfies the boundary conditions at $x = 0$, but $\bar{\chi}_0''(x) \neq 0$ for any finite x. We substitute into Eq. (14) the ansatz $\bar{\chi}(x) = \sum_{l=0}^{\infty} a_l x^{\gamma l+\beta}$, consisting of $\bar{\chi}_0(x)$ times a power series in x^{γ}, with γ a positive power to be determined. This ansatz proves fruitful since both terms on the left-hand side of Eq. (14) give the same powers of x, starting with order x^0 in coincidence with the right-hand side. By power matching one obtains in first order the value for a_0 given above. The second order leaves a_1 as a free coefficient, but fixes the power $\gamma = (8 - 5\beta + \sqrt{-32 + 40\beta + \beta^2})/4$. The coefficients a_l for $l \geq 2$ are then given recursively as a function of a_1, which is finally determined from the condition $\bar{\chi}''(1) = 0$.

In Fig. 1 the profiles of the potential $\bar{\phi} \propto \bar{\chi}$, the electric field $\bar{E} \propto \bar{\chi}'$, and the charge density $\bar{\rho} \propto \bar{\chi}''$ are plotted for various values of α. The coefficient $\bar{\chi}(1)$ appearing in the current-voltage characteristic (15) can be read off from this plot. The behavior at the current source changes qualitatively at $\alpha = -\frac{1}{2}$ (see Section 7).

5 Fluctuations

The rescaled fluctuations $\delta\chi(x,t) = \psi(x,t)$ fulfill the linear differential equation

$$\begin{aligned}\mathcal{L}[\psi] = &-4\bar{\chi}^{\alpha+1}\psi''' + 2\bar{\chi}^{\alpha}\bar{\chi}'\psi'' + 2\bar{\chi}^{\alpha}\bar{\chi}''\psi' \\ &+ \left[2\alpha\bar{\chi}^{\alpha-1}\bar{\chi}'\bar{\chi}'' - 4(\alpha+1)\bar{\chi}^{\alpha}\bar{\chi}'''\right]\psi = \delta i(t).\end{aligned} \tag{16}$$

The solution of the inhomogeneous equation is found with help of the three independent solutions of the homogeneous equation $\mathcal{L}[\psi] = 0$, $\psi_1(x) = \bar{\chi}'(x)$, $\psi_2(x) = \bar{\chi}(x) - (x/\beta)\bar{\chi}'(x)$, and

$$\psi_3(x) = \psi_1(x)\int_x^1 dx'\, \frac{\bar{\chi}^{1/2}(x')\psi_2(x')}{\mathcal{W}^2(x')} - \psi_2(x)\int_x^1 dx'\, \frac{\bar{\chi}^{1/2}(x')\psi_1(x')}{\mathcal{W}^2(x')}, \tag{17}$$

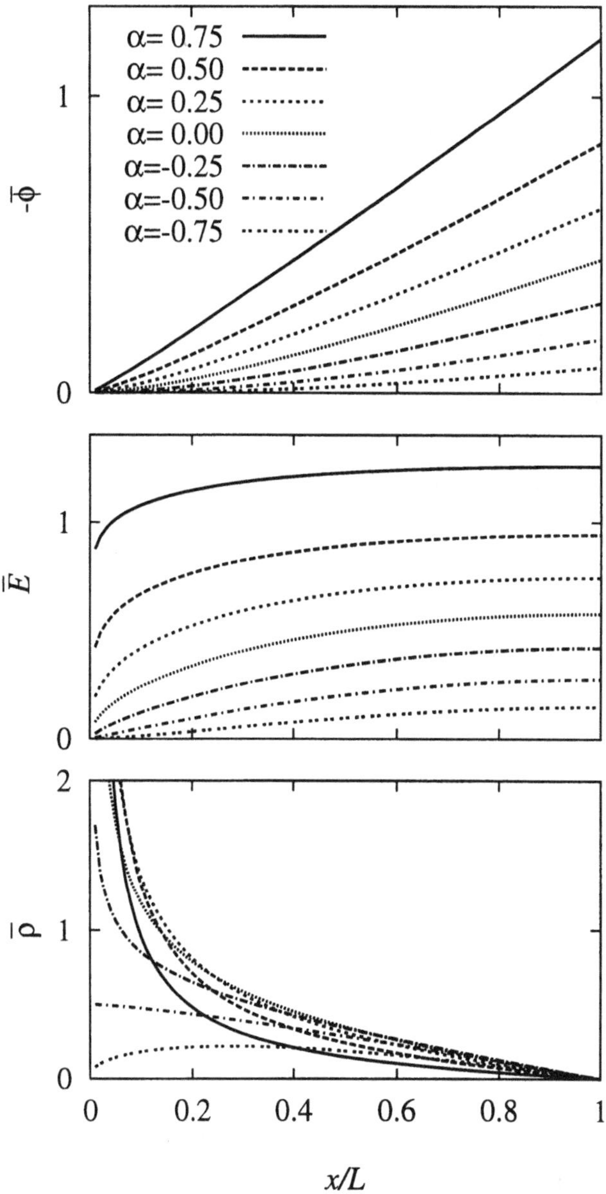

Fig. 1. Profile of the mean electrical potential $\bar{\phi}$ [in units of $(L^3 B\bar{I})^{1/(\alpha+2)}$, with $B = 6m/e^{\alpha+1}\tau_0\kappa A$], the electric field $\bar{E}$ [in units of $(L^3 B\bar{I})^{1/(\alpha+2)}/L$], and the charge density $\bar{\rho}$ [in units of $\kappa(L^3 B\bar{I})^{1/(\alpha+2)}/L^2$], following from Eq. (14) for different values of α.

where we have defined $\mathcal{W}(x) = \psi_1(x)\psi_2'(x) - \psi_1'(x)\psi_2(x)$. The special solution which fulfills $\psi(0,t) = \psi'(0,t) = \psi(1,t) = 0$ is

$$\psi(x,t) = \int_0^1 \mathrm{d}x' \, \frac{\bar{\chi}^{1/2}(x')}{\mathcal{W}^2(x')} \Big[\Theta(x-x')\psi_1(x)\psi_2(x') + \Theta(x'-x)\psi_1(x')\psi_2(x) - \frac{\psi_1(1)}{\psi_2(1)}\psi_2(x)\psi_2(x')\Big] \int_0^{x'} \mathrm{d}x'' \, \frac{\delta I(t) - \delta J(x'',t)}{4\bar{I}} \, \frac{\mathcal{W}(x'')}{\bar{\chi}^{\alpha+3/2}(x'')} . \quad (18)$$

The condition $\psi''(1,t) = 0$ relates the fluctuating current δI to the Langevin current δJ. The resulting expression is of the form

$$\delta I(t) = \mathcal{C}^{-1} \int_0^L \mathrm{d}x \, \delta J(x,t) \mathcal{G}(x), \quad (19)$$

with the definitions $\mathcal{C} = \int_0^1 \mathrm{d}x \, \mathcal{G}(x)$,

$$\mathcal{G}(x) = \frac{\mathcal{W}(x)}{\bar{\chi}^{\alpha+3/2}(x)} \left(1 + \frac{(1-1/\beta)\bar{\chi}'^2(1)}{4\bar{\chi}^{\alpha+1/2}(1)\psi_2(1)} \int_x^1 \mathrm{d}x' \, \frac{\bar{\chi}^{1/2}(x')\psi_2(x')}{\mathcal{W}^2(x')}\right) . \quad (20)$$

The shot-noise power is found by substituting Eq. (19) into Eq. (2) and invoking the correlator (5) for the Langevin current. This results in

$$P = 2\int_0^L \mathrm{d}x \, \left(\frac{\mathcal{G}(x)}{\mathcal{C}}\right)^2 \mathcal{H}(x) \quad (21)$$

with $\mathcal{H}(x) = 2A \int \mathrm{d}\varepsilon \, \sigma(\varepsilon)\bar{\mathcal{F}}(x,\varepsilon) \approx 2\sigma_0[-e\bar{\phi}(x)]^{\alpha+3/2}\bar{\mathcal{F}}(x)$. Eq. (8) gives

$$\mathcal{H}(x) = 2e\bar{I}\bar{\chi}^{\alpha+3/2}(x) \int_x^1 dx' \, \frac{1}{\bar{\chi}^{\alpha+3/2}(x')} = 4P_{\mathrm{Poisson}}\bar{\chi}^{\alpha+1}(x)\bar{\chi}''(x), \quad (22)$$

where we integrated with help of Eq. (14) and used $\bar{\chi}''(1) = 0$.

In Fig. 2 we plot the ratio P/P_{Poisson} as a function of the parameter α (solid curve). The shot-noise suppression factor $P/P_{\mathrm{Poisson}} = 0.3777$ for $\alpha = -\frac{1}{2}$ and goes to zero as $\alpha \to 1$.

6 Drift Approximation

A simple formula for the shot-noise suppression factor can be found when one neglects the diffusion term in Eq. (3) and considers instead of Eq. (13) the corresponding differential equation $(4\alpha+6)\chi^\alpha\chi'\chi'' = 1 + \delta i$. The is the drift approximation of Ref. [6]. The order of the differential equation is reduced by one, so that we also have to drop one of the boundary conditions. The absorbing boundary condition $\chi''(1,t) = 0$ is the most reasonable candidate, because even for the resulting mean profile $\bar{\chi}(x) = b_0 x^\beta$ with $\beta = 3(2+\alpha)^{-1}$ and $b_0 = [\beta^2(\beta-1)]^{-\beta/3}$ most carriers remain concentrated close to the current source. The differential equation for the fluctuations $\alpha\psi/\bar{\chi} + \psi'/\bar{\chi}' +$

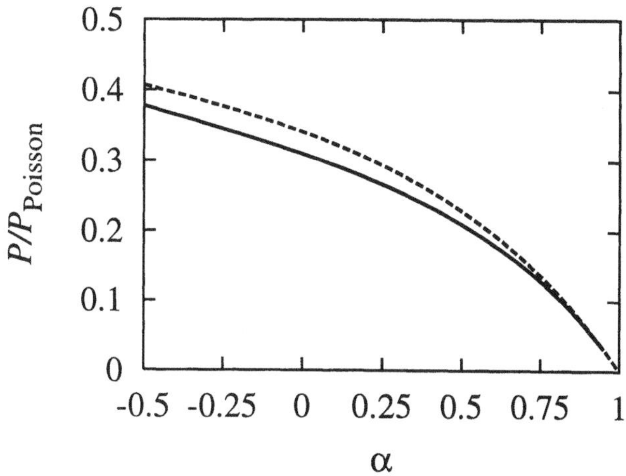

Fig. 2. Shot-noise power P as a function of α. The exact result (solid curve) is compared with the approximate result (24) (dashed curve).

$\psi''/\bar{\chi}'' = \delta i$ can be solved with help of the homogeneous solutions $\psi_1(x) = x^{\beta-1}$ and $\psi_2(x) = x^{3-2\beta}$. The inhomogeneous solution that fulfills $\psi(0,t) = 0$, $\psi'(0,t) = 0$ is

$$\psi(x,t) = b_0 \frac{\beta(\beta-1)}{4-3\beta} \int_0^x dx' \left[x^{3-2\beta} x'^{3\beta-4} - x^{\beta-1} \right] \delta i(x',t). \qquad (23)$$

We demand that the voltage does not fluctuate, $\psi(1,t) = 0$, and obtain Eq. (19) with now $\mathcal{G}(x) = 1 - x^{3\beta-4}$. The shot noise power is finally found from Eq. (21) with $\mathcal{H}(x) = P_{\text{Poisson}} x^{3-\beta/2} \int_x^1 dx' \, x'^{\beta/2-3}$,

$$P/P_{\text{Poisson}} = \frac{6(\alpha-1)(\alpha+2)(16\alpha^2+36\alpha-157)}{5(2\alpha-5)(8\alpha-17)(13+8\alpha)}. \qquad (24)$$

This is the dashed curve in Fig. 2.

7 Discussion

The shot-noise suppression factor P/P_{Poisson} varies from 0.38 to 0 in the range $-\frac{1}{2} < \alpha < 1$, which includes the case of an energy-independent elastic scattering rate ($\alpha = 0$, $P/P_{\text{Poisson}} = 0.3097$) and the case of short-range scattering by uncharged impurities or quasi-elastic scattering by acoustic phonons ($\alpha = -\frac{1}{2}$, $P/P_{\text{Poisson}} = 0.3777$). The results in the drift approximation (24) are about 10% larger. Our values are somewhat smaller than those following from the numerical simulations of González *et al.*, who found $P/P_{\text{Poisson}} = \frac{1}{3}$ for $\alpha = 0$ [4] and $P/P_{\text{Poisson}} = 0.42 - 0.44$ for $\alpha = -\frac{1}{2}$ [12].

Our considerations require the exponent α to be in the range $-\frac{1}{2} < \alpha < 1$. For $\alpha < -\frac{1}{2}$ the mean free path $l \propto \varepsilon^{\alpha+1/2}$ diverges at small kinetic energies. The carriers at the current source therefore enter the conductor ballistically and accumulate only at a finite distance from the injection point. Fig. 1 indicates that the charge density at the current source must be zero if one insists that the electric field vanishes. Nagaev [8] has shown that full shot noise, $P = P_{\text{Poisson}}$, follows for $\alpha = -\frac{3}{2}$. Presumably, P/P_{Poisson} will decrease monotonically from 1 for $\alpha = -\frac{3}{2}$ to 0.38 for $\alpha = -\frac{1}{2}$, but we have no theory for this range of α's. For $\alpha > 1$ the resistance R becomes infinitely large, because the coefficient $\bar{\chi}(1)$ in the current-voltage characteristic (15) diverges. An intuitive understanding can be obtained by equating the potential gain $\phi \sim (Dt)^{3/(2\alpha+4)}$ (acquired by diffusing close to the current source for a time t) with the increase in kinetic energy ε: For $\alpha > 1$ this time $t \propto \varepsilon^{(1-\alpha)/3}$ is seen to diverge for small ε. We found that the shot-noise power vanishes as $\alpha \to 1$. Presumably, a non-zero answer for P would follow for $\alpha > 1$ if the non-zero thermal energy and finite charge density at the current source is accounted for. This remains an open problem.

Discussions with O. M. Bulashenko, T. González, J. M. J. van Leeuwen, and W. van Saarloos are gratefully acknowledged. This work was supported by the European Community (Program for the Training and Mobility of Researchers) and by the Dutch Science Foundation NWO/FOM.

References

[1] M. J. M. de Jong and C. W. J. Beenakker, in: *Mesoscopic Electron Transport*, edited by L. L. Sohn, L. P. Kouwenhoven, and G. Schön, NATO ASI Series E345 (Kluwer, Dordrecht, 1997).
[2] C. W. J. Beenakker and M. Büttiker, Phys. Rev. B **46**, 1889 (1992).
[3] K. E. Nagaev, Phys. Lett. A **169**, 103 (1992).
[4] T. González, C. González, J. Mateos, D. Pardo, L. Reggiani, O. M. Bulashenko, and J. M. Rubí, Phys. Rev. Lett. **80**, 2901 (1998); T. González, J. Mateos, D. Pardo, O. M. Bulashenko, and L. Reggiani, Phys. Rev. B in press (cond-mat/9811069).
[5] R. Landauer, Nature **392**, 658 (1998).
[6] C. W. J. Beenakker, Phys. Rev. Lett. **82**, 2761 (1999).
[7] H. Schomerus, E. G. Mishchenko, and C. W. J. Beenakker, Phys. Rev. B in press (cond-mat/9901346).
[8] K. E. Nagaev, preprint (cond-mat/9812357).
[9] Sh. Kogan, *Electronic Noise and Fluctuations in Solids* (Cambridge University, Cambridge, 1996).
[10] S. V Gantsevich, V. L. Gurevich, and R. Katilius, Rivista Nuovo Cimento **2** (5), 1 (1979).
[11] M. A. Lampert and P. Mark, *Current Injection in Solids* (Academic, New York, 1970).
[12] T. González, C. González, J. Mateos, D. Pardo, L. Reggiani, O. M. Bulashenko, and J. M. Rubí, private communication.

Transport and Noise of Entangled Electrons

Eugene V. Sukhorukov, Daniel Loss, and Guido Burkard

Department of Physics and Astronomy, University of Basel,
Klingelbergstrasse 82, CH-4056 Basel, Switzerland

Abstract. We consider a scattering set-up with an entangler and beam splitter where the current noise exhibits bunching behavior for electronic singlet states and antibunching behavior for triplet states. We show that the entanglement of two electrons in the double-dot can be detected in mesoscopic transport measurements. In the cotunneling regime the singlet and triplet states lead to phase-coherent current contributions of opposite signs and to Aharonov-Bohm and Berry phase oscillations in response to magnetic fields. We analyze the Fermi liquid effects in the transport of entangled electrons.

1 Introduction

The availability of pairwise entangled qubits – Einstein-Podolsky-Rosen (EPR) pairs [1] – is a necessary prerequisite in quantum communication [2]. The prime example of an EPR pair considered here is the singlet/triplet state formed by two electron spins [3], [4]. Its main feature is its non-locality: If we separate the two electrons from each other in real space, their total spin state can still remain entangled. Such non-locality gives rise to striking phenomena such as violations of Bell inequalities and quantum teleportation and has been investigated for photons [5], [6], but not yet for *massive* particles such as electrons, let alone in a solid state environment. In this work we discuss specific properties of transport and noise of entangled electrons as a result of two-particle coherence and nonlocality.

In Sect. (2) we propose and analyze an experimental set-up (see Fig. 1a) by which the entanglement of mobile electrons can be detected in noise measurements with a beam splitter [7]. The entangler is assumed to be a device by which we can generate entangled electron states, a specific realization being the double-dot system [3]. The presence of a beam splitter ensures that the electrons leaving the entangler have a finite amplitude to be interchanged. Thus we can expect that the current-current correlations (noise) measured in leads 3 and/or 4 are sensitive to the symmetry of the *orbital part* of the wave function [8]. Since the spin singlet of two electrons is uniquely associated with a symmetric orbital wave-function, and the triplet with an antisymmetric one we have a means to distinguish singlets from triplets through a bunching or antibunching signature. It is well-known [10] that bosons (fermions) show bunching (antibunching) behavior [11]. Antibunching is so far considered for electrons in the normal state both in theory [12], [13] and in experiments

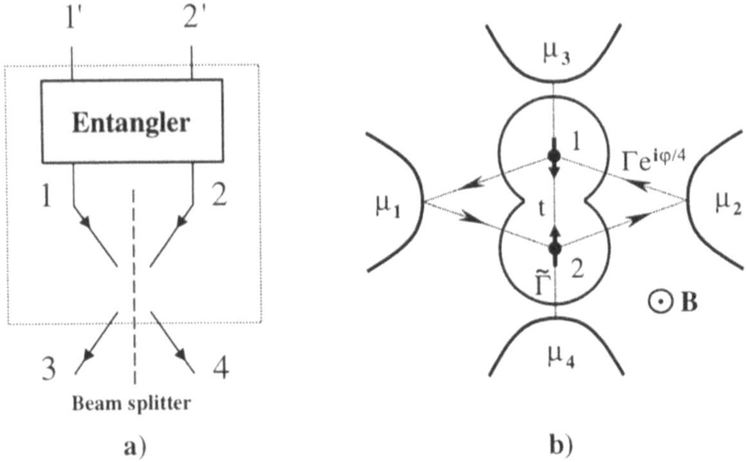

Fig. 1. a) The setup for measuring noise correlations of entangled states. Uncorrelated electrons are fed into the entangler through the Fermi leads $1'$ and $2'$. The entangler is a device that produces pairs of entangled electrons and injects one of the electrons into lead 1 and the other into lead 2. The entanglement can then be detected by performing an interference experiment using a beam splitter. b) Double-dot (DD) system containing two electrons and being weakly coupled to metallic leads 1,...,4, each of which being at the chemical potential $\mu_1,\ldots,\mu_4$. The tunneling amplitudes between dots and leads are denoted by Γ, $\tilde{\Gamma}$. The tunneling (t) between the dots results in a singlet (triplet) ground state. The closed tunneling path between dots and leads 1 and 2 encloses the area A.

[14]. However, this classical effect is independent of phase coherence [15] and should be carefully distinguished from the two-particle phase-coherent effect which we propose here.

The scheme we propose in Sect. (3) [16] consists of two coupled quantum dots (DD) which themselves are weakly coupled in parallel to two leads 1 and 2 (see Fig. 1b). This results in a closed loop, and applying a magnetic field, an Aharonov-Bohm (AB) phase φ will be accumulated by an electron traversing the DD. In the Coulomb blockade (CB) regime we find that due to cotunneling [17] the current depends on the state of the DD: the AB oscillations for singlet and triplets have opposite sign. The amplitude of the AB oscillations provides a measure of the phase coherence of the entangled state, while the period – via the enclosed area A – provides a measure of the non-locality of the EPR pairs. The triplets themselves can be further distinguished by applying a directionally inhomogeneous magnetic field which adds a Berry phase [18] leading to beating.

Finally, in Sect. (4) we would like to address the following question [9]: Is it possible to use mobile electrons, prepared in a entangled spin state, for the purpose of quantum communication? Without spin-dependent interaction we know that the total spin must be conserved even if the two electrons strongly interact with the other electrons in the mesoscopic environment (and among themselves) via Coulomb interaction. It is thus not unreasonable to expect that we still find some spin correlations between initial and final states. But how much is it? And why and how do we loose some of the correlations?

2 Noise of Entangled Electrons: Beam Splitter Set-up

Below, we extend the standard scattering matrix approach [12] to a situation with entanglement. We start by writing the operator for the current carried by electrons with spin σ in lead α of a multiterminal conductor as

$$I_{\alpha\sigma}(t) = \frac{e}{h\nu} \sum_{\epsilon,\epsilon'} \sum_{\beta\gamma} a^{\dagger}_{\beta\sigma}(\epsilon) a_{\gamma\sigma}(\epsilon') A^{\alpha}_{\beta\gamma} \exp\left[\mathrm{i}(\epsilon-\epsilon')t/\hbar\right] \ , \qquad (1)$$

$$A^{\alpha}_{\beta\gamma} = \delta_{\alpha\beta}\delta_{\alpha\gamma} - s^{*}_{\alpha\beta} s_{\alpha\gamma} \ . \qquad (2)$$

where $a^{\dagger}_{\alpha\sigma}(\epsilon)$ creates an incoming electron in lead α with spin σ and energy ϵ, and we assume that the scattering matrix $s_{\alpha\beta}$ is spin- and energy-independent. Note that since we are dealing with discrete energy states here, we normalize the operators $a_{\alpha}(\epsilon)$ such that $[a_{\alpha\sigma}(\epsilon), a_{\beta\sigma'}(\epsilon')^{\dagger}] = \delta_{\sigma\sigma'}\delta_{\alpha\beta}\delta_{\epsilon,\epsilon'}/\nu$, where the Kronecker symbol $\delta_{\epsilon,\epsilon'}$ equals 1 if $\epsilon = \epsilon'$ and 0 otherwise. Therefore we also have included the factor $1/\nu$ in the definition of the current, where ν stands for the density of states in the leads. We will also assume that each lead consists of only a single quantum channel; the generalization to leads with several channels is straightforward but is not needed here.

We restrict ourselves here to unpolarized currents, $I_{\alpha} = \sum_{\sigma} I_{\alpha\sigma}$. The spectral density current fluctuations (noise) $\delta I_{\alpha} = I_{\alpha} - \langle I_{\alpha}\rangle$ between the leads α and β are defined as

$$S_{\alpha\beta}(\omega) = \lim_{T\to\infty} \frac{h\nu}{T} \int_0^T dt \, \exp\left(\mathrm{i}\omega t\right) \langle\Psi|\delta I_{\alpha}(t)\delta I_{\beta}(0)|\Psi\rangle \ , \qquad (3)$$

where $|\Psi\rangle$ is the quantum state of the system to be specified next[1]. We will now investigate the noise for scattering with the entangled incident state

$$|\pm\rangle = \frac{1}{\sqrt{2}} \left(a^{\dagger}_{2\downarrow}(\epsilon_2) a^{\dagger}_{1\uparrow}(\epsilon_1) \pm a^{\dagger}_{2\uparrow}(\epsilon_2) a^{\dagger}_{1\downarrow}(\epsilon_1) \right) |0\rangle \ . \qquad (4)$$

The state $|-\rangle$ is the spin singlet, $|S\rangle$, while $|+\rangle$ denotes the $S_z = 0$ triplet $|T_0\rangle$[2]. Substituting $|\pm\rangle$ defined in (4) for $|\Psi\rangle$ and using the fact that the

[1] Note that since $|\Psi\rangle$ in general does not describe a Fermi liquid state, it is not possible to apply Wick's theorem.

[2] All three triplets, $|T_0\rangle$, $|\uparrow\uparrow\rangle$, $|\downarrow\downarrow\rangle$, give the same result.

unpolarized currents are invariant when all spins are reversed, the expectation value $\langle\pm|\delta I_\alpha \delta I_\beta|\pm\rangle$ can be expressed as the sum of a direct and an exchange term,

$$\langle\pm|\delta I_\alpha \delta I_\beta|\pm\rangle = \langle\uparrow\downarrow|\delta I_\alpha \delta I_\beta|\uparrow\downarrow\rangle \pm \langle\uparrow\downarrow|\delta I_\alpha \delta I_\beta|\downarrow\uparrow\rangle \ , \tag{5}$$

where the upper (lower) sign of the exchange term refers to triplet (singlet). Direct calculation of (5) gives the following result for the zero-frequency ($\omega = 0$) correlations,

$$S_{\alpha\beta} = \frac{e^2}{h\nu}\left[\sum_{\gamma\delta}{}' A^\alpha_{\gamma\delta} A^\beta_{\delta\gamma} \mp \delta_{\epsilon_1,\epsilon_2}\left(A^\alpha_{12}A^\beta_{21} + A^\alpha_{21}A^\beta_{12}\right)\right] \ , \tag{6}$$

where $\sum'_{\gamma\delta}$ denotes the sum over $\gamma = 1, 2$ and all $\delta \neq \gamma$.

We apply formula (6) now to the set-up shown in Fig. 1a involving four leads, described by the single-particle scattering matrix elements, $s_{31} = s_{42} = r$, and $s_{41} = s_{32} = t$, where r and t denote the reflection and transmission amplitudes at the beam splitter, respectively. We assume that there is no backscattering, $s_{12} = s_{34} = s_{\alpha\alpha} = 0$. The unitarity of the s-matrix implies $|r|^2 + |t|^2 = 1$, and $\mathrm{Re}[r^* t] = 0$. Using (2) and (6), we obtain the final result for the noise correlations for the incident states $|\pm\rangle$,

$$\begin{aligned} S_{33} = S_{44} = -S_{34} = eF\,|\langle I\rangle| \ , \\ F = 2eT(1-T)\,(1 \mp \delta_{\epsilon_1,\epsilon_2}) \ , \end{aligned} \tag{7}$$

where $|\langle I\rangle| = e/h\nu$ is the average current in all leads, $T = |t|^2$ is the probability for transmission through the beam splitter, and F is the Fano factor. Note that the total current $\delta I_3 + \delta I_4$ does note fluctuate, i.e. $S_{33} + S_{44} + 2S_{34} = 0$, since we have excluded backscattering.

Above results (7) imply that if two electrons with the same energies, $\epsilon_1 = \epsilon_2$, in the singlet state $|S\rangle = |-\rangle$ are injected into the leads 1 and 2, then the zero frequency noise is *enhanced* by a factor of two, $F = 4eT(1-T)$, compared to the shot noise of uncorrelated particles, $F = 2eT(1-T)$. This enhancement of noise is due to *bunching* of electrons in the outgoing leads, caused by the symmetric orbital wavefunction of the spin singlet $|S\rangle$. On the other hand, all three triplets $|+\rangle$ exhibit an *antibunching* effect, leading to a complete suppression of the zero-frequency noise, $S_{\alpha\alpha} = 0$. The noise enhancement for the singlet $|S\rangle$ is a unique signature for entanglement (there exists no unentangled state with the same symmetry), therefore entanglement can be observed by measuring the noise power of a mesoscopic conductor as shown in Fig. 1a.[3]

[3] These results remain valid in the presence of a Fermi sea.

3 Probing Entanglement of Electrons in a Double-Dot

The DD system (see Fig. 1b) contains 4 metallic leads which are in equilibrium with associated reservoirs kept at the chemical potentials μ_i, $i = 1, \ldots, 4$. The leads are weakly coupled to the dots with tunneling amplitudes Γ and $\tilde{\Gamma}$, and the leads 1, 2 are coupled to *both* dots and play the role of probes where the currents I_i are measured. The leads 3 and 4 are feeding electrodes to manipulate the electron filling in the dots. The quantum dots contain one (excess) electron each, and are coupled to each other by the tunneling amplitude t, which leads to a level splitting [3], [4] $J = E_{\rm t} - E_{\rm s} \sim 4t^2/U$ in the DD, with U being the single-dot Coulomb repulsion energy, and $E_{\rm s/t}$ are the singlet/triplet energies. We recall that for two electrons in the DD (and for weak magnetic fields) the ground state is given by a spin singlet. For convenience we count the chemical potentials μ_i from $E_{\rm s}$. The coupling $\tilde{\Gamma}$ to the feeding leads can be switched off while probing the DD with a current. Here we assume that $\tilde{\Gamma} = 0$.

Using a standard tunneling Hamiltonian approach [19], we write $H = H_0 + V$, where the first term in $H_0 = H_{\rm D} + H_1 + H_2$ describes the DD and $H_{1,2}$ the leads (assumed to be Fermi liquids). The tunneling between leads and dots is described by the perturbation $V = V_1 + V_2$, where

$$V_n = \Gamma \sum_s \left[D_{n,s}^\dagger c_{n,s} + c_{n,s}^\dagger D_{n,s}\right], \quad D_{n,s} = {\rm e}^{\pm {\rm i}\varphi/4} d_{1,s} + {\rm e}^{\mp {\rm i}\varphi/4} d_{2,s} \;, \qquad (8)$$

and where the operators $c_{n,s}$ and $d_{n,s}$, $n = 1, 2$, annihilate electrons with spin s in the nth lead and in the nth dot, resp. The Peierls phase φ in the hopping amplitude accounts for an AB or Berry phase (see below) in the presence of a magnetic field. The upper sign belongs to lead 1 and the lower to lead 2. Finally, we assume that spin is conserved in the tunneling process. For the outgoing currents we have $I_n = {\rm i}e\Gamma \sum_s \left[D_{n,s}^\dagger c_{n,s} - c_{n,s}^\dagger D_{n,s}\right]$. The observable of interest is the average current through the DD system, $I = \langle I_2 \rangle$.

From now on we concentrate on the CB regime where we can neglect double (or higher) occupancy in each dot for all transitions including virtual ones, i.e. we require $\mu_{1,2} < U$. Further we assume that $\mu_{1,2} > J, k_{\rm B}T$ to avoid resonances which might change the DD state. The lead-dot coupling Γ is assumed to be weak so that the state of the DD is not perturbed; this will allow us to retain only the first non-vanishing contribution in Γ to I. Formally, we require $J > 2\pi\nu_{\rm t}\Gamma^2$, where $\nu_{\rm t}$ is the tunneling density of states of the leads. In analogy to the single-dot case [17] we refer to above CB regime as cotunneling regime.

Continuing with our derivation of I, we note that the average $\langle \ldots \rangle \equiv {\rm Tr}\rho\{\ldots\}$ is taken with respect to the equilibrium state of the *entire* system set up in the distant past before V is switched on [19]. Then, in the interaction picture, the current is given by

$$I = \langle U^\dagger I_2(t) U \rangle, \quad U = T\exp\left[-{\rm i}\int_{-\infty}^{t} dt' V(t')\right] \;. \qquad (9)$$

The leading contribution in Γ to the cotunneling current involves the tunneling of one electron from the DD to, say, lead 2 and of a second electron from lead 1 to the DD (see Fig. 1b). This contribution is of order $V_2 V_1^2$, and thus $I \propto \Gamma^4$, as is typical for cotunneling [17]. Taking the trace over Fermi leads, we arrive then at the following compact expression for the cotunneling current

$$I = \frac{1}{2} e\pi\nu_t^2 \Gamma^4 \sum_{i,f,s,s'} \rho_i \, |\langle i|D_{2,s'}^\dagger D_{1,s}|f\rangle|^2 \frac{\Delta_{i,f}\theta(\Delta_{i,f})}{\mu_1\mu_2} \, ,$$
$$\Delta_{i,f} = \mu_1 - \mu_2 + E_i - E_f \, . \tag{10}$$

This equation shows that in the cotunneling regime the initial state $|i\rangle$ (with weight ρ_i) of the DD is changed into a final state $|f\rangle$ by the traversing electron. However, due to the weak coupling Γ, the DD will have returned to its equilibrium state before the next electron passes through it.

For small bias, $|\mu_1 - \mu_2| < J$, only elastic cotunneling is allowed, i.e. $E_i = E_f$. However, this regime is not of interest here since singlet and triplet contributions turn out to be identical and thus indistinguishable. We thus focus on the opposite regime, $|\mu_1 - \mu_2| > J$, where inelastic cotunneling[4] occurs with singlet and triplet contributions being different. In this regime we can neglect the dynamics generated by J compared to the one generated by the bias ("slow spins"), and drop the energies E_i and E_f in (10). Finally, using $1 = \sum_f |f\rangle\langle f|$ we obtain

$$I = e\pi\nu_t^2 \Gamma^4 C(\varphi) \frac{\mu_1 - \mu_2}{\mu_1\mu_2} \, , \tag{11}$$
$$C(\varphi) = \sum_{s,s'} \left[\langle d_{1s'}^\dagger d_{1s} d_{1s}^\dagger d_{1s'} \rangle + \cos\varphi \langle d_{1s'}^\dagger d_{1s} d_{2s}^\dagger d_{2s'} \rangle \right] \, . \tag{12}$$

For the purpose of our analysis we assume that the DD is in its ground state. Equation (12) shows that the cotunneling current depends on the properties of the ground state of the DD through the coherence factor $C(\varphi)$ given in (12). The first term in C is the contribution from the topologically trivial tunneling path which runs from lead 1 through, say, dot 1 to lead 2 and back. The second term (phase-coherent part) in C is the ground state amplitude of the exchange of electron 1 with electron 2 via the leads 1 and 2 such that a closed loop is formed enclosing an area A (see Fig. 1b). Thus, in the presence of a magnetic field B, an AB phase factor $\varphi = ABe/h$ is acquired.

Next, we evaluate $C(\varphi)$ explicitly in the singlet-triplet basis. Note that only the singlet $|S\rangle$ and the triplet $|T_0\rangle$ (see (4)) are entangled EPR pairs while the remaining triplets $|T_+\rangle = |\uparrow\uparrow\rangle$, and $|T_-\rangle = |\downarrow\downarrow\rangle$ are not (they

---

[4] Note that the AB effect is not suppressed by this inelastic cotunneling, since the *entire* cotunneling process involving also leads is elastic: the initial and final states of the *entire* system have the same energy.

factorize). Assuming that the DD is in one of these states we obtain the important result

$$C(\varphi) = \begin{cases} 2 - \cos\varphi\,, & \text{for singlet} \\ 2 + \cos\varphi\,, & \text{for all triplets}\ . \end{cases} \tag{13}$$

Thus, we see that the singlet and the triplets contribute with *opposite sign to the phase-coherent part of the current.* One has to distinguish, however, carefully the entangled from the non-entangled states. The phase-coherent part of the entangled states is a genuine *two-particle* effect, while the one of the product states cannot be distinguished from a phase-coherent *single-particle* effect. Indeed, this follows from the observation that the phase-coherent part in C factorizes for the product states $T_{\pm}$ while it does not so for S, T_0. Also, for states such as $|\uparrow\downarrow\rangle$ the coherent part of C vanishes, showing that two different (and fixed) spin states cannot lead to a phase-coherent contribution since we *know* which electron goes which part of the loop. Finally we note that due to the AB phase the role of the singlet and triplets can be interchanged which is to say that we can continually transmutate the statistics of the entangled pairs S, T_0 from fermionic to bosonic (like in anyons): the symmetric orbital wave function of the singlet S goes into an antisymmetric one at half a flux quantum, and vice versa for the triplet T_0.

We would like to stress that the amplitude of the AB oscillations is a direct measure of the phase coherence of the entanglement, while the period via the enclosed area $A = h/eB_0$ gives a direct measure of the non-locality of the EPR pairs, with B_0 being the field at which $\varphi = 1$. The triplets themselves can be further distinguished by applying a directionally inhomogeneous magnetic field (around the loop) producing a Berry phase Φ^{B} [18], which is positive (negative) for the triplet $m = 1(-1)$, while it vanishes for the EPR pairs S, T_0. Thus, we will eventually see beating in the AB oscillations due to the positive (negative) shift of the AB phase Φ by the Berry phase, $\varphi = \Phi \pm \Phi^{\mathrm{B}}$.

4 Transport of Entangled Electrons

We consider the general scenario of the transport of entangled electrons in a mesoscopic system [9]. In a first step we inject entangled electrons into the leads and create the state $|\mathbf{12}\rangle \equiv |\pm\rangle$ (see (4)) on the top of the Fermi sea (as discussed e.g. in Sect. (2), see Fig. 1a). In a second step, we perform a quantum measurement of the state. As a measure of correlations we consider transition amplitudes between an initial and a final state. We begin with the simplest case given by the wave function overlap of $|\mathbf{12}\rangle$ with $|\mathbf{34}\rangle$,

$$\langle\mathbf{12}|\mathbf{34}\rangle = \delta_{\mathbf{13}}\delta_{\mathbf{24}} \mp \delta_{\mathbf{14}}\delta_{\mathbf{23}}\ , \tag{14}$$

where the upper (lower) sign refers to triplet (singlet). If the quantum numbers coincide, $\mathbf{1} = \mathbf{3}$, and $\mathbf{2} = \mathbf{4}$, the overlap assumes its maximum value 1, reflecting maximum correlation between the two states.

Next we generalize this concept to leads which contain many interacting electrons besides the two entangled electrons. We use a similar overlap as a measure of how much weight remains in the final state $|\mathbf{34}, t\rangle$ when we start from some given initial state $|\mathbf{12}\rangle$. The overlap (14) now becomes a triplet-triplet or singlet-singlet correlation function

$$G^{\mathrm{t/s}}(\mathbf{12}, \mathbf{34}; t) = -G(\mathbf{1}, t)\, G(\mathbf{2}, t)\, [\delta_{\mathbf{13}}\delta_{\mathbf{24}} \mp \delta_{\mathbf{14}}\delta_{\mathbf{23}}] \ , \tag{15}$$

where we have assumed that there is no interactions between lead 1 and 2. Thus the problem is reduced to the evaluation of (time-ordered) single-particle Green's functions $G(\mathbf{1}, t)$, $G(\mathbf{2}, t)$ pertaining to lead 1 and 2, resp. (these leads are still interacting many-body systems though).

For the special case $t = 0$, and no interactions, we have $G = -\mathrm{i}$, and thus $G^{\mathrm{t/s}}$ reduces to the rhs of (14). For the general case, we evaluate G close to the Fermi surface and get the standard result [20]

$$G(\epsilon, t) \approx -\mathrm{i} z_\epsilon \Theta(\epsilon - \epsilon_F) \mathrm{e}^{-\mathrm{i}\epsilon t - \Gamma_\epsilon t} \ , \tag{16}$$

where ϵ is the quasiparticle energy, ϵ_{F} is the Fermi energy, and $1/\Gamma_\epsilon$ is the quasiparticle lifetime. In a 2DEG, $\Gamma_\epsilon \propto (\epsilon - \epsilon_{\mathrm{F}})^2 \log(\epsilon - \epsilon_{\mathrm{F}})$ [21] within the random phase approximation (RPA). Thus, the lifetime becomes infinite when the energy of the added electron approaches ϵ_{F}.

Now, we come to the most important quantity in the present context, the *quasiparticle weight*, $z_{\mathrm{F}} = z_{\epsilon_{\mathrm{F}}}$, evaluated at the Fermi surface; it is defined by

$$z_{\mathrm{F}} = \left[1 - \frac{\partial}{\partial \omega} Re\Sigma(\epsilon_{\mathrm{F}}, \omega = 0)\right]^{-1} \ , \tag{17}$$

where $\Sigma(\epsilon, \omega)$ is the irreducible self-energy occurring in the Dyson equation. The quasiparticle weight, $0 \leq z_\epsilon \leq 1$, describes the weight of the bare electron in the quasiparticle state ϵ, i.e. when we add an electron with energy $\epsilon \geq \epsilon_{\mathrm{F}}$ to the system, some weight (given by $1 - z_\epsilon$) of the original state ϵ will be distributed among all the electrons due to the Coulomb interaction.

Restricting ourselves now to energies close to the Fermi surface we have

$$G^{\mathrm{t/s}}(\mathbf{12}, \mathbf{34}; t) = z_{\mathrm{F}}^2\, [\delta_{\mathbf{13}}\delta_{\mathbf{24}} \mp \delta_{\mathbf{14}}\delta_{\mathbf{23}}] \ , \tag{18}$$

for all times satisfying $0 < t \lesssim 1/\Gamma_\varepsilon$. Thus we see that it is the quasiparticle weight squared, z_{F}^2, which is the measure of our spin correlation function $G^{\mathrm{t/s}}$ we were looking for. It is thus interesting to evaluate z_{F} explicitly. This is indeed possible, again within RPA, and we find after careful calculation [22]

$$z_{\mathrm{F}} = 1 - r_{\mathrm{s}}(\frac{1}{2} + \frac{1}{\pi}) \ , \tag{19}$$

in leading order of the interaction parameter $r_{\mathrm{s}} = 1/q_{\mathrm{F}} a_{\mathrm{B}}$, where $a_{\mathrm{B}} = \epsilon_0 \hbar^2/me^2$ is the Bohr radius. In particular, in a GaAs 2DEG we have $a_{\mathrm{B}} = 10.3$ nm, and $r_{\mathrm{s}} = 0.614$, and thus we obtain from (19) the value $z_{\mathrm{F}} = 0.665$.

We note that a more accurate numerical evaluation of the exact RPA self-energy yields[5] $z_F = 0.691$ [22]. Thus, we see that the spin correlation is reduced by a factor of about two as soon as we inject the two electrons into separate leads consisting of *interacting* Fermi liquids in their ground state.

5 Acknowledgements

This work has been supported by the Swiss National Science Foundation.

References

[1] A. Einstein, B. Podolsky, N. Rosen Phys. Rev. **47**, 777 (1935).
[2] C. H. Bennett and G. Brassard, in Proceedings of the IEEE International Conference on Computers, Systems and Signal Processing, Bangalore, India (IEEE, New York, 1984), p. 175.
[3] D. Loss, D. P. DiVincenzo, Phys. Rev. A **57**, 120 (1998).
[4] G. Burkard, D. Loss, D. P. DiVincenzo, Phys. Rev. B **59** 2070 (1999).
[5] A. Aspect, J. Dalibard, and G. Roger, Phys. Rev. Lett. **49**, 1804 (1982).
[6] A. Zeilinger, in *Physics World*, March 1998.
[7] G. Burkard, D. Loss and E. V. Sukhorukov, preprint (cond-mat/9906071).
[8] R. P. Feynman, R. B. Leighton, and M. Sands, *The Feynman lectures on physics*, Vol. 3 (Addison-Wesley, 1965).
[9] D. P. DiVincenzo, D. Loss, preprint (cond-mat/9901137).
[10] R. Loudon, in *Coherence and Quantum Optics VI*, eds. J. H. Eberly et al. (Plenum, New York, 1990).
[11] R. Hanbury Brown and R. Q. Twiss, Nature (London) **177**, 27 (1956).
[12] M. Büttiker, Phys. Rev. Lett. **65**, 2901 (1990); Phys. Rev. B**46**, 12485 (1992).
[13] T. Martin, R. Landauer, Phys. Rev. B**45**, 1742 (1992).
[14] R. C. Liu et al., Nature **391**, 263 (1998); M. Henny et al., Science **284**, 296 (1999); W. D. Oliver et al., Science **284**, 299 (1999).
[15] E. V. Sukhorukov and D. Loss, Phys. Rev. B**59**, 13054 (1999).
[16] D. Loss and E. V. Sukhorukov, preprint (cond-mat/9907129).
[17] D. V. Averin, Yu. V. Nazarov, in *Single Charge Tunneling*, eds. H. Grabert and M.H. Devoret, NATO ASI Series B: Physics Vol. 294, Plenum Press, New York, 1992; J. Knig, H. Schoeller, and G. Schn, Phys. Rev. Lett. **78**, 4482 (1997).
[18] D. Loss, P. Goldbart, Phys. Rev. B **45**, 13544 (1992).
[19] G. D. Mahan, *Many Particle Physics*, 2nd Ed. (Plenum, New York, 1993).
[20] A.L. Fetter and J.D. Walecka, *Quantum Theory of Many-Particle Systems* (New York, McGraw-Hill, 1971).
[21] G. F. Guiliani and J. J. Quinn, Phys. Rev. B **26**, 4421 (1982).
[22] D. Loss, G. Burkard, and E. Sukhorukov, *Quantum communication with electrons*, unpublished.

---

[5] For 3D metallic leads with, say, $r_s = 2$ (e.g. $r_s^{\mathrm{Cu}} = 2.67$) the loss of correlation is somewhat less strong, since then the quasiparticle weight becomes $z_F = 0.77$ [19].

Shot Noise Suppression in Metallic Quantum Point Contacts

H.E. van den Brom and J.M. van Ruitenbeek

Kamerlingh Onnes Laboratorium, Universiteit Leiden, Postbus 9504, 2300 RA Leiden, The Netherlands

Abstract. Transport properties on atom-size metallic contacts are studied by combining measurements on conductance and shot noise. Results are presented on few-atom contacts of the monovalent metal gold and the trivalent metal aluminum, as well as on larger gold contacts. The experiments are explained in terms of quantum conductance modes. For a single atom gold contact, shot noise tends to be fully suppressed, indicating that only one mode with transmission of nearly unity contributes to the current. In contrast, the trivalent metal aluminum does not show this property. In larger gold contacts the number of partially transmitted modes continuously grows with the contact size.

1 Introduction

Shot noise is a non-equilibrium type of noise, directly resulting from the discreteness of electric charge: the passage of individual electrons causes the current to be a sum of random pulses. These intrinsic current fluctuations were already predicted in 1918 by Schottky, as an artifact of vacuum diodes (Schottky 1918). Shot noise turns out to be present in all kinds of devices, including microscopic conductors. In the last decade, it has become clear that it can be used to obtain information on the electron transport mechanism, e.g. different mesoscopic lengthscales can be distinguished (Steinbach et al. 1996), and the charge in the fractional quantum Hall regime can be determined (Saminadayar et al. 1997, de Picciotto et al. 1997). Recently, from shot noise measurements the quantum nature of the conductance in atom-size gold contacts has been revealed (van den Brom & van Ruitenbeek 1999a). In this paper we present an extension of these measurements to aluminum and to larger gold contacts.

If the passage of each individual electron is completely uncorrelated, the number, N, of such events per unit time obeys Poisson statistics, stating that the fluctuation in this number is given by $\overline{\Delta N^2} \equiv \overline{(N-\overline{N})^2} = \overline{N}$. The spectral density of current fluctuations for such uncorrelated events equals $2eI$ (see, e.g., van der Ziel 1986). This so-called full shot noise can be measured in vacuum diodes or tunneling devices.

Since electrons are fermions, that can be either reflected or transmitted, the occupation number, n, of a given state can be either zero or one, while the average of this number, $\overline{n}$, can be considered as the transmission probability

T_n. Therefore, using $\overline{n^2} = \overline{n}$, for the fluctuations we find $\overline{\Delta n^2} = T_n(1 - T_n)$. Hence, in a ballistic quantum point contact (QPC) with perfect transmission of electrons (i.e., $T_n = 1$ or 0) there are no fluctuations in the occupation numbers of left and right moving electrons, suppressing all shot noise.

For a general multichannel contact at zero temperature, the shot noise spectral density can be expressed in terms of the transmission probabilities T_n of all conducting channels (Lesovik 1989, Büttiker 1990),

$$S_I = 2eV\frac{2e^2}{h}\sum_n T_n(1 - T_n) \ . \tag{1}$$

For a contact with linear current-voltage characteristics, we see that the shot noise power is still linear in bias current. In a ballistic QPC in a two-dimensional electron gas (2DEG), the conductance, G, as a function of contact diameter shows a step-wise increase by integer multiples of the conductance quantum, $G_0 \equiv 2e^2/h$ (van Wees et al. 1988, Wharam et al. 1988), and shot noise was indeed shown to be strongly suppressed at quantized conductance values, where the contributing conductance modes are fully transmitted (Reznikov et al. 1995, Kumar et al. 1996).

From the Landauer formula for conductance, $G = G_0 \sum_n T_n$, we only know the sum of the transmission coefficients. However, from Eq. 1 we also know the sum of the transmission coefficients squared. Hence, the combination of conductance and shot noise measurements can provide us with new knowledge on electron transport properties of the smallest metallic contacts.

2 Experimental techniques

In order to obtain a stable atomic scale contact, we use the mechanically controllable break-junction technique (MCB) (Muller et al. 1992). A notched metal wire of high purity is glued on top of a phosphor bronze substrate, which is insulated with kapton foil. This is mounted into a vacuum can and cooled down to 4.2 K. By bending the substrate, the wire is broken, after which the contact between the fracture surfaces is controlled using a piezoelectric element. An advantage of this technique is its high degree of stability, which is even further improved by careful shielding from external electromagnetic, mechanical and acoustic vibrations. The cryo-pumping action of the low-temperature vacuum can ensures us that two fracture surfaces remain atomically clean for days.

The effect of measuring at low temperature is that the thermal noise is reduced. However, the noise level of the preamplifiers is exceeding the shot noise we are interested in. Using two sets of preamplifiers in parallel and measuring the cross-correlation, this undesired noise is reduced. By subtracting the zero bias thermal noise from the current biased noise measurements, the preamplifier noise, present in both, is further eliminated. For each contact, the electronic transfer characteristics of our wiring is calibrated and corrected for. The result is a white noise spectrum in the frequency interval from a few kHz up to 100 kHz. For currents up to 0.9 μA the shot noise level has the expected linear dependence on current. For further details on the measurement technique, we refer to van den Brom & van Ruitenbeek (1999a).

3 Results

3.1 Atom-size gold contacts

First we investigate the monovalent metal gold, for which a single atom contact is expected to transmit a single conductance mode (Scheer et al. 1998, Cuevas et al. 1998). In Fig. 1 the experimental results for a number of conductance values are shown. All data are strongly suppressed compared to the full shot noise value, with minima close to 1 and 2 times the conductance quantum. In fact, we are considering the shot noise relative to its full value,

$$\frac{S_I}{2eI} = \frac{\sum_n T_n(1-T_n)}{\sum_n T_n} \,. \tag{2}$$

We compare our data to a model that assumes a certain evolution of the values T_n as a function of the total conductance. In the simplest case, the conductance is due to only fully transmitted modes ($T_n = 1$) plus a single partially transmitted mode (full curve). The model gives a measure for the deviation from this ideal case in terms of the contribution x of other partially open channels; the corresponding behavior of the shot noise as a function of conductance is shown as the dashed curves in Fig. 1. This model has no physical basis but merely serves to illustrate the extent to which additional, partially open channels are required to describe the measured shot noise. For a more physical model fitting the data of Fig. 1 see Bürki & Stafford (1999).

We see that for $G < G_0$ the data are very close to the $x = 0\%$ curve, while for $G_0 < G < 2G_0$ the data are closer to the $x = 10\%$ curve. For $G > 2G_0$ the contribution of other partially open channels continues to grow.

The results described above, especially the almost full suppression of shot noise for contacts with conductance close to $G = G_0$, are obtained in gold contacts. We performed similar experiments on aluminum, which showed much weaker suppression of shot noise, as will be discussed in the next section.

3.2 Atom-size aluminum contacts

When the experiment is repeated for aluminum contacts, we observe a different behavior of the shot noise power as a function of conductance compared to gold. For contacts between $0.8\,G_0$ and $2.5\,G_0$ the obtained shot noise values vary from 0.3 to 0.6 $(2eI)$, which is much higher than for gold (see Fig. 1). A systematic dependence of the shot noise power on the conductance seems

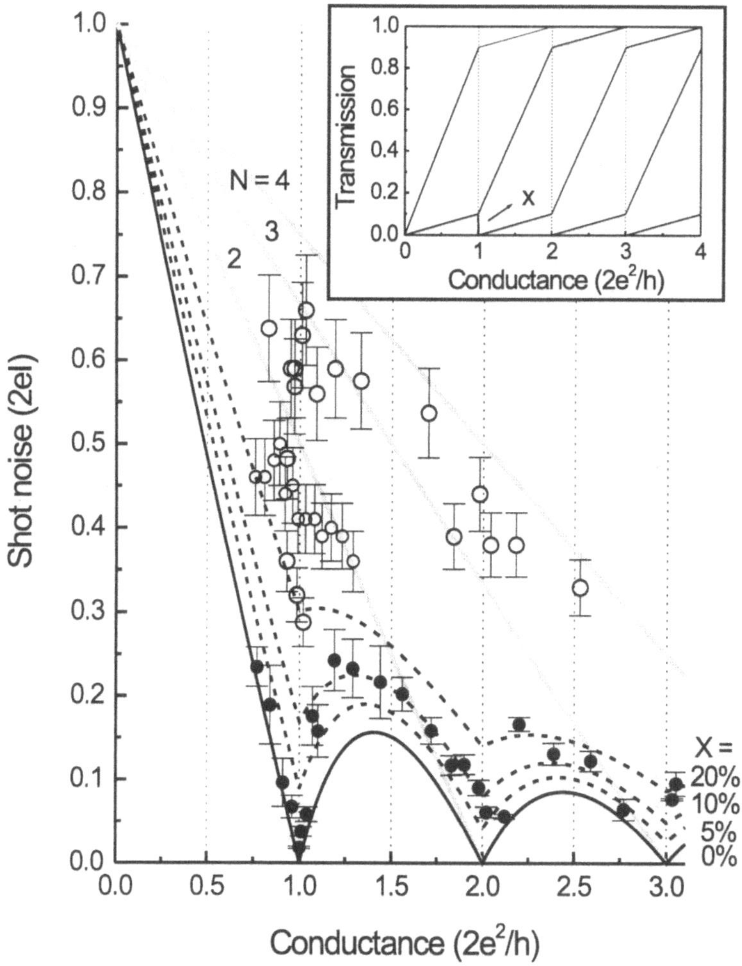

Fig. 1. Measured shot noise values for gold (full circles) and aluminum (open circles) contacts at 4.2 K with a bias current of $0.9\,\mu$A. For gold, comparison is made with calculations described in the text and in the inset (full and dashed black curves). For aluminum, comparison is made with the maximum shot noise that can be produced by N modes (gray curves), as explained in the text. The minimum shot noise is given by the full black curve. The aluminum data measured on the last conductance plateau (see text) are indicated by somewhat smaller circles. Note that in the limit of zero conductance, the theoretical curves all converge to full shot noise. *Inset:* Model for visualizing the effect of contributions of different modes to the conductance and shot noise. The model gives a measure for the deviation from the ideal case of channels opening one by one, by means of a fixed contribution $(1 - T_{n-1}) + T_{n+1} = x$ of the two neighboring modes. As an illustration we show the case of $x = 10\%$ contribution from neighboring modes.

to be absent. The fact that we do see a lowest value at a conductance of $1\,G_0$ is somewhat misleading, since we have much more data-points in this region. The relevant observation is that this lowest shot noise value is still higher than all shot noise values measured for gold, and it cannot be described by a single conductance channel.

A more detailed study of the quantum suppression can be done by following the evolution of the shot noise while stretching a contact, such that the conductance changes only gradually around $1\,G_0$ (i.e. without stepwise behavior due to atomic rearrangements). This is possible for aluminum because of the anomalous slope of the last conductance plateau before jumping to tunneling, showing a conductance *increasing* with contact elongation (Krans et al. 1993). In contrast, for gold the last plateau has a nearly constant conductance of about $1\,G_0$. Due to this anomalous behavior of aluminum, we are able to start the experiment at a conductance of $0.76\,G_0$, and continue measuring without an abrupt change in the conductance, until the contact breaks, at $G = 1.29\,G_0$. Simultaneously the shot noise is measured, showing a value changing gradually from 0.50 to 0.36 $(2eI)$, with no response to the crossing of the unit of conductance (van den Brom & van Ruitenbeek 1999b).

From the two measured parameters, the conductance, G, and the shot noise, S_I, one can not determine the full set of transmission probabilities. However, the shot noise values found for aluminum, especially the ones at conductance values close to G_0, can be fit to Eq. 1 only if we assume that more than one mode is transmitted. The maximum shot noise generated by two, three or four modes respectively as a function of conductance is plotted as the gray curves in Fig. 1; the minimum shot noise in all cases is given by the full black curve. Hence, for a contact with shot noise higher than indicated by the gray N-mode maximum shot noise curve, at least $N+1$ modes are contributing to the conductance. From this simple analysis we can see that for a considerable number of contacts with a conductance close to $1\,G_0$, the number of contributing modes is at least three.

3.3 Larger gold contacts

The measurements on gold contacts, discussed in section 3.1, are performed on the smallest contacts only. We already noted that the contribution of partially open channels grows with increasing conductance. This raises the question, what happens at higher conductance values. For long conductors a crossover is expected to the diffusive regime, where the shot noise suppression is $1/3$, or to the interacting hot-electron regime, where the shot noise grows to $\sqrt{3}/4\,(2eI)$ (Steinbach et al. 1996). In the limit of a macroscopic conductor, shot noise is commonly believed to be absent (Horowitz & Hill 1989) because of inelastic scattering (Beenakker & Büttiker 1992). For larger ballistic point contacts, with resistance values of approximately $1\,\Omega$, it is reported (Akimenko et al. 1982) that the measured noise contains spectroscopic features

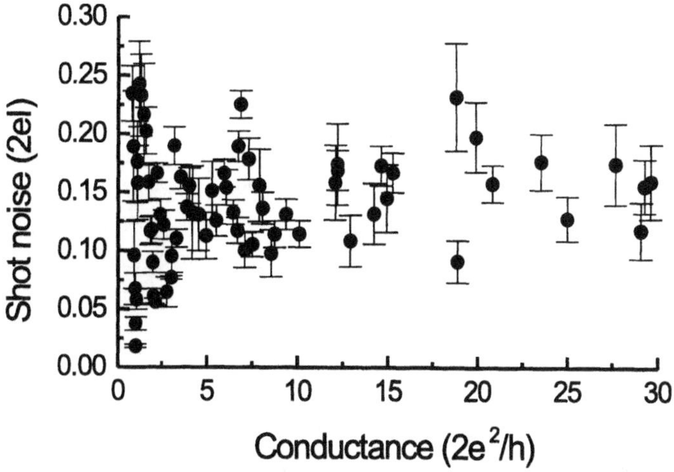

Fig. 2. Measured shot noise values for larger gold contacts at 4.2 K. The gold data from Fig. 1, with conductance below $3\,G_0$, are also shown. Note that above this value, nearly all data are in the range from $0.10-0.20\,(2eI)$, without an apparent systematic dependence on the conductance.

showing the influence of phonons on the fluctuations. However, this noise was attributed to the same mechanism that is responsible for $1/f$-noise.

We repeated the experiments for larger contacts, up to a conductance of $G = 30\,G_0$. Since determining the contact conductance is done by DC current bias, the measured signal, i.e. the voltage across the contact, decreases with increasing contact size. Therefore, for contacts of conductance $G \gtrsim 10\,G_0$, it turned out to be more convenient to raise the bias current. Down to $G \approx 4\,G_0$ this higher bias did to influence the results, as would have been the case when local heating raises the thermal noise.

As can be seen in Fig. 2, nearly all shot noise data for contacts above $G = 3\,G_0$ are in the range from $0.10-0.20\,(2eI)$, without a systematic dependence on the conductance. This means that the number of partially transmitted modes continues to grow with contact size. This can be seen by realizing that according to Eq. 2 the single mode contribution to the shot noise decreases with increasing conductance.

4 Discussion

Let us first focus on the results for the monovalent metal gold, which has been studied most intensively. Starting with a tunneling contact, we found that for increasing contact size, the transmission of the first mode entering the contact goes to unity, before the second channel opens, after which the second channel tends to go fully open, before the third one opens, and so on. This property was also inferred from measurements of conductance fluctuations (Ludoph et al. 1999) and thermopower (Ludoph & van Ruitenbeek 1999), and was referred to as 'saturation of channel transmission'. From Fig. 1 we can even quantify the accuracy to which this rule is obeyed; deviations increase to around 20% admixture of the next channels already at $G = 3\,G_0$. Where the interpretation of Ludoph et al. depends on ensemble averaged properties, the present results show that saturation of channel transmission is observed for *individual* contacts, and is independent of any adjustable parameters.

For atom-size contacts of superconductors Scheer et al. (1997, 1998) performed experiments on current-voltage characteristics in the superconducting state. This was used to demonstrate that the number of conducting channels in a single atom contact is determined by the number of valence orbitals (Scheer et al. 1998, Cuevas et al. 1998). However, reservations were made about the results on gold, since the proximity effect was used to induce superconductivity in a gold QPC, causing a modified quasiparticle density of states. Our results for $G < G_0$ show unambiguously that in the monovalent metal gold, the current through a single atom contact is indeed almost exclusively carried by one single conductance channel.

Using an independent-electron model of a disordered nanocontact in a 2DEG, Bürki and Stafford (1999) quantitatively reproduce our experimental shot noise data for the smallest gold contacts. They justify the use of their model by comparing the histogram it produces with experimental histograms.

In model simulations for gold, Brandbyge et al. (1997) find nearly full transmission of the channels for G near 1 and $3\,G_0$. However, they report two half opened channels around $G = 2\,G_0$, which is at variance with the experimental results of section 3.1. For larger contacts, up to $G = 12\,G_0$, they find an increasing number of partially transmitted modes, in agreement with the results described here in section 3.3.

In the semiclassical limit, shot noise in a ballistic point contact is predicted to be suppressed below its full value by a factor proportional to the contact diameter, and inversely proportional to the electron mean free path (Kulik & Omel'yanchuk 1984). The physical interpretation of this prediction is, that for larger contacts, the amount of backscattering from defects in the banks increases with increasing contact size, enlarging the current fluctuations. However, quantum mechanically, apart from an increasing number of partially transmitted modes, also the number of fully open conductance channels could be increasing. The way these two effects are canceling each other in Eq. 2 is very sensitive to the exact transmission values. From our experiments we see that for contacts between 3 and $30\,G_0$, the contribution of partially transmitted modes continues to grow. A transition to the diffusive regime is not observed. This is to be expected, for the electron mean free path is estimated to be around 5 nm (Ludoph et al. 1999), while the contact diameter is on the same order of magnitude or smaller.

Our results for a single atom contact of aluminum show that several modes are transmitted, confirming the results of Scheer et al. (1997), who found three conducting channels. These results are fully consistent with recent tight binding calculations for single atom junctions, in which the conductance modes were attributed to the atomic valence orbitals (Cuevas et al. 1998). Gold has a single s-like orbital, giving rise to a single conductance channel. Aluminum has one s and three p orbitals, resulting in only three partially open conductance channels, for the anti-symmetric s-p_z hybridized orbital does not give any significant transmission.

From conductance versus interelectrode distance curves, the quantum conductance properties cannot be identified as unambiguously in metals as they are in 2DEG contacts (van Wees et al. 1988, Wharam et al. 1988). Histograms built up from many of these conductance curves display the statistical preference of certain conductance values, which may be influenced by both quantization effects (including symmetry effects, Krans et al. 1995) and preferred atomic configurations. This can be best illustrated by the conductance histogram of aluminum (Yanson & van Ruitenbeek 1997), which shows pronounced peaks near the first four integer multiples of G_0. Neglecting the effect of preferred atomic configurations, one would be tempted to attribute these peaks to conductance quantization. However, using the knowledge that a single atom contact of conductance $G \approx 1\,G_0$ carries several conductance modes, for which we find strong evidence in our shot noise data, a natural interpretation for the first peak in the histogram would be, that it arises from a reproducible last contact configuration of a single atom. In analogy, also the other peaks in the histogram are expected to have their origin in a certain preferred atomic configuration.

References

Akimenko A.I., Verkin A.B., and Yanson I.K. (1982): J. Low Temp. Phys. **54**, 247
Beenakker C.W.J., and Büttiker M. (1992): Phys. Rev. B **46**, 1889
Brandbyge M., Schiøtz J., Sørensen M.R., Stoltze P., Jacobsen K.W., Nørskov J.K., Olesen L., Lægsgaard E., Stensgaard I., and Besenbacher F. (1995): Phys. Rev. B **52**, 8499
Brandbyge M., Sørensen M.R., and Jacobsen K.W. (1997): Phys. Rev. B **56**, 14956
Bürki J., and Stafford C.A. (1999): Phys. Rev. Lett. **83**, in print
Büttiker M. (1990): Phys. Rev. Lett. **65**, 2901
Cuevas J.C., Levy Yeyati A., and Martín-Rodero A. (1998): Phys. Rev. Lett. **80**, 1066
de Picciotto R., Reznikov M., Heiblum M., Umansky V., Bunin G., and Mahalu D. (1997): Nature **389**, 162
Gai Z., He Y., Yu H., and Yang W.S. (1996): Phys. Rev. B **53**, 1042

Horowitz P., and Hill W. (1989), *The art of Electronics* (Cambridge University Press, Cambridge), p. 432
Krans J.M., Muller C.J., Yanson I.K., Govaert Th.C.M., Hesper R., and van Ruitenbeek J.M. (1993): Phys. Rev. B **48**, 14721
Krans J.M., van Ruitenbeek J.M., Fisun V.V., Yanson I.K., and de Jongh L.J. (1995): Nature **375**, 767
Kulik I.O., and Omel'yanchuk A.N. (1984): Sov. J. Low Temp. Phys. **10**, 158
Kumar A., Saminadayar L., Glattli D.C., Jin Y., and Etienne B. (1996): Phys. Rev. Lett. **76**, 2778
Lesovik G.B. (1989): Sov. Phys. JETP Lett. **49**, 592
Ludoph B., Esteve D., Urbina C., Devoret M.H., and van Ruitenbeek J.M. (1999): Phys. Rev. Lett. **82**, 1530
Ludoph B., and van Ruitenbeek J.M. (1999): Phys. Rev. B **59**, 12290
Muller C.J., van Ruitenbeek J.M., and de Jongh L.J. (1992): Physica C **191**, 485; Phys. Rev. Lett. **69**, 140
Reznikov M., Heiblum M., Shtrikman H, and Mahalu D. (1995): Phys. Rev. Lett. **75**, 3340
Saminadayar L., Glattli D.C., Jin Y., and Etienne B. (1997): Phys. Rev. Lett. **79**, 2526
Scheer E., Joyez P., Esteve D., Urbina C., and Devoret M.H. (1997): Phys. Rev. Lett. **78**, 3535
Scheer E., Agraït N., Cuevas J.C., Levy Yeyati A., Ludoph B., Martín-Rodero A., Rubio G., van Ruitenbeek J.M., and Urbina C. (1998): Nature **394**, 154
Schottky W. (1918): Ann. Phys. (Leipzig) **57**, 541
Steinbach A.H., Martinis J.M., and Devoret M.H. (1996): Phys. Rev. Lett. **76**, 3806
van den Brom H.E., and van Ruitenbeek J.M. (1999a): Phys. Rev. Lett. **82**, 1526
van den Brom H.E., and van Ruitenbeek J.M. (1999b), in *Quantum physics at the mesoscopic scale*, D.C. Glattli and M. Sanquer, eds. (Editions Frontières, Gif-sur-Yvette), in print
van der Ziel A. (1986), *Noise in Solid State Devices and Circuits* (Wiley, New York)
van Wees B.J., van Houten H., Beenakker C.W.J., Williamson J.G., Kouwenhoven L.P., van der Marel D., and Foxon C.T. (1988): Phys. Rev. Lett. **60**, 848
Wharam D.A., Thornton T.J., Newsbury R., Pepper M., Ahmed H., Frost J.E.F., Hasko D.G., Peacock D.C., Ritchie D.A., and Jones G.A.C. (1988): J. Phys. C **21**, L209
Yanson A.I., and van Ruitenbeek J.M. (1997): Phys. Rev. Lett. **79**, 2157

Part IV

Driven Tunneling

Driven Tunneling: Chaos and Decoherence

Peter Hänggi[1], Sigmund Kohler[2], and Thomas Dittrich[3]

[1] Institut für Physik, Universität Augsburg, D-86135 Augsburg, Germany
[2] Depto. de Física Teórica de la Materia Condensada, Universidad Autónoma, E-28049 Madrid, Spain
[3] Depto. de Física, Universidad de los Andes, Santafé de Bogotá, Colombia

Abstract. Chaotic tunneling in a driven double-well system is investigated in absence as well as in the presence of dissipation. As the constitutive mechanism of chaos-assisted tunneling, we focus on the dynamics in the vicinity of three-level crossings in the quasienergy spectrum. They are formed when a tunnel doublet, located on a pair of symmetry-related tori in the classical phase space, approaches a chaotic singlet in energy. The coherent quantum dynamics near the crossing, in particular the enhanced tunneling that involves the chaotic singlet state as a "step stone", is described satisfactorily by a three-state model. It fails, however, for the corresponding dissipative dynamics, because incoherent transitions due to the interaction with the environment indirectly couple the three states in the crossing to the remaining quasienergy states. We model dissipation by coupling the double well, the driving included, to a heat bath. The time dependence of the central system, with a quasienergy spectrum containing exponentially small tunnel splittings, requires special considerations when applying the Born-Markov and rotating-wave approximations to derive a master equation for the density operator. We discuss the effect of decoherence on the now transient chaos-assisted tunneling: While decoherence is accelerated practically independent of temperature near the center of the crossing, it can be stabilzed with increasing temperature at a chaotic-state induced exact crossing of the ground-state quasienergies. Moreover the asymptotic amount of coherence left within the vicinity of the crossing is enhanced if the temperature is below the splitting of the avoided crossing; but becomes diminished when temperature raises above the splitting (chaos-induced coherence or incoherence, respectively). The asymptotic state of the driven dissipative quantum dynamics partially resembles the, possibly strange, attractor of the corresponding damped driven classical dynamics, but also exhibits characteristic quantum effects.

1 Introduction

The interplay of classical chaos and dissipation in a quantum system bears interesting effects at the border between classical and quantum mechanics like, e.g., the suppression of classical chaos by quantum interference [1] or its restauration by dissipation [2]. While the mutual influence of quantum coherence and classical chaos is under investigation since many years, the additional effects caused by coupling the chaotic system to an environment, namely dissipation and decoherence, have been studied only rarely. One reason is that by including dissipation, the computational effort grows drastically, since one has to deal with density matrices instead of wave functions.

In classical Hamiltonian systems, the transition from regular motion to chaos is most clearly visible in the change of the phase-space structure: With increasing nonlinearity, regular tori successively dissolve in adjacent chaotic layers which grow in size and merge until the whole phase space is uniformly covered by a chaotic sea where the dynamics is locally hyperbolic and globally diffusive [3]. Research in quantum chaos has initially been concentrated on this limiting case of "hard chaos", because the absence of structure in phase space facilitates the description.

Closer to the generic situation, however, is the intermediate regime with an extremely intricate interweaving of regular and chaotic areas, as described by the Kolmogorov-Arnol'd-Moser (KAM) theorem, with self-similar hierarchies of regular islands. It is in this regime that we expect the most interesting, but at the same time least tractable, phenomena of chaos-coherence interplay to occur. A prominent example is chaotic tunneling, the coherent exchange of probability between symmetry-related regular regions that are separated dynamically by a chaotic layer, instead of a static potential barrier [4–20]. Chaotic tunneling comes about by the simultaneous action of classical nonlinear dynamics and quantum coherence. Tunneling is extremely sensitive to any disruption of coherence as it occurs due to the unavoidable coupling to the environment: In presence of dissipation, coherent tunneling becomes a transient that fades out on the way to an asymptotic state [21, 22]. This is just one instance of the general rule that decoherence tends to restore classical behaviour, other examples being the partial lifting, by dissipation, of the quantum suppression of chaos, and the appearance of quantum stationary states that show a close resemblance to corresponding classical strange attractors [2]. However, particularly for weak dissipation, more complicated cross effects occur, such as the strong modification of the decoherence time by chaotic tunneling.

In this contribution, we investigate the mutual influence of chaotic tunneling and dissipation for a specific, but nevertheless generic case: a periodically forced bistable system. The quartic double well with a harmonic driving will serve as our working model. In Section 2 we introduce its Hamiltonian and the underlying symmetries. To provide the necessary background, we also briefly review other important features of this system, in particular driven tunneling and its coherent suppression and modification in the presence of classical chaos without damping.

Dealing with a driven system, its quantum dynamics is adequately analyzed in terms of the Floquet or quasienergy spectrum, also introduced in Section 2. The quasispectrum associated with chaotic tunneling exhibits a characteristic feature: Quasienergies of chaotic singlets frequently intersect tunnel doublets which are supported by regular tori. As the basic mechanism of chaotic tunneling we study, in Sections 3 and 4, the coherent and dissipative quantum dynamics in the vicinity of such singlet-doublet crossings. While in the coherent case the dynamics is well described in a three-state approximation, the coupling to the environment indirectly couples the three states to all other states. On the basis of numerical results for the full driven double well with dissipation, we reveal the limitations of the three-level approximation and identify additional features of the full dynamics not covered by it. In particular, we consider the long-time asymptotics and the phase-space structure associated with it.

Also on the classical level, the presence of friction has profound consequences for the phase-space structure: Due to the net contraction of phase-space volume, stationary states are restricted to manifolds of lower dimensionality than the underlying phase space. Depending on friction strength and details of the system, this attractor may be consist of fixed points, of limit cycles, or, if the classical dynamics is chaotic, of a strange attractor with self-similar, fractal geometry. On a quantum level, the structures associated with classical attractors are smeared out on a scale $\hbar$, yet leave clear traces in the asymptotic state of the corresponding dissipative quantum dynamics [23]. We study the classical-quantum correspondence of the asymptotic state in Section 5.

2 The model

We consider the quartic double well with a spatially homogeneous driving force, harmonic in time. It is defined by the Hamiltonian

$$H(t) = H_{\mathrm{DW}} + H_F(t), \tag{1}$$

$$H_{\mathrm{DW}} = \frac{p^2}{2m} - \frac{1}{4}m\omega_0^2 x^2 + \frac{m^2\omega_0^4}{64E_{\mathrm{B}}}x^4, \tag{2}$$

$$H_F(t) = Sx\cos(\Omega t). \tag{3}$$

The potential term of the static bistable Hamiltonian H_{DW} possesses two minima at $x = \pm x_0$, $x_0 = (8E_{\mathrm{B}}/m\omega_0^2)^{1/2}$, separated by a barrier of height E_{B} (cf. Fig. 1). The parameter ω_0 denotes the (angular) frequency of small oscillations near the bottom of each well. Apart from mere scaling, the classical phase space of H_{DW} only depends on the presence or absence, and the signs, of the x^2 and the x^4 term. Besides that, it has no free parameter. This is obvious from the scaled form of the classical equations of motion,

$$\dot{\bar{x}} = \bar{p}, \tag{4}$$

$$\dot{\bar{p}} = \frac{1}{2}\bar{x} - \frac{1}{2}\bar{x}^3 - F\cos(\bar{\Omega}\bar{t}), \tag{5}$$

where the dimensionless quantities $\bar{x}$, $\bar{p}$ and $\bar{t}$ are given by x/x_0, $p/m\omega_0 x_0$ and $\omega_0 t$, respectively. The influence of the driving on the classical phase-space

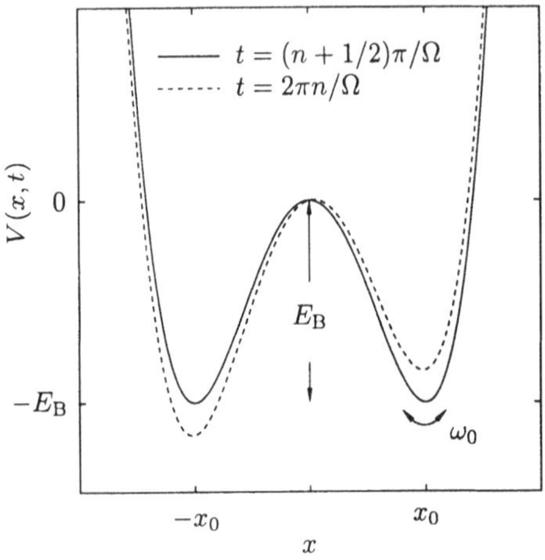

Fig. 1. Sketch of the driven double well potential described by the time-dependent Hamiltonian (1) at two different phases.

structure is fully characterized by the rescaled amplitude and frequency of the driving,

$$F = \frac{S}{\sqrt{8m\omega_0^2 E_B}}, \qquad \bar{\Omega} = \frac{\Omega}{\omega_0}. \tag{6}$$

This implies that the classical dynamics is independent of the barrier height E_B.

In the quantum-mechanical case, this is no longer true: The finite size of Planck's constant results in a finite number

$$D = \frac{E_B}{\hbar\omega_0} \tag{7}$$

of doublets with energy below the barrier top. The classical limit amounts to $D \to \infty$. This is evident from the classical scales for position, x_0, and momentum, $m\omega_0 x_0$, introduced above: The corresponding action scale is $m\omega_0 x_0^2$ and therefore, the position-momentum uncertainty relation in the scaled phase space $(\bar{x}, \bar{p})$ reads

$$\Delta\bar{x}\,\Delta\bar{p} \geq \frac{\hbar_{\text{eff}}}{2}, \tag{8}$$

where

$$\hbar_{\text{eff}} = \frac{\hbar}{m\omega_0 x_0^2} = \frac{1}{8D} \tag{9}$$

denotes the effective quantum of action.

In the following, we restrict the driving amplitude to moderate values, such that the difference between the potential minima remains much smaller than the barrier height. This implies that the bistable character of the potential is never lost.

2.1 Symmetries

Time periodicity. The Hamiltonian (1) is P-periodic, with $P = 2\pi/\Omega$. As a consequence of this discrete time-translational invariance of $H(x,p;t)$, the relevant generator of the quantum dynamics is the Floquet operator [24–28]

$$U = \mathcal{T} \exp\left(-\frac{\mathrm{i}}{\hbar} \int_0^P \mathrm{d}t \, H(t) \right) , \tag{10}$$

where $\mathcal{T}$ denotes time ordering. According to the Floquet theorem, the adiabatic states of the system are the eigenstates of U. They can be written in the form

$$|\psi_\alpha(t)\rangle = \mathrm{e}^{-\mathrm{i}\epsilon_\alpha t/\hbar} |\phi_\alpha(t)\rangle, \tag{11}$$

with

$$|\phi_\alpha(t+P)\rangle = |\phi_\alpha(t)\rangle.$$

Expanded in these Floquet states, the propagator of the driven system reads

$$U(t',t) = \sum_\alpha \mathrm{e}^{-\mathrm{i}\epsilon_\alpha (t'-t)/\hbar} |\phi_\alpha(t')\rangle\langle\phi_\alpha(t)|. \tag{12}$$

The associated eigenphases ϵ_α, referred to as quasienergies, come in classes, $\epsilon_{\alpha,n} = \epsilon_\alpha + n\hbar\Omega$, $n = 0, \pm1, \pm2, \ldots$. This is suggested by a Fourier expansion of the $|\phi_\alpha(t)\rangle$,

$$\begin{aligned} |\phi_\alpha(t)\rangle &= \sum_n |c_{\alpha,n}\rangle \, \mathrm{e}^{-\mathrm{i}n\Omega t}, \\ |c_{\alpha,n}\rangle &= \frac{1}{P} \int_0^P \mathrm{d}t \, |\phi_\alpha(t)\rangle \, \mathrm{e}^{\mathrm{i}n\Omega t}. \end{aligned} \tag{13}$$

The index n counts the number of quanta in the driving field. Otherwise, the members of a class α are physically equivalent. Therefore, the quasienergy spectrum can be reduced to a single "Brillouin zone", $-\hbar\Omega/2 \le \epsilon < \hbar\Omega/2$.

Since the quasienergies have the character of phases, they can be ordered only locally, not globally. A quantity that is defined on the full real axis and therefore does allow for a complete ordering, is the mean energy [22, 27, 28]

$$E_\alpha = \frac{1}{P} \int_0^P \mathrm{d}t \, \langle\psi_\alpha(t)| \, H(t) \, |\psi_\alpha(t)\rangle \equiv \langle\langle\phi_\alpha(t)| \, H(t) \, |\phi_\alpha(t)\rangle\rangle. \tag{14}$$

It is related to the corresponding quasienergy by

$$E_\alpha = \epsilon_\alpha + \langle\langle\phi_\alpha(t)| \, \mathrm{i}\hbar \frac{\partial}{\partial t} \, |\phi_\alpha(t)\rangle\rangle, \tag{15}$$

where the outer angle brackets denote the time average over one period of the driving, as indicated by Eq. (14). The second term on the right-hand side plays the rôle of a geometric phase accumulated over this period [22,29]. Without the driving, $E_\alpha = \epsilon_\alpha$, as it should be. By inserting the Fourier expansion (13), the mean energy takes the form

$$E_\alpha = \sum_n (\epsilon_\alpha + n\hbar\Omega)\,\langle c_{\alpha,n}|c_{\alpha,n}\rangle. \tag{16}$$

It shows that the nth Floquet channel gives a contribution $\epsilon_\alpha + n\hbar\Omega$ to the mean energy, weighted by the Fourier coefficient $\langle c_{\alpha,n}|c_{\alpha,n}\rangle$ [28].

Quasienergies and Floquet states are obtained numerically by solving the matrix eigenvalue equation [24,27,28]

$$\sum_{n'}\sum_{k'} \mathcal{H}_{n,k;n',k'}\, c_{n',k'} = \epsilon c_{n,k}, \tag{17}$$

equivalent to the time-dependent Schrödinger equation. It is derived by inserting the eigenstates (11) into the Schrödinger equation, Fourier expanding, and using the representation in the eigenbasis of the unperturbed Hamiltonian, $H_0|\Psi_k\rangle = E_k|\Psi_k\rangle$. We introduced the abbreviations

$$\mathcal{H}_{n,k;n',k'} = (E_k - n\hbar\Omega)\delta_{n-n'}\delta_{k-k'} + \frac{1}{2}\, S\, x_{k,k'}\,(\delta_{n-1-n'} + \delta_{n+1-n'}), \tag{18}$$

$$c_{n,k} = \langle \Psi_k|c_n\rangle, \tag{19}$$

$$x_{k,k'} = \langle \Psi_k|\, x\, |\Psi_{k'}\rangle. \tag{20}$$

Time-reversal symmetry. The energy eigenfunctions of an autonomous Hamiltonian with time-reversal symmetry,

$$\mathsf{T}: \quad x \to x, \quad p \to -p, \quad t \to -t \tag{21}$$

can be chosen as real [30,31]. Time-reversal invariance is generally broken by a magnetic field or by an explicit time-dependence of the Hamiltonian. However, for the sinusoidal shape of the driving together with the initial phase chosen above, T invariance is retained and the Schrödinger operator $\mathcal{H}(t) = H(t) - i\hbar\partial_t$ obeys $\mathcal{H}(t) = \mathcal{H}^*(-t)$. If now $\phi(x,t)$ is a Floquet state in position representation with quasienergy ϵ, then $\phi^*(x,-t)$ also is a Floquet state with the same quasienergy. This means that we can always find a linear combination of these Floquet states such that $\phi(x,t) = \phi^*(x,-t)$, or in the frequency domain, $\phi(x,\Omega) = \phi^*(x,\Omega)$, i.e., the Fourier coefficients of the Floquet states can be chosen real.

Generalized parity. The invariance of H_{DW} under parity P: $x \to -x$, $p \to -p$, $t \to t$ is destroyed by any spatially constant driving force. With the above choice of $H_F(t)$, however, a more general, dynamical symmetry remains [32–34]. It is defined by the operation

$$\mathsf{P}_P: \quad x \to -x, \quad p \to -p, \quad t \to t + P/2 \tag{22}$$

and represents a generalized parity acting in the extended phase space spanned by x, p, and phase, i.e., time $t \bmod P$. While such a discrete symmetry is of minor importance in classical physics, its influence on the quantum mechanical quasispectrum $\{\epsilon_\alpha(F)\}$ is profound: It devides the Hilbert space in an even and an odd sector, thus allowing for a classification of the Floquet states as even or odd. Quasienergies from different symmetry classes may intersect, while quasienergies with the same symmetry typically form avoided crossings [31]. The fact that P_P acts in the phase space extended by time $t \bmod P$, results in a particularity: If, e.g., $|\phi(t)\rangle$ is an even Floquet state, then $\exp(\mathrm{i}\Omega t)|\phi(t)\rangle$ is odd, and vive versa. Thus, two equivalent Floquet states from neighboring Brillouin zones have opposite generalized parity. This means that a classification of the corresponding solutions of the Schrödinger equation, $|\psi(t)\rangle = \exp(-\mathrm{i}\epsilon t/\hbar)|\phi(t)\rangle$, as even or odd is meaningful only with respect to a given Brillouin zone.

The invariance of the system under P_P is also of considerable help in the numerical treatment of the Floquet matrix (18) [11, 12]. To obtain a complete set of Floquet states, it is sufficient to compute all eigenvectors of the Floquet Hamiltonian in the even subspace whose eigenvalues lie in the first two Brillouin zones. The even Floquet states are given by the eigenvectors of $\mathcal{H}_\mathrm{e}$ from the first Brillouin zone; the odd Floquet states are obtained by shifting the (even) ones from the second to the first Brillouin zone, which changes their generalized parity. Thus, in the even subspace, we have to diagonalize the matrix

$$\mathcal{H}_\mathrm{e} = \begin{pmatrix} \ddots & \vdots & \vdots & \vdots & \vdots & \vdots & \\ \cdots & E_\mathrm{e} + 2\hbar\Omega & X_\mathrm{eo} & 0 & 0 & 0 & \cdots \\ \cdots & X_\mathrm{eo} & E_\mathrm{o} + \hbar\Omega & X_\mathrm{oe} & 0 & 0 & \cdots \\ \cdots & 0 & X_\mathrm{oe} & E_\mathrm{e} & X_\mathrm{eo} & 0 & \cdots \\ \cdots & 0 & 0 & X_\mathrm{eo} & E_\mathrm{o} - \hbar\Omega & X_\mathrm{oe} & \cdots \\ \cdots & 0 & 0 & 0 & X_\mathrm{oe} & E_\mathrm{e} - 2\hbar\Omega & \cdots \\ & \vdots & \vdots & \vdots & \vdots & \vdots & \ddots \end{pmatrix} . \tag{23}$$

For the same number of Floquet channels, it has only half the dimension of the original Floquet matrix (18). The entries in $\mathcal{H}_\mathrm{e}$ are themselves blocks of infinite dimension, in principle. They read explicitly

$$E_\mathrm{e} = \begin{pmatrix} E_0 & 0 & 0 & \cdots \\ 0 & E_2 & 0 & \cdots \\ 0 & 0 & E_4 & \cdots \\ \vdots & \vdots & \vdots & \ddots \end{pmatrix}, \quad E_\mathrm{o} = \begin{pmatrix} E_1 & 0 & 0 & \cdots \\ 0 & E_3 & 0 & \cdots \\ 0 & 0 & E_5 & \cdots \\ \vdots & \vdots & \vdots & \ddots \end{pmatrix}, \tag{24}$$

$$X_{\rm eo} = \frac{S}{2} \begin{pmatrix} x_{0,1} & x_{0,3} & x_{0,5} & \cdots \\ x_{2,1} & x_{2,3} & x_{2,5} & \cdots \\ x_{4,1} & x_{4,3} & x_{4,5} & \cdots \\ \vdots & \vdots & \vdots & \ddots \end{pmatrix}, \quad X_{\rm oe} = \frac{S}{2} \begin{pmatrix} x_{1,0} & x_{1,2} & x_{1,4} & \cdots \\ x_{3,0} & x_{3,2} & x_{3,4} & \cdots \\ x_{5,0} & x_{5,2} & x_{5,4} & \cdots \\ \vdots & \vdots & \vdots & \ddots \end{pmatrix}. \tag{25}$$

Here, $E_{\rm e}$, $E_{\rm o}$, represent the undriven Hamiltonian, and $X_{\rm eo}$, $X_{\rm oe}$ the driving field $H_1 = Sx/2$, decomposed in the even and odd eigenstates $|\Psi_k\rangle$ of $H_{\rm DW}$, with E_k denoting an eigenvalue of $H_{\rm DW}$, and $x_{k,k'}$ a matrix element of the position operator, see Eq. (20).

2.2 Tunneling, driving, and dissipation

With the driving $H_F(t)$ switched off, the classical phase space generated by $H_{\rm DW}$ exhibits the constituent features of a bistable Hamiltonian system. A separatrix at $E = 0$ forms the border between two sets of trajectories: One set, with $E < 0$, comes in symmetry-related pairs, each partner of which oscillates in either one of the two potential minima. The other set consists of unpaired, spatially symmetric trajectories, with $E > 0$, which encircle both wells.

Torus quantization of the integrable undriven double well, Eq. (2), implies a simple qualitative picture of its eigenstates: The unpaired tori correspond to singlets with positive energy, whereas the symmetry-related pairs below the top of the barrier correspond to degenerate pairs of eigenstates. Due to the almost harmonic shape of the potential near its minima, neighboring pairs are separated in energy approximately by $\hbar\omega_0$. Exact quantization, however, predicts that the partners of these pairs have small but finite overlap. Therefore, the true eigenstates come in doublets, each of which consists of an even and an odd state, $|\Phi_n^+\rangle$ and $|\Phi_n^-\rangle$, respectively. The energies of the nth doublet are separated by a finite tunnel splitting Δ_n. We can always choose the global relative phase such that the superpositions

$$|\Phi_n^{\rm R,L}\rangle = \frac{1}{\sqrt{2}} \left(|\Phi_n^+\rangle \pm |\Phi_n^-\rangle\right) \tag{26}$$

are localized in the right and the left well, respectively. As time evolves, the states $|\Phi_n^+\rangle$, $|\Phi_n^-\rangle$ acquire a relative phase $\exp(-\mathrm{i}\Delta_n t/\hbar)$ and $|\Phi_n^{\rm R}\rangle$, $|\Phi_n^{\rm L}\rangle$ are transformed into one another after a time $\pi\hbar/\Delta_n$. Thus, the particle tunnels forth and back between the wells with a frequency $\Delta_n/\hbar$. This introduces an additional, purely quantum-mechanical frequency scale, the tunnel rate $\Delta_0/\hbar$ of a particle residing in the ground-state doublet. Typically, tunnel rates are extremely small compared to the frequencies of the classical dynamics, all the more in the semiclassical regime we are interested in.

A driving of the form (3), even if its influence on the classical phase space is minor, can entail significant consequences for tunneling: It may enlarge the tunnel rate by orders of magnitude or even suppress tunneling altogether.

For adiabatically slow driving, $\Omega \ll \Delta_0/\hbar$, tunneling is governed by the instantaneous tunnel splitting, which is always larger than its unperturbed value Δ_0 and results in an enhancement of the tunneling rate [33]. If the driving is faster, $\Delta_0/\hbar \lesssim \Omega \ll \omega_0$, cf. Fig. 2, the opposite holds true: The relevant time scale is now given by the inverse of the quasienergy splitting of the ground-state doublet $\hbar/|\epsilon_1 - \epsilon_0|$. It has been found [33, 35] that in this case, for finite driving amplitude, $|\epsilon_1 - \epsilon_0| < \Delta_0$. Thus tunneling is always decelerated. Where the quasienergies of the ground-state doublet (which are of different generalized parity) intersect as a function of F, the splitting vanishes and tunneling is brought to a complete standstill by the purely coherent influence of the driving [32].

The small energy scales associated with tunneling make it extremely sensitive to any loss of coherence. As a consequence, the symmetry underlying the formation of tunnel doublets is generally broken, and an additional energy scale is introduced, the effective finite width attained by each discrete level. Tunneling and related coherence phenomena thus fade out on a time scale t_{decoh}. In general, this time scale gets shorter for higher temperatures, reflecting the growth of the transition rates (53) [36]. However, there exist counterintuitive effects: For driven tunneling in the vicinity of an exact crossing of the ground-state doublet, the coherent suppression of tunneling [22,32,33] can be stabilized with higher temperatures [37–39] until levels outside the doublet start to play a rôle.

So far, we have considered only driving frequencies much smaller than the frequency scale ω_0 of the relevant classical resonances, i.e., a parameter regime where classical motion is predominantly regular. In this regime, coherent tunneling is well described within a two-state approximation [33, 35]. In the dissipative case, however, a two-state approximation must fail for temperatures $k_B T \gtrsim \hbar\omega_0$, where thermal activation to higher doublets becomes relevant.

2.3 The onset of chaos

Driving with a frequency $\Omega \approx \omega_0$ affects also the dynamics of the classical bistable system, as small oscillations near the bottom of the wells become resonant with the driving and classical chaos comes into play (cf. Fig. 2). In a quantum description, this amounts to resonant multiple excitation of interdoublet transitions until levels near the top of the barrier are significantly populated.

In this frequency regime, switching on the driving has two principal consequences for the classical dynamics: The separatrix is destroyed as a closed curve and replaced by a homoclinic tangle [40] of stable and unstable manifolds. A chaotic layer forms in the vicinity and with the topology of the former separatrix (cf. Fig. 6, below). This opens the way for diffusive transport between the two potential wells. Due to the nonlinearity of the potential,

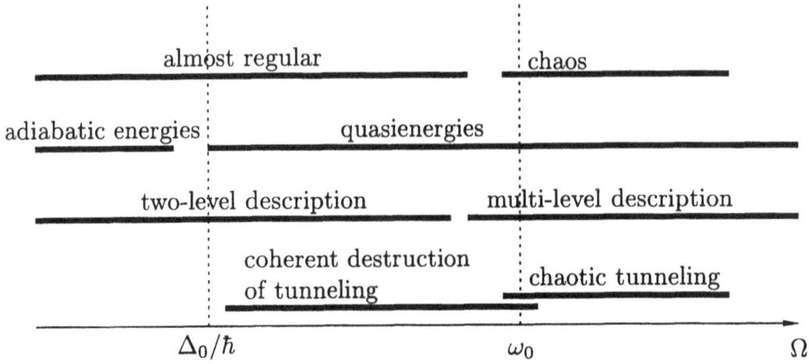

Fig. 2. Tunneling phenomena and the according appropriate levels of description for the non-dissipative driven double-well potential, Eq. (1). The bars depict the corresponding regimes of the driving frequency Ω. See Section 2 for a detailed discussion.

there is an infinite set of resonances of the driving with the unperturbed motion, both inside as well as outside the wells [41,42]. Since the period of the unperturbed, closed trajectories diverges for $E \to 0$, both from below and above, the resonances accumulate towards the separatrix of the unperturbed system. By its large phase-space area, the first resonance (the one for which the periods of the driving and of the unperturbed oscillation are in a 1-to-1 ratio) is prominent among the others and soon (in terms of increasing amplitude F) exceeds in size the "order-zero" regular areas near the bottom of each well [11].

Both major tendencies in the evolution of the classical phase space, extension of the chaotic layer and growth of the first resonance, leave their specific traces in the quasienergy spectrum. The tunnel doublets characterizing the unperturbed spectrum for $E < 0$ pertain to states located on pairs of symmetry-related quantizing tori in the regular regions within the wells. With increasing size of the chaotic layer, the quantizing tori one by one resolve in the chaotic sea. The corresponding doublets disappear as distinct structures in the spectrum as they attain a splitting of the same order as the mean level separation. The gradual widening of the doublets proceeds as a smooth function of the driving amplitude [11,12], which roughly obeys a power law [43–45]. As soon as a pair of states is no longer supported by any torus-like manifold, including fractal [46] and vague tori [47], the corresponding eigenvalues detach themselves from the regular ladder to which they formerly belonged. They can then fluctuate freely in the spectrum and thereby "collide" with other chaotic singlets or regular doublets.

The appearance of a regular region, large enough to accommodate several eigenstates, around the first resonance introduces a second ladder of doublets into the spectrum. Size and shape of the first resonance vary in a way different

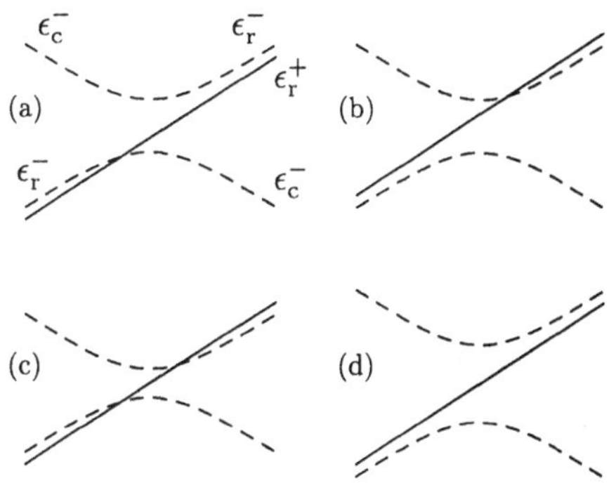

Fig. 3. Possible configurations of quasienergy crossings between a chaotic singlet and a regular doublet. Different line types signify different parity. See Section 3.1 for the labeling of the levels. Note that only for configurations (a),(b), the order of the regular doublet is restored in passing through the crossing. In configurations (c),(d), it is reversed.

from the main regular region. The corresponding doublet ladder therefore moves in the spectrum independently of the doublets that pertain to the main regular region, and of the chaotic singlets. This gives rise to additional singlet-doublet and even to doublet-doublet encounters.

3 Chaotic tunneling near singlet-doublet crossings

Near a crossing, level separations deviate vastly, in both directions, from the typical tunnel splitting (cf. Fig. 8, below). This is reflected in time-domain phenomena ranging from the suppression of tunneling to a strong increase in its rate and to complicated quantum beats [13–15]. Singlet-doublet crossings, in turn, drastically change the non-dissipative quasienergy scales and replace the two-level by a three-level structure. As a consequence, the familiar way tunneling fades out in the presence of dissipation is also significantly altered. Near a crossing, the coherent dynamics can last much longer than for the unperturbed doublet, despite the presence of the same dissipation as outside the crossing, establishing "chaos-induced coherence." Depending on temperature, it can also be destroyed on a much shorter time scale.

For the parameters chosen in our numerical studies, higher resonances are negligible in size. Accordingly, the "coastal strip" between the chaotic layer along the former separatrix and the regular regions within and outside the wells, formed by hierarchies of regular islands around higher resonances, remains narrow (cf. Fig. 6). For the tunneling dynamics, the rôle of states located in the border region [17, 18] is therefore not significant in our case.

3.1 Three-level crossings

Among the various types of quasienergy crossings that occur according to the above scenario, those involving a regular doublet and a chaotic singlet are the most common. In order to give a quantitative account of such crossings and the associated coherent dynamics, and for later reference in the context of the incoherent dynamics, we shall now discuss them in terms of a simple three-state model, devised much in the spirit of Ref. [7].

Far to the left of the crossing, we expect the following situation: There is a doublet of Floquet states

$$|\psi_{\mathrm{r}}^{+}(t)\rangle = \mathrm{e}^{-\mathrm{i}\epsilon_{\mathrm{r}}^{+}t/\hbar}|\phi_{\mathrm{r}}^{+}(t)\rangle, \tag{27}$$

$$|\psi_{\mathrm{r}}^{-}(t)\rangle = \mathrm{e}^{-\mathrm{i}(\epsilon_{\mathrm{r}}^{+}+\Delta)t/\hbar}|\phi_{\mathrm{r}}^{-}(t)\rangle, \tag{28}$$

with even (superscript +) and odd (−) generalized parity, respectively, residing on a pair of quantizing tori in one of the regular (subscript r) regions. We have assumed the quasienergy splitting $\Delta = \epsilon_{\mathrm{r}}^{-} - \epsilon_{\mathrm{r}}^{+}$ (as opposed to the unperturbed splitting) to be positive. The global relative phase is chosen such that the superpositions

$$|\phi_{\mathrm{R,L}}(t)\rangle = \frac{1}{\sqrt{2}}\left(|\phi_{\mathrm{r}}^{+}(t)\rangle \pm |\phi_{\mathrm{r}}^{-}(t)\rangle\right) \tag{29}$$

are localized in the right and the left well, respectively, and tunnel back and forth with a frequency $\Delta/\hbar$.

As the third player, we introduce a Floquet state

$$|\psi_{\mathrm{c}}^{-}(t)\rangle = \mathrm{e}^{-\mathrm{i}(\epsilon_{\mathrm{r}}^{+}+\Delta+\Delta_{\mathrm{c}})t/\hbar}|\phi_{\mathrm{c}}^{-}(t)\rangle, \tag{30}$$

located mainly in the chaotic (subscript c) layer, so that its time-periodic part $|\phi_{\mathrm{c}}^{-}(t)\rangle$ contains a large number of harmonics. Without loss of generality, its parity is fixed to be odd. For the quasienergy, we assume that $\epsilon_{\mathrm{c}}^{-} = \epsilon_{\mathrm{r}}^{+} + \Delta + \Delta_{\mathrm{c}} = \epsilon_{\mathrm{r}}^{-} + \Delta_{\mathrm{c}}$, where $|\Delta_{\mathrm{c}}|$ can be regarded as a measure of the distance from the crossing.

The structure of the classical phase space then implies that the mean energy of the chaotic state should be close to the top of the barrier and far above that of the doublet. We assume, like for the quasienergies, a small splitting of the mean energies pertaining to the regular doublet, $|E_{\mathrm{r}}^{-} - E_{\mathrm{r}}^{+}| \ll E_{\mathrm{c}}^{-} - E_{\mathrm{r}}^{\pm}$.

In order to model an avoided crossing between $|\phi_{\mathrm{r}}^{-}\rangle$ and $|\phi_{\mathrm{c}}^{-}\rangle$, we suppose that there is a non-vanishing fixed matrix element

$$b = \langle\langle\phi_{\mathrm{r}}^{-}|H_{\mathrm{DW}}|\phi_{\mathrm{c}}^{-}\rangle\rangle > 0. \tag{31}$$

For the singlet-doublet crossings under study, we typically find that $\Delta \ll b \ll \hbar\Omega$. Neglecting the coupling with all other states, we model the system by the three-state (subscript 3s) Floquet Hamiltonian

$$\mathcal{H}_{3\mathrm{s}} = \epsilon_{\mathrm{r}}^{+} + \begin{pmatrix} 0 & 0 & 0 \\ 0 & \Delta & b \\ 0 & b & \Delta+\Delta_{\mathrm{c}} \end{pmatrix} \tag{32}$$

in the three-dimensional Hilbert space spanned by $\{|\phi_{\mathrm{r}}^{+}(t)\rangle, |\phi_{\mathrm{r}}^{-}(t)\rangle, |\phi_{\mathrm{c}}^{-}(t)\rangle\}$. Its Floquet states are

$$
\begin{aligned}
|\phi_0^+(t)\rangle &= |\phi_r^+(t)\rangle, \\
|\phi_1^-(t)\rangle &= \left(|\phi_r^-(t)\rangle \cos\beta - |\phi_c^-(t)\rangle \sin\beta\right), \\
|\phi_2^-(t)\rangle &= \left(|\phi_r^-(t)\rangle \sin\beta + |\phi_c^-(t)\rangle \cos\beta\right).
\end{aligned} \tag{33}
$$

with quasienergies

$$
\epsilon_0^+ = \epsilon_r^+, \quad \epsilon_{1,2}^- = \epsilon_r^+ + \Delta + \frac{1}{2}\Delta_c \mp \frac{1}{2}\sqrt{\Delta_c^2 + 4b^2}, \tag{34}
$$

and mean energies, neglecting contributions of the matrix element b,

$$
\begin{aligned}
E_0^+ &= E_r^+, \\
E_1^- &= E_r^- \cos^2\beta + E_c^- \sin^2\beta, \\
E_2^- &= E_r^- \sin^2\beta + E_c^- \cos^2\beta.
\end{aligned} \tag{35}
$$

The angle β describes the mixing between the Floquet states $|\phi_r^-\rangle$ and $|\phi_c^-\rangle$ and is an alternative measure of the distance to the avoided crossing. By diagonalizing the Hamiltonian (32), we obtain

$$
2\beta = \arctan\left(\frac{2b}{\Delta_c}\right), \quad 0 < \beta < \frac{\pi}{2}. \tag{36}
$$

For $\beta \to \pi/2$, corresponding to $-\Delta_c \gg b$, we retain the situation far left of the crossing, as outlined above, with $|\phi_1^-\rangle \approx |\phi_c^-\rangle$, $|\phi_2^-\rangle \approx |\phi_r^-\rangle$. To the far right of the crossing, i.e., for $\beta \to 0$ or $\Delta_c \gg b$, the exact eigenstates $|\phi_1^-\rangle$ and $|\phi_2^-\rangle$ have interchanged their phase-space structure [13–15]. Here, we have $|\phi_1^-\rangle \approx |\phi_r^-\rangle$ and $|\phi_2^-\rangle \approx |\phi_c^-\rangle$. The mean energy is essentially determined by this phase-space structure, so that there is also an exchange of E_1^- and E_2^- in an exact crossing, cf. Eq. (35), while E_0^+ remains unaffected (Fig. 4b). The quasienergies ϵ_0^+ and ϵ_1^- must intersect close to the avoided crossing of ϵ_1^- and ϵ_2^- (Fig. 4a). Their crossing is exact, since they pertain to states with opposite parity (cf. Fig. 3a,b).

In order to illustrate the above three-state model and to demonstrate its adequacy, we have numerically studied a singlet-doublet crossing that occurs for the double-well potential, Eq. (1), with $D = 4$, at a driving frequency $\Omega = 0.982\,\omega_0$ and amplitude $F = 0.015029$ (Fig. 5). The phase-space structure of the participating Floquet states (Figs. 6, 7) meets the assumptions of our three-state theory. A comparison of the appropriately scaled three-state theory (Fig. 4) with this real singlet-doublet crossing (Fig. 5) shows satisfactory agreement. Note that in the real crossing, the quasienergy of the chaotic singlet *decreases* as a function of F, so that the exact crossing occurs to the left of the avoided one. This numerical example also shows a deficiency of the idealized three-state model. Following the global tendency of widening of the splittings with increasing driving amplitude [11, 44, 45], it may happen that even far away from a crossing, the doublet splitting does not return to its value on the opposite side (see Fig. 8). It is even possible that an exact

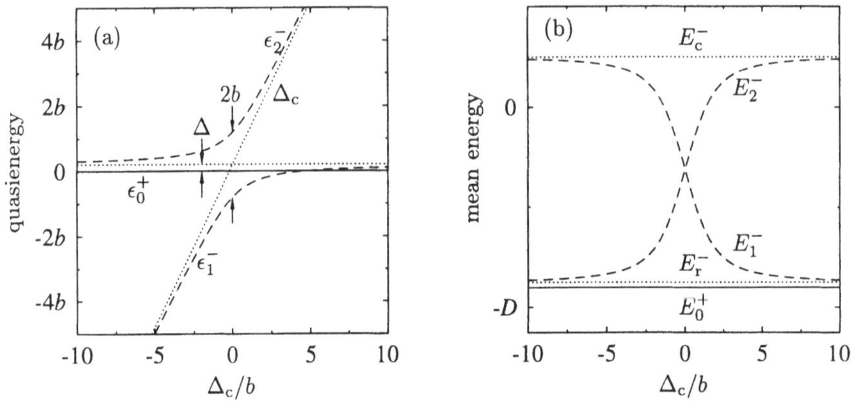

Fig. 4. A singlet-doublet crossing, according to the three-state model (32), in terms of the quasienergies (a) and the mean energies (b) as functions of the coupling parameter Δ_c/b. Energies for a corresponding exact crossing (i.e., with the crossing states uncoupled) are marked by dotted lines, the energies in the presence of coupling by full and dashed lines for even and odd states, respectively.

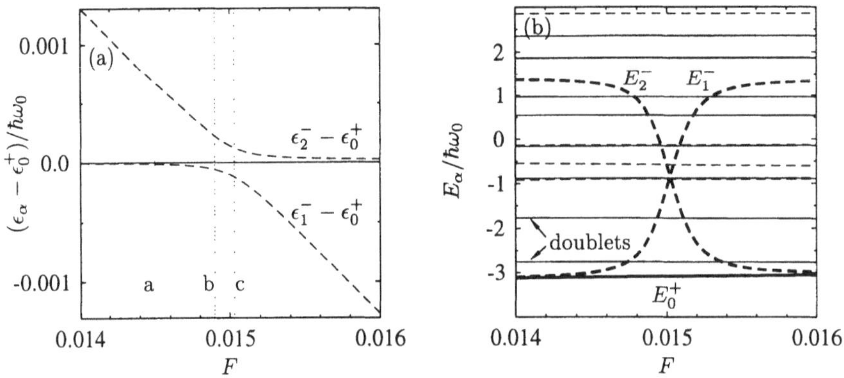

Fig. 5. Singlet-doublet crossing found numerically for the driven double well, Eq. (1), at $D = 4$ and $\Omega = 0.982\,\omega_0$, in terms of the dependence of the quasienergies (a) and the mean energies (b) on the driving amplitude F. Values of the driving amplitude used in Fig. 9 are marked by dotted vertical lines. Full and dashed lines indicate energies of even and odd states, respectively. Bold lines give the mean energies of the chaotic singlet and the ground-state doublet depicted in panel (a).

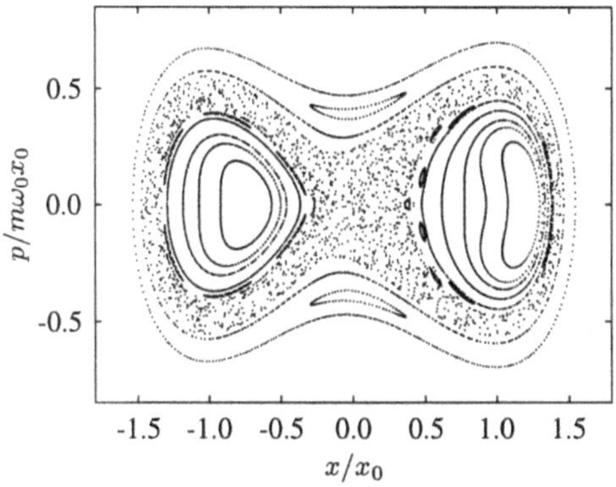

Fig. 6. Classical stroboscopic phase space portrait, at $t = 2\pi n/\Omega$, of the harmonically driven quartic double well, Eq. (1). The driving parameters $F = 0.015$, $\Omega = 0.982\,\omega_0$, are at the center of the singlet-doublet crossing shown in Fig. 5.

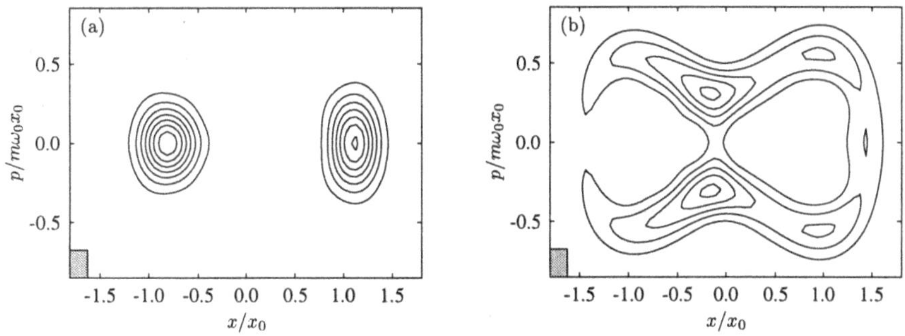

Fig. 7. Contour plots of the Husimi functions for the Floquet states $|\phi_1^-\rangle \approx |\phi_r^-\rangle$ (a) and $|\phi_2^-\rangle \approx |\phi_c^-\rangle$ (b) of the harmonically driven quartic double well, Eq. (1), at stroboscopic times $t = nP$. The driving parameters $F = 0.014$, $\Omega = 0.982\,\omega_0$, are in sufficient distance to the singlet-doublet crossing such that the admixture from the chaotic singlet state is negligible. The rectangle in the lower left corner has the size of the effective quantum of action $\hbar_{\text{eff}}$.

crossing of ϵ_0^+ and ϵ_1^- never takes place in the vicinity of the crossing. In that case, the relation of the quasienergies in the doublet gets reversed via the crossing (Fig. 3c,d). Nevertheless, the three-state scenario captures the essential features.

To study the dynamics of the tunneling process, we focus on the state

$$|\psi(t)\rangle = \frac{1}{\sqrt{2}}\left(e^{-i\epsilon_0^+ t/\hbar}|\phi_0^+(t)\rangle + e^{-i\epsilon_1^- t/\hbar}|\phi_1^-(t)\rangle \cos\beta + e^{-i\epsilon_2^- t/\hbar}|\phi_2^-(t)\rangle \sin\beta\right). \tag{37}$$

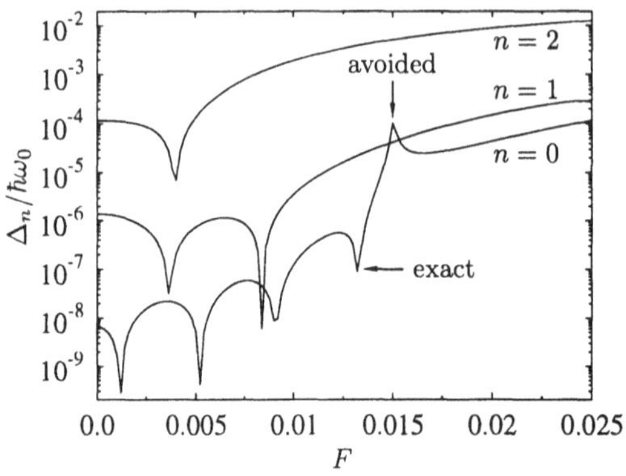

Fig. 8. Splitting of the lowest doublets for $D = 4$ and $\Omega = 0.982\,\omega_0$. The arrows indicate the locations of the exact and the avoided crossing within a three-level crossing of the type sketched in Fig. 3a.

It is constructed such that at $t = 0$, it corresponds to the decomposition of $|\phi_{\mathrm{R}}\rangle$ in the basis (33) at finite distance from the crossing. Therefore, it is initially localized in the regular region in the right well and follows the time evolution under the Hamiltonian (32). From Eqs. (29), (33), we find the probabilities for its evolving into $|\phi_{\mathrm{R}}\rangle$, $|\phi_{\mathrm{L}}\rangle$, or $|\phi_{\mathrm{c}}\rangle$, respectively, to be

$$\begin{aligned} P_{\mathrm{R,L}}(t) &= |\langle\phi_{\mathrm{R,L}}(t)|\psi(t)\rangle|^2 \\ &= \frac{1}{2}\left(1 \pm \left[\cos\frac{(\epsilon_1^- - \epsilon_0^+)t}{\hbar}\cos^2\beta + \cos\frac{(\epsilon_2^- - \epsilon_0^+)t}{\hbar}\sin^2\beta\right]\right. \\ &\quad \left. + \left[\cos\frac{(\epsilon_1^- - \epsilon_2^-)t}{\hbar} - 1\right]\cos^2\beta\sin^2\beta\right), \qquad (38) \\ P_{\mathrm{c}}(t) &= |\langle\phi_{\mathrm{c}}(t)|\psi(t)\rangle|^2 = \left[1 - \cos\frac{(\epsilon_1^- - \epsilon_2^-)t}{\hbar}\right]\cos^2\beta\sin^2\beta. \end{aligned}$$

We discuss the coherent dynamics of the three-state model for different distances to the crossing and illustrate it by numerical results for the real crossing introduced above.

At sufficient distance from the crossing, there is only little mixing between the regular and the chaotic states, i.e., $\sin\beta \ll 1$ or $\cos\beta \ll 1$. The tunneling process then follows the familiar two-state dynamics involving only $|\phi_{\mathrm{r}}^+\rangle$ and $|\phi_{\mathrm{r}}^-\rangle$, with tunnel frequency $\Delta/\hbar$ (Fig. 9a). Close to the avoided crossing, $\cos\beta$ and $\sin\beta$ are of the same order of magnitude, and $|\phi_1^-\rangle$, $|\phi_2^-\rangle$ become very similar to one another. Both now have support in the chaotic layer as well as in the symmetry-related regular regions, they are of a hybrid nature. Here, the tunneling involves all the three states and must be described at least by a three-level system. The exchange of probability between the two regular regions proceeds via a "stop-over" in the chaotic region [7,8,13–15].

The three quasienergy differences that determine the time scales of this process are in general all different, leading to complicated beats (Fig. 9b).

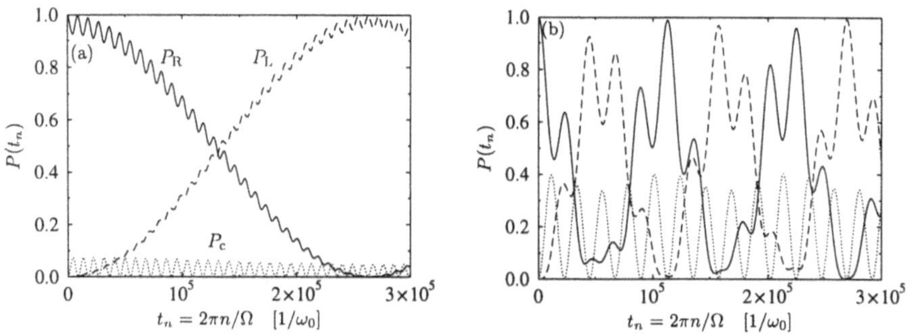

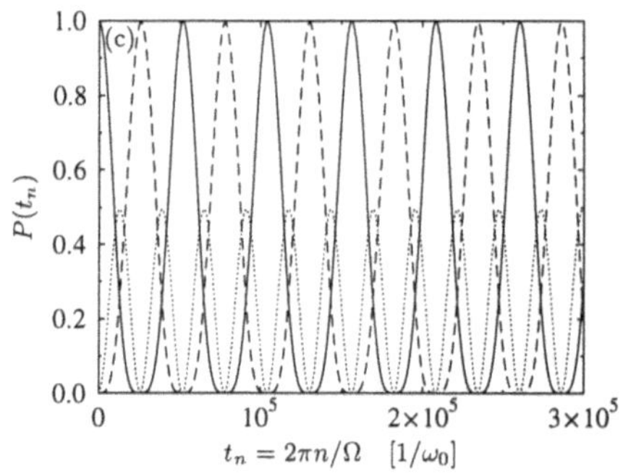

Fig. 9. Stroboscopic time evolution of a state initially localized in the right well, in the vicinity of the singlet-doublet crossing shown in Fig. 5, in terms of the probabilities to be in the right well (which here is identical to the return probability, marked by full lines), in the reflected state in the left well (dashed), or in the chaotic state $|\psi_c\rangle$ (dotted). Parameter values are as in Fig. 5, and $F = 0.0145$ (a), 0.0149 (b), 0.015029 (c).

However, for $\Delta_c = -2\Delta$, the two quasienergies $\epsilon_1^- - \epsilon_0^+$ and $\epsilon_0^+ - \epsilon_2^-$ are degenerate. At this point, the center of the crossing, the number of different frequencies in the three-level dynamics reduces to two again. This restores the familiar coherent tunneling in the sense that there is again a simple periodic exchange of probability between the regular regions [13–15]. However, the rate is much larger if compared to the situation far off the crossing, and the chaotic region is now temporarily populated during each probability transfer, twice per tunneling cycle (Fig. 9c).

4 Incoherent quantum dynamics

4.1 Master equation

System-bath model. To achieve a microscopic model of dissipation, we couple the system (1) bilinearly to a bath of non-interacting harmonic oscillators [48, 49]. The total Hamiltonian of system and bath is then given by

$$H(t) = H_{\mathrm{DW}}(t) + \sum_{\nu=1}^{\infty} \left(\frac{p_\nu^2}{2m_\nu} + \frac{m_\nu}{2}\omega_\nu^2 \left(x_\nu - \frac{g_\nu}{m_\nu \omega_\nu^2} x \right)^2 \right) . \qquad (39)$$

The position x of the system is coupled, with coupling strength g_ν, to an ensemble of oscillators with masses m_ν, frequencies ω_ν, momenta p_ν, and coordinates x_ν. The bath is fully characterized by the spectral density of the coupling energy,

$$J(\omega) = \pi \sum_{\nu=1}^{\infty} \frac{g_\nu^2}{2m_\nu\omega_\nu}\delta(\omega - \omega_\nu). \tag{40}$$

For the time evolution we choose an initial condition of the Feynman-Vernon type: at $t = t_0$, the bath is in thermal equilibrium and uncorrelated to the system, i.e.,

$$\rho(t_0) = \rho_{\mathrm{S}}(t_0) \otimes \rho_{\mathrm{B,eq}}, \tag{41}$$

where $\rho_{\mathrm{B,eq}} = \exp(-\beta H_{\mathrm{B}})/\mathrm{tr_B}\exp(-\beta H_{\mathrm{B}})$ is the canonical ensemble of the bath and $1/\beta = k_{\mathrm{B}}T$.

Due to the bilinearity of the system–bath coupling, one can always eliminate the bath variables to get an exact, closed integro-differential equation for the reduced density matrix $\rho_{\mathrm{S}}(t) = \mathrm{tr_B}\rho(t)$. It describes the dynamics of the central system, subject to dissipation [50].

Born-Markov approximation. In most cases, however, the integro-differential equation for $\rho_{\mathrm{S}}(t)$ can be solved only approximately. In particular, in the limit of weak coupling,

$$\gamma \ll k_{\mathrm{B}}T/\hbar, \tag{42}$$

$$\gamma \ll |\epsilon_\alpha - \epsilon_{\alpha'}|/\hbar, \tag{43}$$

it is possible to truncate the time-dependent perturbation expansion in the system–bath interaction after the second-order term. The quantity γ, to be defined below, denotes the effective damping of the dissipative system, and $|\epsilon_\alpha - \epsilon_{\alpha'}|/\hbar$ are the transition frequencies of the central system. In the present case, *the central system is understood to include the driving* [51–54], so that the transition frequencies are given by quasienergy differences. The autocorrelations of the bath decay on a time scale $\hbar/k_{\mathrm{B}}T$ and thus in the present limit, instantaneously on the time scale $1/\gamma$ of the system correlations. With the initial preparation (41), the equation of motion for the reduced density matrix in this approximation is given by [54]

$$\begin{aligned}\dot\rho_{\mathrm{S}}(t) = &-\frac{\mathrm{i}}{\hbar}\left[H_{\mathrm{S}}(t), \rho_{\mathrm{S}}(t)\right] + \frac{1}{\pi\hbar}\int_{-\infty}^{\infty} \mathrm{d}\omega\, J(\omega) n_{\mathrm{th}}(\hbar\omega) \\ &\times \int_0^\infty \mathrm{d}\tau \left(\mathrm{e}^{\mathrm{i}\omega\tau}\left[\tilde{x}(t-\tau, t)\rho_{\mathrm{S}}(t), x\right] + \mathrm{H.c.}\right),\end{aligned} \tag{44}$$

where $\tilde{x}(t', t)$ denotes the position operator in the interaction picture defined by

$$\tilde{x}(t', t) = U^\dagger(t', t)\, x\, U(t', t), \tag{45}$$

with $U(t',t)$, the propagator of the conservative driven double well, given in Eq. (12). 'H.c.' means 'Hermitian conjugate', and

$$n_{\rm th}(\epsilon) = \frac{1}{\mathrm{e}^{\epsilon/k_{\rm B}T} - 1} = -n_{\rm th}(-\epsilon) - 1 \qquad (46)$$

is the thermal occupation of the bath oscillator with energy ϵ. To achieve a more compact notation, we require $J(-\omega) = -J(\omega)$. In the following, we shall restrict ourselves to an Ohmic bath, $J(\omega) = m\gamma\omega$. This defines the effective damping constant γ.

We use the time-periodic components $|\phi_\alpha(t)\rangle$ of the Floquet states as a basis to expand the density operator, Eq. (44). Expressing the matrix elements

$$X_{\alpha\beta}(t) = \langle\phi_\alpha(t)|x|\phi_\beta(t)\rangle \qquad (47)$$

of the position operator by their Fourier coefficients

$$X_{\alpha\beta,n} = \langle\langle\phi_\alpha(t)|x\,\mathrm{e}^{-\mathrm{i}n\Omega t}|\phi_\beta(t)\rangle\rangle = X^*_{\beta\alpha,-n}\,, \qquad (48)$$

$$X_{\alpha\beta}(t) = \sum_n \mathrm{e}^{\mathrm{i}n\Omega t} X_{\alpha\beta,n}\,, \qquad (49)$$

yields the equation of motion for the elements $\rho_{\alpha\beta}$ of the reduced density matrix $\rho_{\rm S}$ [37, 39, 52, 54],

$$\begin{aligned} \dot\rho_{\alpha\beta}(t) &= \frac{\mathrm{d}}{\mathrm{d}t}\langle\phi_\alpha(t)|\rho_{\rm S}(t)|\phi_\beta(t)\rangle \\ &= -\frac{\mathrm{i}}{\hbar}(\epsilon_\alpha - \epsilon_\beta)\rho_{\alpha\beta}(t) \\ &\quad + \sum_{\alpha'\beta' nn'} \big(N_{\alpha\alpha',n}X_{\alpha\alpha',n}\rho_{\alpha'\beta'}X_{\beta'\beta,n'} \\ &\quad - N_{\alpha'\beta',n}X_{\alpha\alpha',n'}X_{\alpha\alpha',n}\rho_{\beta'\beta}\big)\mathrm{e}^{\mathrm{i}(n+n')\omega t} + \mathrm{H.c.}\,. \end{aligned} \qquad (50)$$

The coefficients of this differential equation are periodic in time with the period of the driving. The $N_{\alpha\beta,n}$ are given by

$$N_{\alpha\beta,n} = N(\epsilon_\alpha - \epsilon_\beta + n\hbar\Omega), \quad N(\epsilon) = \frac{m\gamma\epsilon}{\hbar^2} n_{\rm th}(\epsilon). \qquad (51)$$

For $\epsilon \gg k_{\rm B}T$, $N(\epsilon)$ approaches zero.

Since the position operator x is odd under P_P (cf. Eq. (22)), the master equations (44) and (50) are invariant under P_P. Therefore, both the conservative and the dissipative dynamics preserve the parity of the operator $|\phi_\alpha\rangle\langle\phi_\beta|$. If $|\phi_\alpha\rangle$ and $|\phi_\beta\rangle$ belong to the same parity class, it is even, and odd otherwise. In particular, the projectors $|\phi_\alpha\rangle\langle\phi_\alpha|$ and thus all density matrices diagonal in the Floquet basis are even under P_P.

Rotating-wave approximation. Assuming that dissipative effects are relevant only on a time scale much larger than the period P of the driving, we average the coefficients of the master equation (50) over P to obtain the equation of motion

$$\dot{\rho}_{\alpha\beta}(t) = -\frac{\mathrm{i}}{\hbar}(\epsilon_\alpha - \epsilon_\beta)\rho_{\alpha\beta}(t) + \sum_{\alpha'\beta'} \mathcal{L}_{\alpha\beta,\alpha'\beta'}\rho_{\alpha'\beta'}, \tag{52}$$

with the time-independent dissipative part

$$\begin{aligned}\mathcal{L}_{\alpha\beta,\alpha'\beta'} = &\sum_n (N_{\alpha\alpha',n} + N_{\beta\beta',n})\, X_{\alpha\alpha',n} X_{\beta'\beta,-n} \\ &-\delta_{\beta\beta'} \sum_{\beta'',n} N_{\beta''\alpha',n} X_{\alpha\beta'',-n} X_{\beta''\alpha',n} \\ &-\delta_{\alpha\alpha'} \sum_{\alpha''n} N_{\alpha''\beta',n} X_{\beta'\alpha'',-n} X_{\alpha''\beta,n}.\end{aligned} \tag{53}$$

This step amounts to a rotating-wave approximation which is, however, less restrictive than the one introduced in [51, 52] where dissipative effects are averaged over the generally longer time scale $\max_{\alpha,\beta,n}(2\pi\hbar/(\epsilon_\alpha - \epsilon_\beta + n\hbar\Omega))$.

4.2 Chaos-assisted dissipative tunneling

The crucial effect of dissipation on a quantum system is the disruption of coherence: a coherent superposition evolves into an incoherent mixture. Thus, phenomena based on coherence, such as tunneling, are rendered transients that fade out on a finite time scale t_{decoh}. In general, for driven tunneling in the weakly damped regime, this time scale gets shorter for higher temperatures, as transition rates grow [36]. However, in the vicinity of an exact crossing of the ground-state quasienergies, the coherent suppression of tunneling [22, 32, 33] can be stabilized with higher temperatures [37–39] and increasing friction [55,56] until levels outside the doublet start to play a rôle. We have studied dissipative chaos-assisted tunneling, at the particular real singlet-doublet crossing introduced in Sec. 3.1 (see Fig. 5). The time evolution has been computed numerically by iterating the dissipative quantum map for the improved master equation in moderate rotating-wave approximation, Eq. (52). As initial condition, we have chosen the density operator $\rho(0) = |\phi_{\mathrm{R}}\rangle\langle\phi_{\mathrm{R}}|$, a pure state located in the right well.

In the vicinity of a singlet-doublet crossing, the tunnel splitting increases significantly—the essence of chaos-assisted tunneling. During the tunneling, the chaotic singlet $|\phi_{\mathrm{c}}\rangle$ becomes populated periodically with frequency $|\epsilon_2^- - \epsilon_1^-|/\hbar$, cf. Eq. (38) and Fig. 9. The high mean energy of this singlet results in an enhanced decay of coherence at times when it is well populated (Fig. 10). For the relaxation towards the asymptotic state, also the slower transitions within doublets are relevant. Therefore, the corresponding time scale t_{relax} can be much larger than t_{decoh} (Fig. 11).

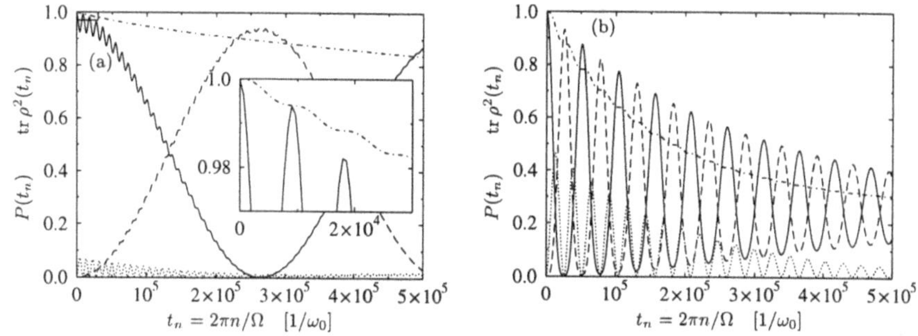

Fig. 10. Occupation probabilities as in Fig. 9a,c, but in the presence of dissipation. The dash-dotted line shows the time evolution of $\operatorname{tr}\rho^2$. The parameter values are $D = 4$, $\Omega = 0.982\,\omega_0$, $\gamma = 10^{-6}\omega_0$, $k_B T = 10^{-4}\hbar\omega_0$, and $F = 0.0145$ (a), 0.015029 (b). The inset in (a) is a blow up of the rectangle in the upper left corner of that panel.

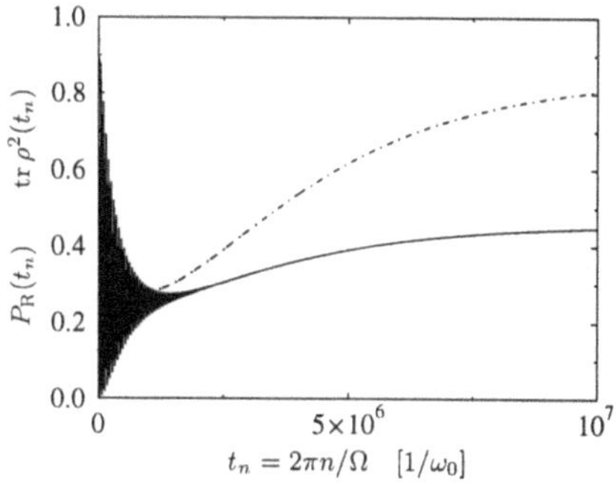

Fig. 11. Time evolution of the return probability P_R (full line) and the coherence function $\operatorname{tr}\rho^2$ (dash-dotted) during loss and regain of coherence. The parameter values are as in Fig. 10b.

To obtain quantitative estimates for the dissipative time scales, we approximate t_decoh by the decay rate of $\operatorname{tr}\rho^2$, as a measure of coherence, averaged over a time t_p,

$$\frac{1}{t_\mathrm{decoh}} = -\frac{1}{t_\mathrm{p}}\int_0^{t_\mathrm{p}} \mathrm{d}t' \frac{\mathrm{d}}{\mathrm{d}t'} \operatorname{tr}\rho^2(t') \tag{54}$$

$$= \frac{1}{t_\mathrm{p}}\Big(\operatorname{tr}\rho^2(0) - \operatorname{tr}\rho^2(t_\mathrm{p})\Big)\,. \tag{55}$$

Because of the stepwise loss of coherence (Fig. 10), we have chosen the propagation time t_p as an $n$fold multiple of the duration $2\pi\hbar/|\epsilon_2^- - \epsilon_1^-|$ of the chaotic beats. For this procedure to be meaningful, n should be so large that the coherence decays substantially during the time t_p (in our numerical studies to a value of approximately 0.9). The time scale t_relax of the approach to the asymptotic state is given by the reciprocal of the smallest real part of the eigenvalues of the dissipative kernel.

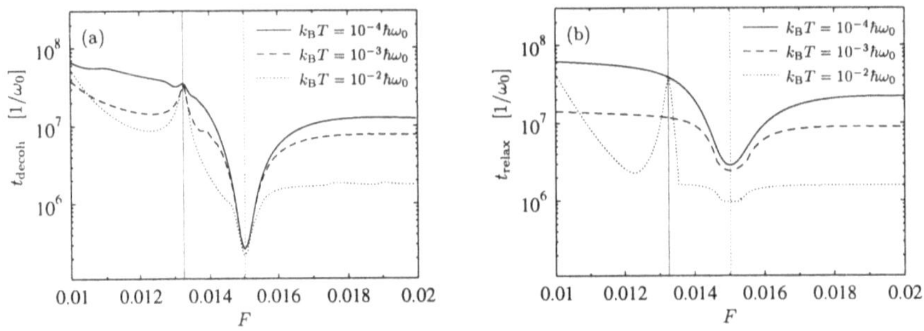

Fig. 12. Time scales of the decay of the coherence measure $\operatorname{tr}\rho^2$ (a) and of the relaxation towards the asymptotic solution (b) near the singlet-doublet crossing. Near the exact crossing ($F \approx 0.013$, full vertical line) coherence is stabilized, whereas at the center of the avoided crossing ($F \approx 0.015$, dashed vertical line) the decay of coherence is accelerated. The parameter values are $D = 4$, $\Omega = 0.982\,\omega_0$, $\gamma = 10^{-6}\omega_0$, temperature as given in the legend.

Outside the singlet-doublet crossing we find that the decay of coherence and the relaxation take place on roughly the same time scale (Fig. 12). At $F \approx 0.013$, the chaotic singlet induces an exact crossing of the ground-state quasienergies (see Fig. 8), resulting in a *stabilization of coherence* with increasing temperature. At the center of the avoided crossing, the decay of coherence becomes much faster and is essentially independent of temperature. This indicates that transitions from states with mean energy far above the ground state play a crucial rôle.

4.3 Asymptotic state

As the dynamics described by the master equation (52) is dissipative, it converges in the long-time limit to an asymptotic state $\rho_\infty(t)$. In general, this attractor remains time dependent but shares the symmetries of the central system, i.e. here, periodicity and generalized parity. However, the coefficients (53) of the master equation for the matrix elements $\rho_{\alpha\beta}$, valid within a moderate rotating-wave approximation, are time independent and so the asymptotic solution also is. The explicit time dependence of the attractor has been effectively eliminated by representing it in the Floquet basis and introducing a mild rotating-wave approximation.

To gain some qualitative insight into the asymptotic solution, we focus on the diagonal elements

$$\mathcal{L}_{\alpha\alpha,\alpha'\alpha'} = 2\sum_n N_{\alpha\alpha',n}|X_{\alpha\alpha',n}|^2, \quad \alpha \neq \alpha', \tag{56}$$

of the dissipative kernel. They give the rates of direct transitions from $|\phi_{\alpha'}\rangle$ to $|\phi_\alpha\rangle$. Within a full rotating-wave approximation [51, 52], these were the only non-vanishing contributions to the master equation to affect the diagonal elements $\rho_{\alpha\alpha}$ of the density matrix.

In the case of zero driving amplitude, the Floquet states $|\phi_\alpha\rangle$ reduce to the eigenstates of the undriven Hamiltonian H_{DW}. The only non-vanishing Fourier component is then $|c_{\alpha,0}\rangle$, and the quasienergies ϵ_α reduce to the corresponding eigenenergies E_α. Thus $\mathcal{L}_{\alpha\alpha,\alpha'\alpha'}$ only consists of a single term proportional to $N(\epsilon_\alpha - \epsilon_{\alpha'})$. It describes two kinds of thermal transitions: decay to states with lower energy and, if the energy difference is less than $k_{\mathrm{B}}T$, thermal activation to states with higher energy. The ratio of the direct transitions forth and back then reads

$$\frac{\mathcal{L}_{\alpha\alpha,\alpha'\alpha'}}{\mathcal{L}_{\alpha'\alpha',\alpha\alpha}} = \exp\left(-\frac{\epsilon_\alpha - \epsilon_{\alpha'}}{k_{\mathrm{B}}T}\right). \tag{57}$$

We have detailed balance and therefore the steady-state solution is

$$\rho_{\alpha\alpha'}(\infty) \sim \mathrm{e}^{-\epsilon_\alpha/k_{\mathrm{B}}T}\,\delta_{\alpha\alpha'}. \tag{58}$$

In particular, the occupation probability decays monotonically with the energy of the eigenstates. In the limit $k_{\mathrm{B}}T \to 0$, the system tends to occupy the ground state only.

For a strong driving, each Floquet state $|\phi_\alpha\rangle$ contains a large number of Fourier components and $\mathcal{L}_{\alpha\alpha,\alpha'\alpha'}$ is given by a sum over contributions with quasienergies $\epsilon_\alpha - \epsilon_{\alpha'} + n\hbar\Omega$. Thus decay to states with "higher" quasienergy (recall that quasienergies do not allow for a global ordering) becomes possible due to terms with $n < 0$. Physically, it amounts to an incoherent transition under absorption of driving-field quanta. Correspondingly, the system tends to occupy Floquet states comprising many Fourier components with low index n. According to Eq. (16), these states have a low mean energy.

The effects under study are found for a driving with a frequency of the order of unity. Thus for a quasienergy doublet, not close to a crossing, we have $|\epsilon_\alpha - \epsilon_{\alpha'}| \ll \hbar\Omega$, and $\mathcal{L}_{\alpha'\alpha',\alpha\alpha}$ is dominated by contributions with $n < 0$, where the splitting has no significant influence. However, up to the tunnel splitting, the two partners in the quasienergy doublet are almost identical. Therefore, with respect to dissipation, both should behave similarly. In particular, one expects an equal population of the doublets even in the limit of zero temperature (Fig. 13a), in contrast to the time-independent case.

In the vicinity of a singlet-doublet crossing the situation is more subtle. Here, the odd partner, say, of the doublet mixes with a chaotic singlet, cf. Eq. (33), and thus acquires components with higher energy. Due to the high mean energy E_{c}^- of the chaotic singlet, close to the top of the barrier, the decay back to the ground state can also proceed indirectly via other states with mean energy below E_{c}^-. Thus $|\phi_1^-\rangle$ and $|\phi_2^-\rangle$ are depleted and mainly $|\phi_0^+\rangle$ will be populated. However, if the temperature is significantly above the splitting $2b$ at the avoided crossing, thermal activation from $|\phi_0^+\rangle$ to $|\phi_{1,2}^-\rangle$, accompanied by depletion via the states below E_{c}^-, becomes possible.

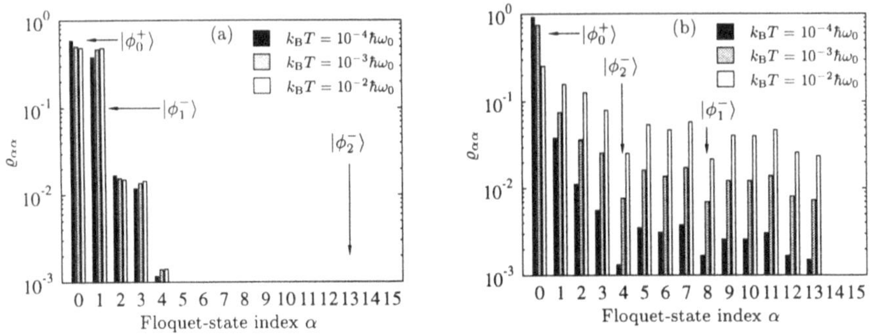

Fig. 13. Occupation probability $\rho_{\alpha\alpha}$ of the Floquet states $|\phi_\alpha\rangle$ in the long-time limit. The parameter values are $D = 4$, $\Omega = 0.982\,\omega_0$, $\gamma = 10^{-6}\omega_0$, and $F = 0.013$ (a), 0.015029 (b), temperature as given in the legend.

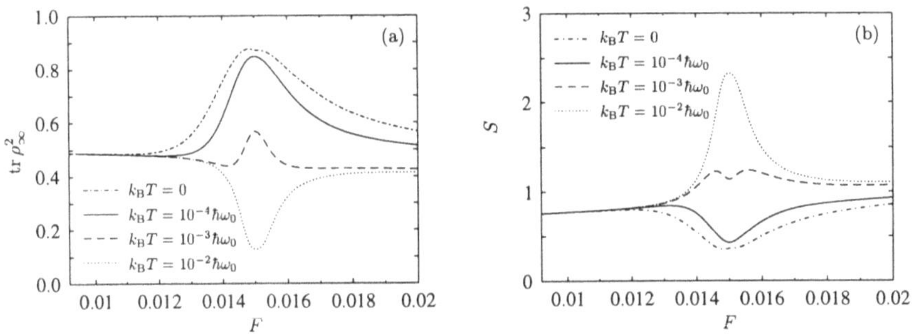

Fig. 14. Coherence (a) and Shannon entropy (b) of the asymptotic state in the vicinity of a singlet-doublet crossing for different temperatures as given in the legend. The other parameter values are $D = 4$, $\Omega = 0.982\,\omega_0$, and $\gamma = 10^{-6}\omega_0$.

Thus asymptotically, all these states become populated in a steady closed flow (Fig. 13b). The long-time limit of the corresponding classical dynamics converges to one of two limit cycles, each of which is located close to one of the potential minima. In a stroboscopic map they correspond to two isolated fixed points. This behavior is qualitatively different from the asymptotic limit of the dissipative quantum dynamics near the center of the crossing and shows that the occupation of levels outside the singlet and the doublet for $t \to \infty$ is a pure quantum effect.

Important global characteristics of the asymptotic state, measuring its degree of spreading over phase space, are the Shannon entropy $S = -\mathrm{tr}\,(\rho_\infty \ln \rho_\infty)$ or, alternatively, $\mathrm{tr}\,\rho_\infty^2$. The latter gives approximately the inverse number of incoherently occupied states and can be considered an "incoherent inverse participation ratio" [57]. It equals unity only if the

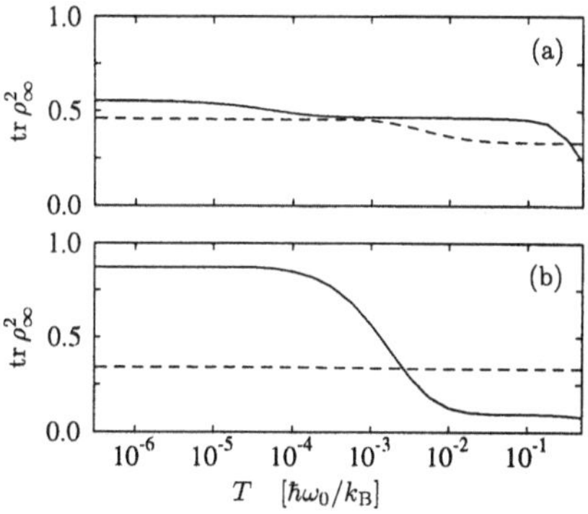

Fig. 15. Coherence of the asymptotic state in the vicinity of a singlet-doublet crossing for $F = 0.013$ (a) and $F = 0.015029$ (b): exact calculation (full line) compared to the result of a three-level description (dashed) of the dissipative dynamics. The other parameter values are $D = 4$, $\Omega = 0.982\,\omega_0$, and $\gamma = 10^{-6}\omega_0$.

attractor is a pure state. According to the above scenario, we expect tr ρ_∞^2 to assume the value 1/2, in a regime with strong driving but preserved doublet structure, reflecting the incoherent population of the ground-state doublet. In the vicinity of the singlet-doublet crossing where the doublet structure is dissolved, its value should be close to unity for temperatures $k_B T \ll 2b$ and much less than unity for $k_B T \gg 2b$ (Figs. 14a, 15). This means that the crossing of the chaotic singlet with the regular doublet leads to an improvement of coherence if the temperature is below the splitting of the avoided crossing, and to a loss for temperatures above the splitting. This phenomenon amounts to a *chaos-induced* coherence or incoherence, respectively. The corresponding Shannon entropy (Fig. 14b) assumes approximately the value $\ln n$ for n incoherently populated states. Thus outside the crossing, we have $S \approx \ln 2$ and at the center of the crossing the entropy exhibits a significant temperature dependence.

The crucial rôle of the decay via states not involved in the three-level crossing can be demonstrated by comparing it to the dissipative dynamics including only these three levels (plus the bath). At the crossing, the three-state model results in a completely different type of asymptotic state (Fig. 15). The failure of the three-state model in the presence of dissipation clearly indicates that in the vicinity of the singlet-doublet crossing, it is important to take a large set of levels into account.

5 Signatures of chaos in the asymptotic state

Phase-space representations of quantum mechanics, like the Husimi or the Wigner distributions, help to reveal the structures of the corresponding classical phase space [7,58–62]. In particular, for the case of regular classical dynamics, the Husimi function of a (quasi)energy eigenstate is localized along the corresponding quantizing torus; for chaotic motion, it is spread over the entire chaotic layer. If the classical dynamics is mixed, quantum-mechanical states can be classified as regular or chaotic according to their distribution in

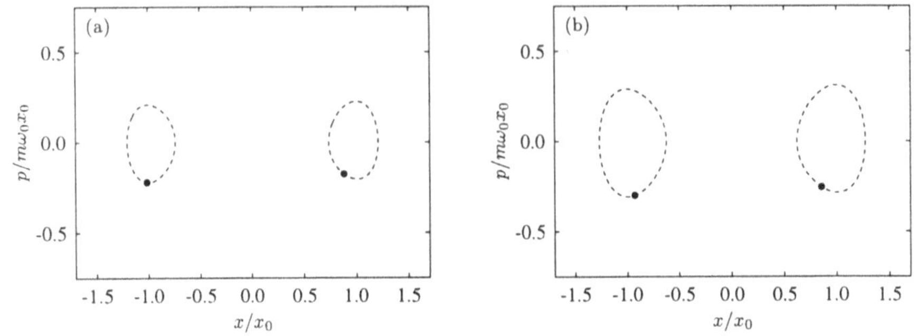

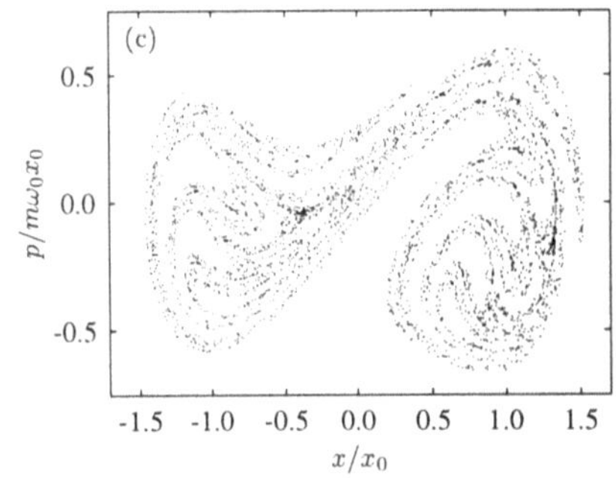

Fig. 16. Stroboscopic classical phase space portrait at $t = nP$, of the dissipative harmonically driven quartic double well, Eqs. (59), (60), for the driving amplitude $F = 0.09$ and frequency $\Omega = 0.9\,\omega_0$. The friction strength is $\gamma = 0.3\,\omega_0$ (a), $0.2\,\omega_0$ (b), $0.03\,\omega_0$ (c). In panels (a) and (b) the stroboscopic portrait is marked by a full dot and the broken lines show the corresponding limit cycles.

phase space [62]. Moreover, the phase-space representation of the asymptotic state of a dissipative quantum map exhibits the structures of the corresponding classical attractor [23]. However, these analogies find their limit in the Heisenberg uncertainty principle. It does not allow for arbitrarily fine phase-space structures and results in smearing on action scales below h.

5.1 Classical attractor

To describe the classical dissipative dynamics of the driven double well, we add an Ohmic friction force $F_\gamma = -\gamma p$ to the conservative equations (4), (5),

$$\dot{x} = \frac{1}{m} p, \tag{59}$$

$$\dot{p} = -\gamma p - \frac{\partial V(x,t)}{\partial x}. \tag{60}$$

Friction destroys the time-reversal symmetry (21) of the conservative system. Accordingly, dissipation breaks the reflection symmetry of the phase-space portrait with respect to the x-axis, found at zero phase of the driving (cf. Fig. 6).

A constituent feature of dissipative flows is the net exponential contraction of phase-space volume. Therefore, the dynamics is asymptotically confined to

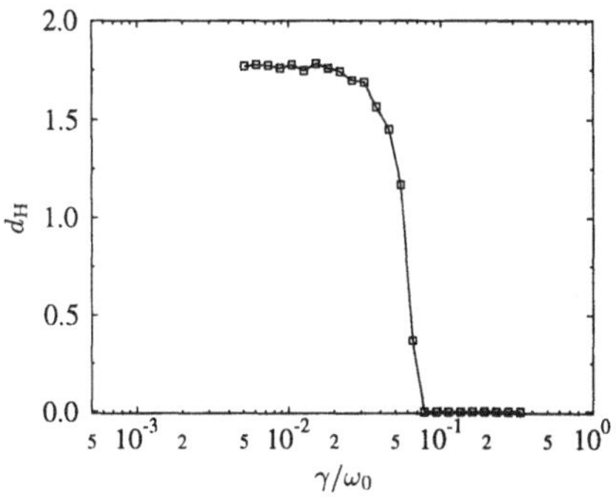

Fig. 17. Hausdorff dimension of the classical attractor of the dissipative harmonically driven quartic double well, Eqs. (59), (60), for $F = 0.09$, $\Omega = 0.9\,\omega_0$.

an attractor, a measure-zero manifold in phase space to which all trajectories starting from within the surrounding basin of attraction converge. For periodically driven dissipative systems, the attractor is in general also time-dependent with the period of the driving and is adequately described by a stroboscopic map [63–65].

Depending on the parameters that control the dissipative flow, an attractor can consist of fixed points, limit cycles, or manifolds of fractal dimension, less than that of phase space. An adequate concept to characterize the geometry of an attractor is the Hausdorff dimension d_{H} defined, for example, in Ref. [3]. It typically increases with decreasing contraction rate, so that strange attractors are expected to occur in the regime of weak dissipation of a system that in absence of friction, is chaotic.

The Hausdorff dimension of the classical attractor for the driven double well with dissipation, Eqs. (59), (60), at the parameter values $F = 0.09$ and $\Omega = 0.9\,\omega_0$, is shown in Fig. 17 for various values of the friction γ. Although the attractor itself is periodically time dependent with the period of the driving, its Hausdorff dimension d_{H} remains nearly constant [63]. Near $\gamma \approx 0.06\,\omega_0$, with decreasing γ, the classical dynamics undergoes a transition from regular motion to chaos, manifest in a corresponding transition from limit cycles (Fig. 16a,b) to a strange attractor (Fig. 16c) and a concomitant jump in d_{H}. At the same values of F and Ω, the regular islands near the potential minima of the corresponding undamped dynamics have already completely dissolved in the chaotic sea.

5.2 Quantum attractor

The "quantum attractors", i.e., the asymptotic states of the dissipative quantum dynamics, for example in a Husimi representation, resemble the corresponding classical attractors up to coarse graining (Fig. 18). Correspondingly, the qualitative transformation from limit cycles to a strange attractor is visible in the asymptotic quantum distribution, but proceeds continuously. Although the asymptotic state for $\gamma = 0.2\,\omega_0$ (Fig. 18b,e), is still concentrated

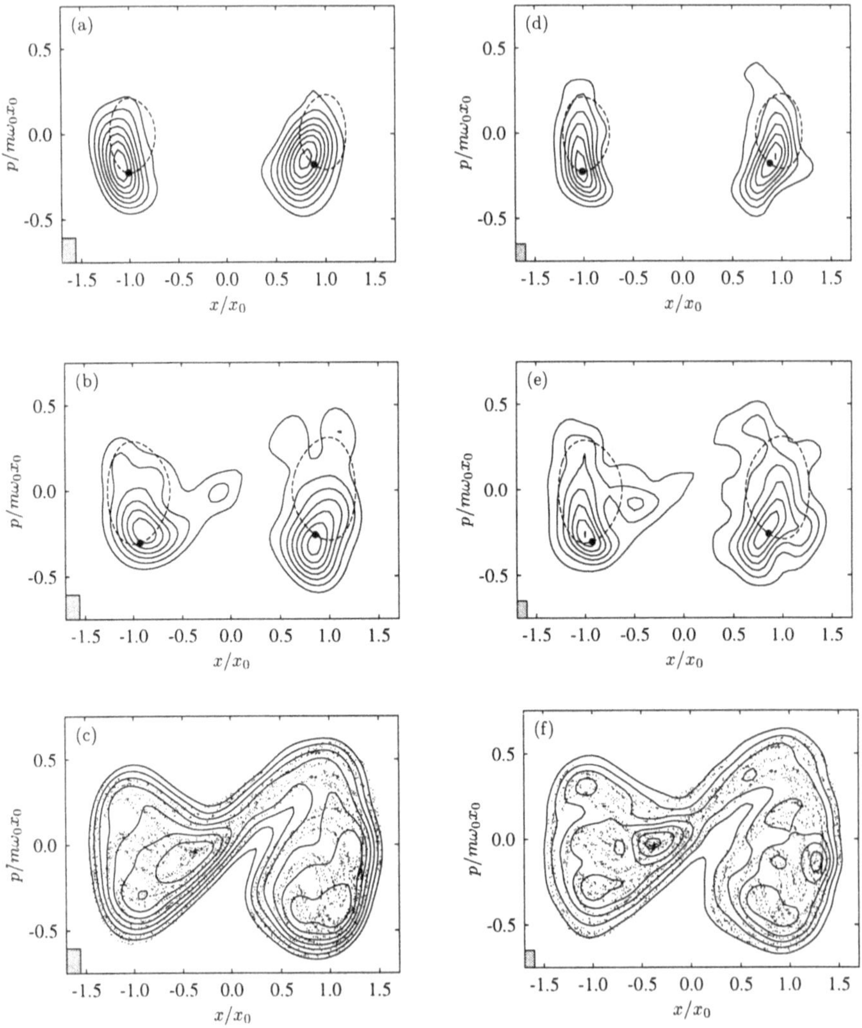

Fig. 18. Contour plot of the Husimi function of the quantum attractor (full lines) at $t = nP$, $n \to \infty$, superposed on the corresponding classical phase-space portrait, Fig. 16. The parameter values $F = 0.09$, $\Omega = 0.9\,\omega_0$, $\gamma = 0.3\,\omega_0$ (a,d), $0.2\,\omega_0$ (b,e), $0.03\,\omega_0$ (c,f) are as in Fig. 16. The effective action is $D = 6$ (a–c) and $D = 12$ (d–f). The rectangle in the lower left corner has the size of the effective quantum of action $\hbar_{\text{eff}} = 1/8D$.

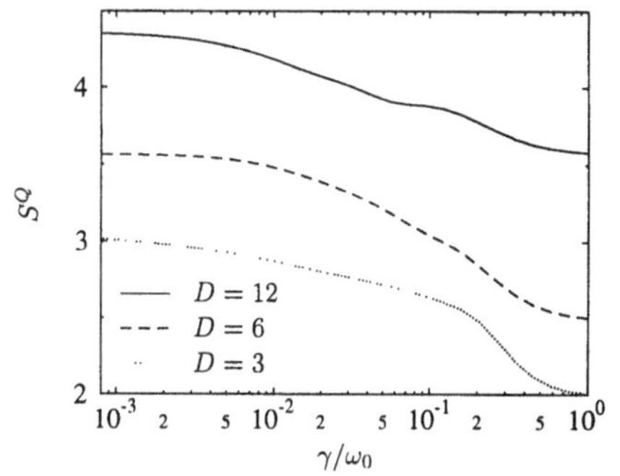

Fig. 19. Wehrl entropy of the asymptotic state of the dissipative quantum map for different values of the effective quantum of action $\hbar_{\text{eff}} = 1/8D$. Other parameters as in Fig. 17.

near the fixed points of the classical stroboscopic map, it covers a broader phase-space area that already anticipates the shape of the strange attractor. Reducing the effective quantum of action $\hbar_{\text{eff}} = 1/8D$ allows for a sharper resolution of the underlying classical structures in the Husimi functions, as expected.

Like the attractors of the dissipative classical dynamics (Fig. 16), their quantum-mechanical counterparts have lost the reflection symmetry with respect to the x-axis, in contrast to the Husimi representations of the Floquet states in absence of dissipation (cf. Fig. 7). This symmetry breaking is caused by finite off-diagonal elements of the asymptotic density matrix in Floquet representation, since diagonal representations share the symmetries of the basis. Thus, off-diagonal matrix elements play a significant rôle for the asymptotic state. This demonstrates that a description within a full rotating-wave approximation is insufficient, since it would result in an asymptotic state diagonal in the Floquet representation [51–53].

Because the quantum attractor, in contrast to the classical one, has support all over phase space (or a region of finite measure), we cannot characterize it by a Hausdorff dimension. A more suitable measure for the extension of the quantum attractor is a phase-space version of the Shannon entropy, the Wehrl entropy [62, 66, 67]

$$S^Q = -\int \frac{\mathrm{d}x\,\mathrm{d}p}{2\pi\hbar}\, Q(x,p)\ln[Q(x,p)] \tag{61}$$

of its Husimi representation

$$Q(x,p) = \langle x,p|\rho_{\mathrm{S}}|x,p\rangle, \tag{62}$$

where $|x,p\rangle$ denotes a coherent state centered at (x,p) in phase space. Its exponential, $\exp(S^Q)$, gives approximately the number of coherent states covered by the Husimi function. Thus, the occupied phase-space area is $2\pi\hbar\exp(S^Q)$. The Wehrl entropy of the asymptotic state for our numerical example is depicted in Fig. 19 for various values of $\hbar_{\text{eff}} = 1/8D$. It grows with decreasing friction γ, reflecting the increasing dispersion of the Husimi functions. In the semiclassical regime, i.e., for a sufficiently large value of the effective action D, we observe a kink-like behavior of the entropy near $\gamma \approx 0.06\,\omega_0$, where the classical attractor undergoes the transition mentioned above, from a set of isolated fixed points to a strange attractor.

Note that for $\gamma \gtrsim 0.1\,\omega_0$, the Markov approximation becomes inaccurate, since γ is then of the order of the mean level spacing and the condition (43) is violated for at least part of the transitions between Floquet states. Nevertheless, we obtain the qualitative behavior which we expected from classical considerations.

6 Conclusion

For the generic situation of the dissipative quantum dynamics of a particle in a driven double-well potential, classical chaos plays a significant rôle for the coherent dynamics. Even for arbitrarily small driving amplitude, the separatrix is replaced by a chaotic layer, while the motion near the bottom of the wells remains regular. Nevertheless, the influence of states located in the chaotic region alters the splittings of the regular doublets and thus the tunnel rates, which is the essence of chaotic tunneling. We have studied chaotic tunneling in the vicinity of crossings of chaotic singlets with tunnel doublets under the influence of an environment. As a simple intuitive model to compare against, we have constructed a three-state system which in the case of vanishing dissipation, provides a faithful description of an isolated singlet-doublet crossing. Dissipation introduces new time scales to the system: one for the loss of coherence and a second one for the relaxation to an asymptotic state. Well outside the crossing, both time-scales are of the same order, reflecting an effective two-state behavior. The center of the crossing is characterized by a strong mixing of the chaotic state with one state of the tunnel doublet. The high mean energy of the chaotic state introduces additional decay channels to states outside the three-state system. Thus, decoherence becomes far more effective and, accordingly, tunneling fades out much faster.

The study of the asymptotic state, the quantum attractor, demonstrates clearly that a three-state model of the singlet-doublet crossing is insufficient once dissipation is effective. This is so because the coupling to the heat bath enables processes of decay and thermal activation that connect the states in the crossing with other, "external" states of the central system. In the presence of driving, the asymptotic state is no longer literally a state of equilibrium. Rather, incoherent processes create a steady flow of probability involving states within as well as outside the crossing. As a result, the composition of the asymptotic state, expressed for example by its coherence $\mathrm{tr}\,\rho_\infty^2$, is markedly different at the center of the crossing as compared to the asymptotic state far away from the crossing, even if that is barely visible in the corresponding classical phase-space structure.

With increasing driving amplitude, in absence of dissipation, even the dynamics near the bottom of the wells becomes fully chaotic. This has striking consequences for the corresponding dissipative classical dynamics: For sufficiently weak dissipation, it remains chaotic, but for strong friction it becomes regular. Accordingly, the geometry of the classical attractor is fractal or regular, respectively. We have observed the signatures of this qualitative difference in the asymptotic state of the corresponding quantum dynamics. However, in contrast to the sudden change of the classical behavior, the quantum attractor undergoes a smooth transition: The structure of the strange attractor is already felt by the Husimi function for parameter values where the classical attractor consists only of two isolated fixed points. For the observation of these semiclassical structures, off-diagonal matrix elements of the asymptotic state in Floquet basis proved crucial. This clearly indicates that a full rotating-wave approximation must fail.

Many more phenomena at the overlap of chaos, tunneling, and dissipation await being unraveled. They include four-state crossings formed when two doublets intersect, chaotic Bloch tunneling along extended potentials with a large number of unit cells instead of just two, and the influence of decoherence on a multi-step mechanism of chaotic tunneling. These phenomena are typically observed in the far semiclassical regime, which requires to take very many levels into account. A semiclassical description of the dissipative quantum system may circumvent this problem.

References

[1] G. Casati, B. V. Chirikov, F. M. Izrailev, and J. Ford, in *Stochastic Behavior in Classical and Quantum Hamiltonian Systems*, Vol. 93 of *Lecture Notes in Physics*, edited by G. Casati and J. Ford (Springer, Berlin, 1979), p. 334.
[2] T. Dittrich and R. Graham, Ann. Phys. (N.Y.) **200**, 363 (1990).
[3] E. Ott, *Chaos in Dynamical Systems*, Cambridge University Press (Cambridge 1993).
[4] M.J. Davis and E.J. Heller, J. Chem. Phys. 75, 246 (1981).
[5] O. Bohigas, S. Tomsovic, and D. Ullmo, Phys. Rev. Lett. **64**, 1479 (1990).
[6] O. Bohigas, S. Tomsovic, and D. Ullmo, Phys. Rev. Lett. **65**, 5 (1990).
[7] O. Bohigas, S. Tomsovic, and D. Ullmo, Phys. Rep. **223**, 43 (1993).
[8] S. Tomsovic and D. Ullmo, Phys. Rev. E **50**, 145 (1994).
[9] W. A. Lin and L. E. Ballentine, Phys. Rev. Lett. **65**, 2927 (1990).
[10] W. A. Lin and L. E. Ballentine, Phys. Rev. A **45**, 3637 (1992).
[11] R. Utermann, T. Dittrich, and P. Hänggi, Phys. Rev. E **49**, 273 (1994).
[12] P. Hänggi, R. Utermann, and T. Dittrich, Physica B **194-196**, 1013 (1994).
[13] M. Latka, P. Grigolini, and B. J. West, Phys. Rev. E **50**, 596 (1994).
[14] M. Latka, P. Grigolini, and B. J. West, Phys. Rev. A **50**, 1071 (1994).
[15] M. Latka, P. Grigolini, and B. J. West, Phys. Rev. E **50**, R3299 (1994).
[16] E. M. Zanardi, J. Gutiérrez, and J. M. Gomez Llorente, Phys. Rev. E **52**, 4736 (1995).
[17] E. Doron and S. D. Frischat, Phys. Rev. Lett **75**, 3661 (1995).

[18] S. D. Frischat and E. Doron, Phys. Rev. E **57**, 1421 (1998).
[19] F. Leyvraz and D. Ullmo, J. Phys. A **29**, 2529 (1996).
[20] R. Roncaglia, L. Bonci, F. M. Izrailev, B. J. West, and P. Grigolini, Phys. Rev. Lett. **73**, 802 (1994).
[21] A. O. Caldeira and A. L. Leggett, Ann. Phys. (N.Y.) **149**, 374 (1983); erratum: Ann. Phys. (N.Y.) **153**, 445 (1984).
[22] M. Grifoni and P. Hänggi, Phys. Rep. **304**, 219 (1998).
[23] T. Dittrich and R. Graham, Europhys. Lett. **4**, 263 (1987).
[24] J. H. Shirley, Phys. Rev. **138**, B979 (1965).
[25] H. Sambe, Phys. Rev. A **7**, 2203 (1973).
[26] N. L. Manakov, V. D. Ovsiannikov, and L. P. Rapoport, Phys. Rep. **141**, 319 (1986).
[27] S.-I. Chu, Adv. Chem. Phys. **73**, 739 (1989).
[28] T. Dittrich, P. Hänggi, G.-L. Ingold, B. Kramer, G. Schön, and W. Zwerger, *Quantum Transport and Dissipation* (Wiley-VCH, Weinheim, 1998).
[29] D. J. Moore, Helv. Phys. Acta **66**, 3 (1993).
[30] M. L. Mehta, *Random matrices and the statistical theory of energy levels* (Academic Press, New York, 1967).
[31] F. Haake, *Quantum Signatures of Chaos*, Vol. 54 of *Springer Series in Synergetics* (Springer, Berlin, 1991).
[32] F. Grossmann, T. Dittrich, P. Jung, and P. Hänggi, Phys. Rev. Lett. **67**, 516 (1991).
[33] F. Grossmann, P. Jung, T. Dittrich, and P. Hänggi, Z. Phys. B **84**, 315 (1991).
[34] A. Peres, Phys. Rev. Lett. **67**, 158 (1991).
[35] F. Grossmann and P. Hänggi, Europhys. Lett. **18**, 571 (1992).
[36] P. Hänggi, P. Talkner, and M. Borkovec, Rev. Mod. Phys. **62**, 251 (1990).
[37] T. Dittrich, B. Oelschlägel, and P. Hänggi, Europhys. Lett. **22**, 5 (1993).
[38] B. Oelschlägel, T. Dittrich, and P. Hänggi, Acta Physica Polonica B **24**, 845 (1993).
[39] T. Dittrich, P. Hänggi, B. Oelschlägel, and R. Utermann, in *25 Years of Non-Equilibrium Statistical Mechanics*, Vol. 445 of *Lecture Notes in Physics*, edited by J. J. Brey (Springer, Berlin, 1995), p. 269.
[40] A. J. Lichtenberg and M. A. Liebermann, *Regular and Stochastic Motion*, Vol. 38 of *Applied Mathical Sciences* (Springer, New York, 1983).
[41] D. F. Escande, Phys. Rep. **121**, 165 (1985).
[42] L. E. Reichl and W. M. Zheng, in *Directions in Chaos*, edited by H. B. Lin (World Scientific, Singapore, 1987), Vol. 1, p. 17.
[43] M. Wilkinson, Physica D **21**, 341 (1986).
[44] M. Wilkinson, J. Phys. A **20**, 635 (1987).
[45] S. Kohler, R. Utermann, P. Hänggi, and T. Dittrich, Phys. Rev. E **58**, 7219 (1998).
[46] L. E. Reichl, *The Transition to Chaos: In Conservative and Classical Systems: Quantum Manifestations* (Springer, New York, 1992).
[47] R. B. Shirts and W. P. Reinhardt, J. Chem. Phys. **77**, 5204 (1982).
[48] A first quantum Langevin formulation has been given by: V. B. Magalinskiĭ, Zh. Eksp. Teor. Fiz. **36**, 1942 (1959), [Sov. Phys. JETP **9**, 1381 (1959)].
[49] Classical formulations of the system-harmonic oscillator bath coupling scheme have been put forward by: N. N. Bogoliubov, in *On some statistical methods*

in mathematical physics (Ukr. S.S.R. Acad. of Science Press, 1945), p. 115 (in russian); R. Zwanzig, J. Stat. Phys. **9**, 215 (1973).
[50] F. Haake, in *Quantum Statistics in Optics and Solid-State Physics*, Vol. 66 of *Springer Tracts in Modern Physics*, edited by G. Höhler (Springer, Berlin, 1973).
[51] R. Blümel *et al.*, Phys. Rev. Lett. **62**, 341 (1989).
[52] R. Blümel *et al.*, Phys. Rev. A **44**, 4521 (1991).
[53] R. Graham and R. Hübner, Ann. Phys. (N.Y.) **234**, 300 (1994).
[54] S. Kohler, T. Dittrich, and P. Hänggi, Phys. Rev. E **55**, 300 (1997).
[55] D. E. Makarov and N. Makri, Phys. Rev. E **52**, 5863 (1995).
[56] N. Makri, J. Chem. Phys. **106**, 2286 (1997).
[57] T. Dittrich and U. Smilansky, Nonlinearity **4**, 59 (1991).
[58] K. Takahashi and N. Saitô, Phys. Rev. Lett. **55**, 645 (1985).
[59] S.-J. Chang and K.-J. Shi, Phys. Rev. Lett. **55**, 269 (1985).
[60] S.-J. Chang and K.-J. Shi, Phys. Rev. A **34**, 7 (1986).
[61] B. Mirbach and H. J. Korsch, J. Phys. A **27**, 6579 (1994).
[62] T. Gorin, H. J. Korsch, and B. Mirbach, Chem. Phys. **217**, 145 (1997).
[63] F. C. Moon and G.-X. Li, Physica D **17**, 99 (1985).
[64] F. C. Moon and G.-X. Li, Phys. Rev. Lett. **55**, 1439 (1985).
[65] W. Szemplinska-Stupnicka, Nonlinear Dynamics **3**, 225 (1992).
[66] A. Anderson and J. J. Halliwell, Phys. Rev. D **48**, 2753 (1993).
[67] A. Wehrl, Reps. Math. Phys. **16**, 353 (1979); Reps. Math. Phys. **30**, 119 (1991).

A Fermi Pump

Mathias Wagner[1] and Fernando Sols[2]

[1] Hitachi Cambridge Laboratory, Madingley Road, Cambridge CB3 0HE, United Kingdom
[2] Dpto. de Física Teórica de la Materia Condensada, C-V and Instituto de Ciencia de Materiales "Nicolás Cabrera", Universidad Autónoma de Madrid, E-28049 Madrid, Spain

Abstract. The use of a band offset between the two leads of an electron pump driven by a locally applied oscillating gate voltage is shown to increase the pump current dramatically. A spectral analysis reveals that the bulk of the pump current flows deep within the Fermi sea and not at the Fermi surface, especially in higher spatial dimensions, thereby rendering this effect insensitive to temperature.

The physics of mesoscopic systems has taught us that the spatial coherence of electronic wave functions has a profound impact on the transport characteristics. Equally, recent studies of *ac driven* nanostructured systems show clear evidence that the spatio-temporal coherence of the wave function brings about a wealth of new effects, such as the experimental realization [1] of the adiabatic Thouless pump [2, 3, 4, 5, 6], which is the mesoscopic analog of Archimedes' water pump, or the prediction of Rabi oscillations in driven quantum-dot pumps [7, 8]. Frequency-locked turnstiles [9, 10] clocking electrons through the system one by one are now promising candidates for future current standards, and in ratchets [11, 12, 13] the combination of asymmetry, nonlinearity, and noise, also results in current rectification. It is not necessary, though, to employ very sophisticated operational schemes: Standing-wave resonances above a single driven asymmetric barrier have also been predicted to yield a pump action [14]. In all these cases, to observe a pump current at zero applied bias it is instrumental that the inversion symmetry be broken, either in real or in k space — a fact which has long been known in the theory of photogalvanic and photovoltaic effects in ferroelectrics [15].

In this contribution we want to put forward a new idea on how to achieve pumping in a driven system which, at the same time, also illustrates a perhaps more general mechanism that underlies some of the other pumps mentioned above. The point we wish to make is that under fairly general conditions the pump current is a genuine Fermi-sea effect, with the bulk of the current being carried *deep within* the Fermi sea in spatial dimensions $D > 1$, thus rendering it resilient against temperature. This is in stark contrast to what is found in dc transport, where only the states within a few kT of the Fermi surface contribute to the current.

We consider a very simple model system as depicted in Fig. 1. In this model, a quantum well is driven harmonically by an external ac potential

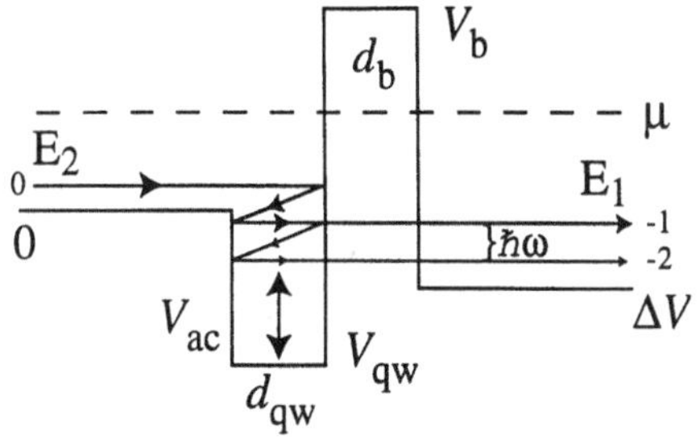

Fig. 1. Schematic potential profile for a Fermi-sea pump. The chemical potential μ is the same in the left and right contacts.

$V_{ac} \cos \omega t$ provided by some gates, for instance. Adjacent to the well is a static barrier, and the overall potential profile features a band offset ΔV between the left and right leads. The spatial asymmetry of the model, and in particular the band offset ΔV, is vital for a large pump current to exist. In the following, we will discuss this model in 1, 2 and 3 spatial dimensions, assuming separability of the Hamiltonian $H(t) = -(\hbar^2/2m)\Delta + V_0(z) + V_{ac}(z,t)$.

The central point to realise is that the driving force leads to *inelastic* transport channels opening up, as schematically shown in Fig. 1, which in the following we will call "pipelines". In particular, we are interested in pipelines close to the band edge, where the energy of the electron is comparable to the photon energy $\hbar\omega$ of the driving. Consider, for instance, an electron incident from the left with longitudinal energy $E_z = E_2$ less than $\hbar\omega$. Upon reflection at the barrier, it can emit or absorb one or more photon quanta. In the case of emission, its longitudinal energy E_z will be below the band edge of the left side, resulting in a reflection at the left wall of the quantum well and, with a good probability, a subsequent "capturing" of the electron in the quantum well. Once captured, it can tunnel to the right through the barrier at an energy E_1, which is one or more photon quanta less than the energy E_2 it had initially. In this way, a range of pipelines, or inelastic transmission channels, are set up which have one common entry point to the left at energy E_2, and a range of lower exit points to the right at energies $E_1 = E_2$ - $\hbar\omega$, E_2 - $2\hbar\omega$, ... Obviously, each pipeline can be traversed in both directions and, due to time-reversal symmetry, $V_{ac}(z,t) = V_{ac}(z,-t)$, microreversibility holds: The probability for an electron to go from E_2 to E_1 can be proved to be the same as for the reverse direction, i.e., $T_{\rightarrow}(E_1, E_2) \equiv T_{\leftarrow}(E_2, E_1)$. It is therefore convenient to study the *net* total transmission probability $T_{net}(E_z)$, which is defined as the difference between the total right- and leftward transmission probabilities at a given longitudinal energy E_z of the incident electron. T_{net} can, for instance, be calculated using a transfer-matrix formalism [16, 17]. In Fig. 2 we provide a contour plot of T_{net} as a function of driving V_{ac} and longitudinal energy E_z. The parameters used were $\hbar\omega = 0.1\,\text{meV} \approx 24\,\text{GHz}$,

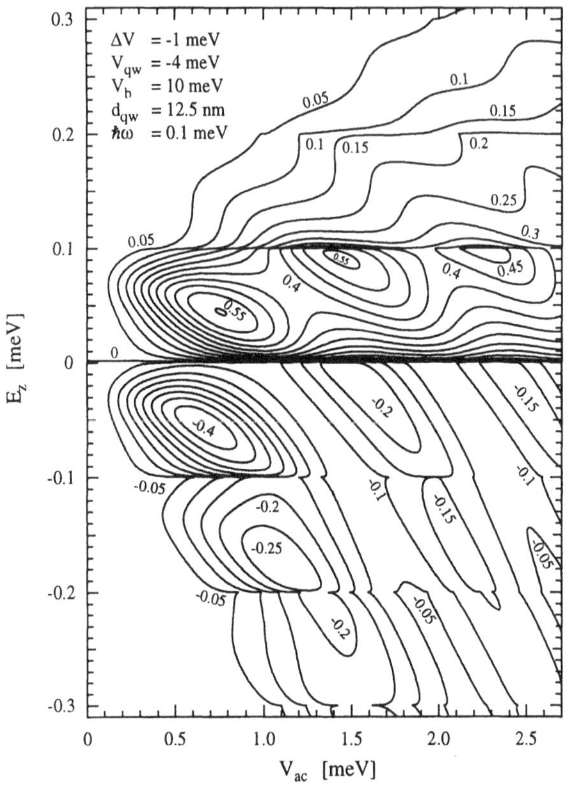

Fig. 2. Contour plot of the net transmission probability T_{net} as function of driving V_{ac} and longitudinal energy E_z of the incident electron. $E_z = 0$ is the band edge on the left-hand side of Fig. 1. Parameters: $\hbar\omega = 0.1\,\text{meV} \approx 24$ GHz, $V_{\text{b}} = 10$, $V_{\text{qw}} = -4$, $\Delta V = -1$ (meV), $d_{\text{b}} = 10$, $d_{\text{qw}} = 12.5$ (nm).

$V_{\text{b}} = 10$, $V_{\text{qw}} = -4$, $\Delta V = -1$ (all measured in meV with respect to the band edge in the left lead), $d_{\text{b}} = 10\,\text{nm}$, $d_{\text{qw}} = 12.5\,\text{nm}$, and an effective mass of $0.067\,m_0$. Clearly, the energy range most important for transport is located within a few $\hbar\omega$ of the band edge ($\equiv E_z = 0$) in the left-hand lead, as T_{net} tends to zero for $E_z \gg \hbar\omega$. For $0 < E_z < \hbar\omega$ pronounced maxima are seen at $V_{\text{ac}} \approx 0.7$, 1.5 and 2.3 meV, with corresponding minima for $E_z < 0$ at roughly the same driving strengths. These maxima and minima are generated by dominant pipelines. For instance, the maximum at $E_z \approx 0.04\,\text{meV}$ and $V_{\text{ac}} \approx 0.7\,\text{meV}$ is the "entrance" (E_2 in Fig. 1) of a pipeline on the left of the structure, whilst the series of minima at energies -0.06, $-0.16\,\text{meV}$, etc. correspond to the "exits" at the right (E_1 in Fig. 1).

The next step is to calculate the dc current response to the ac driving. For this we take advantage of the fact that in the absence of an external dc bias the chemical potential μ characterizing the electrons incident from the (thermal) reservoirs at the far left and right is the same on either side of the structure. Hence, a Tsu-Esaki type of current formula valid for coherent transport can be employed,

$$I = \int_{\Delta V}^{\infty} dE f(E-\mu) J(E) \ , \tag{1}$$

where $E = E_z + E_\perp$ is the *total* energy of the incident electron consisting of its longitudinal and transversal energies (with respect to the direction of transport), $f(E-\mu)$ is the Fermi function describing electrons incident from reservoirs assumed in thermal equilibrium, and $J(E)$ is the spectral current density defined as [18, 19]

$$J(E) = \frac{2e}{h} \int_{\Delta V}^{E} dE_z D_\perp (E - E_z) T_{\text{net}}(E_z) \ . \tag{2}$$

In Eq. (2) $D_\perp$ is the density of states relating to the spatial dimensions *perpendicular* to the current transport. In the absence of any ac driving, T_{net} vanishes on account of the principle of detailed balance, but for $V_{\text{ac}} \neq 0$ the spatial asymmetry of the model system induces a substantial net transmission probability, as seen in Fig. 2.

Figure 3 shows the results for the dc pump current at a low temperature of 100 mK (i.e. $< \hbar\omega$) as a function of the driving potential V_{ac} and the chemical potential μ, for spatial dimensions 1, 2, and 3. In all three cases we find a strong increase of the pump current with V_{ac}, typical values being of the order of a few nA in one, and a few 100 nA/μm^2 in three dimensions. Except for the 3D case, the current exhibits a pronounced maximum as a function of the chemical potential μ. In 1D this is somewhere between 0 and $\hbar\omega$, whereas in 2D it is shifted to slightly larger chemical potentials. Note that the current is *negative*, i.e., it is an "uphill" particle current from right to left in Fig. 1. In Fig. 4 we present the same data, but now for a much higher temperature of 4.2 K, which is *larger* than $\hbar\omega$. Naively, at such elevated temperatures, one would expect the pump current to be diminished, as now electrons with kinetic energies larger than $\hbar\omega$ should contribute to the current — and this is indeed what is observed in the 1D case. However, in 2D, and in particular in 3D the current is still rather strong. In fact, the only difference in the 3D case is a smaller slope $\partial I/\partial\mu$ for small μ and large V_{ac}, but the limiting current for large μ is virtually the same as in Fig. 3. This is, at first, a rather surprising result: The pump current is only little, if anything, affected by temperature in the 3D geometry.

In order to understand the essence of the relevant physics involved, we analyse a simplified model, which allows for analytical results at finite temperatures. Let us consider a single pipeline connecting two energies E_2 and E_1 on the left and right, respectively. In this model we can write

$$T_{\text{net}}(E_z) = T_{\text{p}} \left[\delta(E_z - E_2) - \delta(E_z - E_1) \right] \ , \tag{3}$$

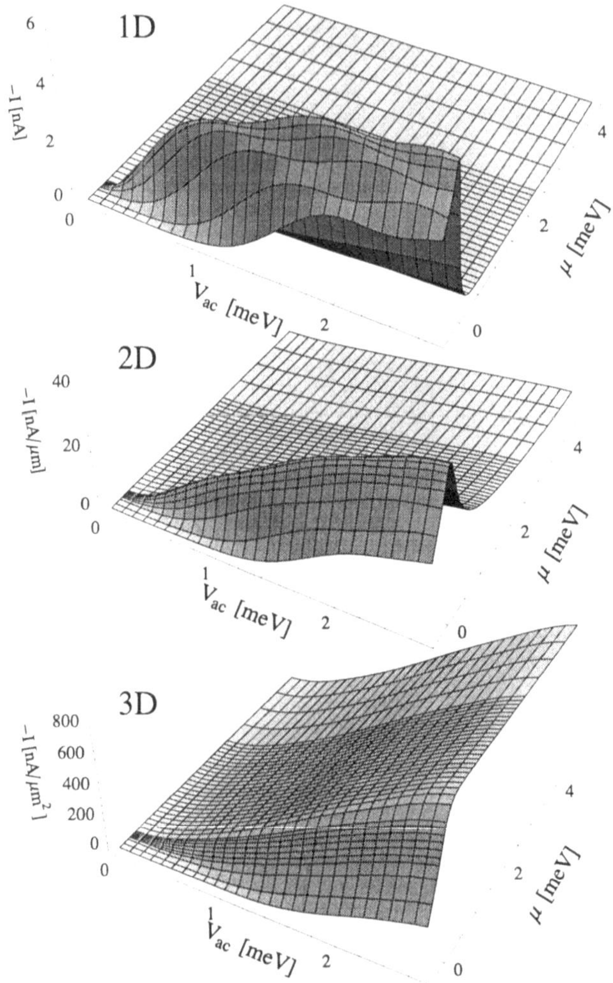

Fig. 3. Net pump current for 1, 2, and 3D at $T = 100\,\mathrm{mK}$ as a function of the driving amplitude V_{ac} and chemical potential μ. Parameters as in Fig. 2.

where T_{p} is the strength of the pipeline. With Eqs. (1) and (2) the pump current then becomes

$$I = \frac{2e}{h} T_{\mathrm{p}} \int_0^{\infty} dE_{\perp} D_{\perp}(E_{\perp}) \times [f(E_{\perp} + E_2 - \mu) - f(E_{\perp} + E_1 - \mu)] \quad (4)$$

and has a straight-forward physical interpretation: As the chemical potential μ raises, the lower end of the pipeline at E_1 will get submerged first, resulting in a particle current being pumped from right to left in Fig. 1. Once the left-

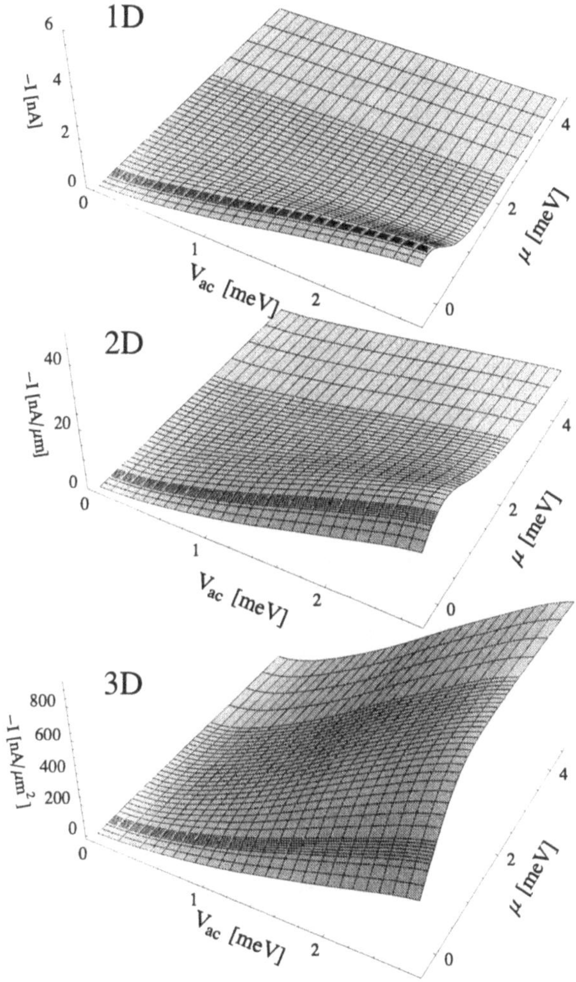

Fig. 4. Net pump current for 1, 2, and 3D at $T = 4.2\,\mathrm{K}$ as a function of the driving amplitude V_{ac} and chemical potential μ. Parameters as in Fig. 2.

hand side of the pipeline is also submerged below μ, current will be pumped in the opposite direction as well, and whether or not these two currents cancel depends, as we will see in the following, on $D_\perp$, i.e., the dimensionality in transversal direction.

For a 1D pump where $D_\perp(E_\perp) = \delta(E_\perp)$, Eq. (4) yields

$$I_{1\mathrm{D}} = \frac{2e}{h} T_{\mathrm{p}} \left[f(E_2 - \mu) - f(E_1 - \mu) \right] \;, \tag{5}$$

which has a peak at $\mu = (E_1 + E_2)/2$, and an exponential decay for $\mu \gg kT$: $I_{1\mathrm{D}} = -(2e/h) T_{\mathrm{p}} \mathrm{e}^{-\beta\mu} \beta (E_2 - E_1)$. In 2D we have $D_\perp(E_\perp) = \sqrt{2m/E_\perp}/h$, and Eq. (4) gives

$$I_{2D} \approx \frac{2e}{h^2} T_p \sqrt{2\pi m \beta} \Big[\mathrm{Li}_{-\frac{1}{2}}(-e^{\beta\mu})(E_2 - E_1) - \frac{\beta}{2} \mathrm{Li}_{-\frac{3}{2}}(-e^{\beta\mu})(E_2^2 - E_1^2) \Big] , \quad (6)$$

where Li is the polylogarithm function [20]. We see that in 2D the current decays only *algebraically* for $\mu \gg kT$, $I_{2D} = -(2e/h^2)T_p\sqrt{2m/\mu}(E_2 - E_1)$, independent of temperature, and has a maximum at $\mu \approx E_2 + kT$. Finally, in 3D we have $D_\perp = 2\pi m/h^2$ independent of energy, and

$$I_{3D} \approx \frac{4\pi me}{h^3} T_p \left[f(-\mu)(E_1 - E_2) + \frac{f'(-\mu)}{2}(E_1^2 - E_2^2) \right] . \quad (7)$$

For $\mu \gg kT$ one has $f(-\mu) \approx 1$ and $f'(-\mu) \approx 0$, i.e., the current becomes *independent* of μ in this limit, $I_{3D} = -(4\pi me/h^3)T_p(E_2 - E_1)$. This result can be interpreted as $I = D_\perp \times$ *strength* $\times$ *lift* of the pump. The maxima and asymptotic behaviour for large μ predicted by Eqs. (6) and (7) for the 2D and 3D cases agree within 5% with the numerical results of Figs. 3 and 4. However, to get quantitative agreement with the asymptotic behaviour in 1D, or to describe the dependence on V_{ac} in any spatial dimension, a distribution of pipelines is required.

The 3D result can also be understood in the following way: Consider a single pipeline in 3D as in Fig. 1 and electrons incident from both sides at *total* energy $E = E_z + E_\perp > \max(E_1, E_2)$. For electrons from the left only those with $E_z = E_2$ will be able to use the pipeline, whereas on the right $E_z = E_1$ must hold. Since $D_\perp$ does not depend on energy in 3D, it follows that the number of electrons fulfilling these conditions is the same on either side and, with microreversibility, $T_\rightarrow(E_1, E_2) = T_\leftarrow(E_2, E_1)$, the net current carried by states in the energy shell $(E, E + dE)$ of the *incident* electrons is zero regardless of temperature. Any pump current, therefore, must stem from states with total energy E between E_1 and E_2. We stress that it is the *total* energy E of the states we are referring to here, and not merely their longitudinal energy E_z! This distinction is important when asking the question where, in the Fermi sea, does the pump current flow.

To make this point more quantitative, we have plotted in Fig. 5 the spectral net current density $J(E)$, as defined in Eq. (2), as a function of the total energy E of the incident electron. For clarity, we have offset the curves for 2D and 3D by 1 and 2 meV, respectively. In all three dimensions, $J(E)$ shows a pronounced structure at $E \approx 0$, and quickly decays to zero for $E > \hbar\omega$. This confirms that for $\mu \gg \hbar\omega$ the pump current is indeed carried by states deeply submerged in the Fermi sea. In 1D and 2D $J(E)$ changes sign, leading to a strong cancellation once μ exceeds $\hbar\omega$, which partially negates this effect. However, for 3D pumps, $J(E)$ has a single, pronounced peak at $E < \hbar\omega$ that carries the bulk of the pump current. The fact that the current-carrying states are far away from the Fermi surface also explains why in Fig. 4 (where

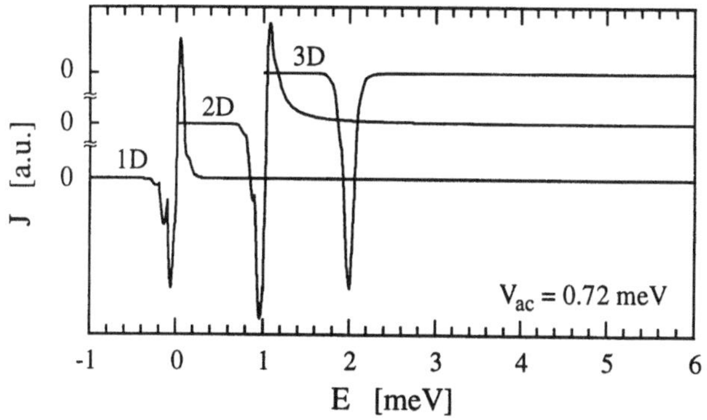

Fig. 5. Spectral current density $J(E)$ [see Eq.(1)] as function of the *total* energy of the incident electron, for 1, 2, and 3D. Curves are displaced for clarity. Parameters as in Fig. 2.

$kT \gg \hbar\omega$) the current, in particular for 3D and to some extend for 2D, is so insensitive to temperature, which is a "melting" of the Fermi surface.

We have made extensive studies on the dependence of the pump current on the structural parameters ΔV, V_{qw}, V_b, and d_{qw}, and have found the pumping action to be very robust against changes. In particular, we found that ΔV needs to exceed $\hbar\omega$ only by a few times to achieve an increase of the pump current by typically three orders of magnitude, which is in agreement with the fact that the most dominant pipelines are within a few $\hbar\omega$ of the band edge.

Finally, as a hint at the more general nature and robustness of the effect, we present in Fig. 6 the dc current through our Fermi pump when operated in a *two-terminal* mode (at 100 mK), where we have assumed the driving potential to extend throughout the left lead such that $V_{\mathrm{ac}}^{\mathrm{lead}} = 2V_{\mathrm{ac}}^{\mathrm{qw}}$, in a crude attempt to model screening. Again, a substantial pump current was found.

Before concluding, it is instructive to compare these results with very recent work of the group of Büttiker, where a statement has been made in Ref. [21] to the effect that there is *no* dc pump current at all, if the chemical potential is the same in the left and right leads. At first, this seems like a complete contradiction to our findings, but on closer inspection one finds that their conclusion is essentially based on the assumption that all the current-carrying electrons have a kinetic energy much *larger* than the photon energy, in which case one can neglect the differences in the phase velocities of the different sidebands [see their comment after Eq. (13)]. This result is very interesting, as it strongly supports our findings that it is exactly the *low-energetic* electrons in the pipelines of Fig. 1, where the phase velocities do depend strongly on energy, which dominate the dc current transport in the Fermi pump presented in this work.

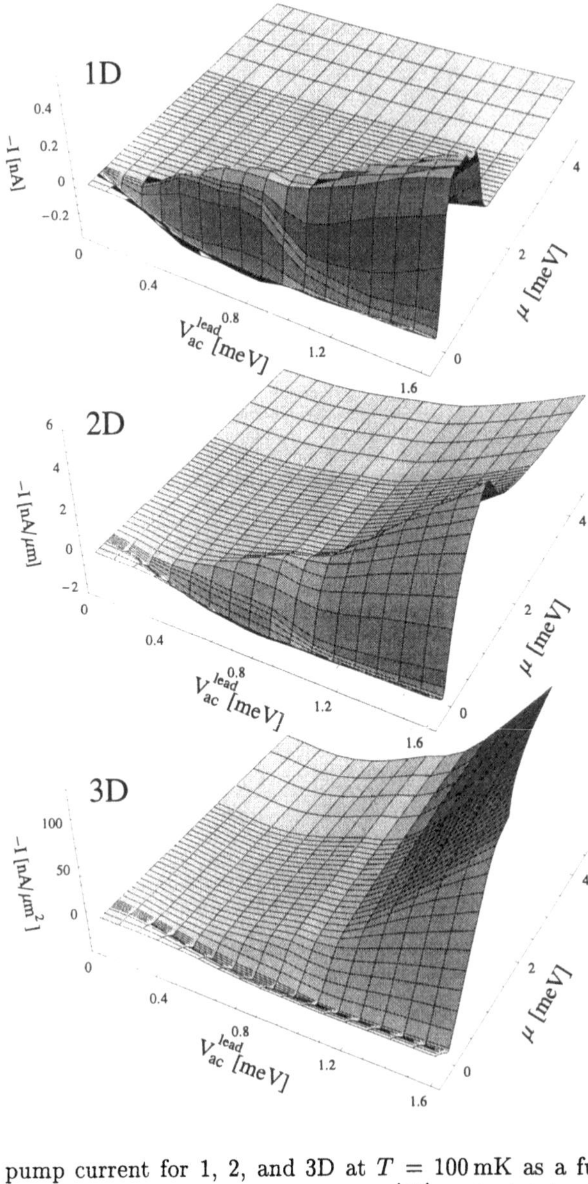

Fig. 6. Net pump current for 1, 2, and 3D at $T = 100\,\text{mK}$ as a function of the chemical potential μ and the driving amplitude $V_{\text{ac}}^{\text{lead}}$ in the left lead of Fig. 1 (for two-terminal operation). Parameters as in Fig. 2.

In summary, we have described a simple and efficient electron pump where the current-carrying states are deeply submerged in the Fermi sea. The resulting "subsea" electron current is resilient against temperature — especially in 3D, but also to some extent in 2D. In 3D the current is independent of the chemical potential μ provided $\mu \gg kT$, while in 2D the pump current decays as $\mu^{-1/2}$. The effect relies mostly on the existence of a band offset between the leads and thus is very stable against imperfections of the model parameter. The pump current predicted is large and should be easily measurable with today's technologies. Due to its fundamental character, the mechanism of inelastic transmission deep below the Fermi surface should be applicable in a wide range of structures. The authors appreciate helpful discussions with S. Kohler. This work has been supported by the EU via TMR contract FMRX-CT98-0180, and by DGICyT (PB96-0080-C02).

References

[1] M. Switkes, C. M. Marcus, K. Campman, and A. C. Gossard, Science **283**, 1905 (1999).
[2] D. J. Thouless, Phys. Rev. B **27**, 6083 (1983).
[3] Q. Niu, Phys. Rev. Lett. **64**, 1812 (1990).
[4] I. L. Aleiner and A. V. Andreev, Phys. Rev. Lett. **81**, 1286 (1998).
[5] F. Zhou, B. Spivak, and B. L. Altshuler, Phys. Rev. Lett. **82**, 608 (1999).
[6] B. L. Altshuler and L. I. Glazman, Science **283**, 1864 (1999).
[7] C. A. Stafford and N. S. Wingreen, Phys. Rev. Lett. **76**, 1916 (1996).
[8] T. H. Stoof and Y. V. Nazarov, Phys. Rev. B **53**, 1050 (1996).
[9] L. J. Geerligs *et al.*, Phys. Rev. Lett. **64**, 2691 (1990).
[10] L. P. Kouwenhoven *et al.*, Phys. Rev. Lett. **67**, 1626 (1991).
[11] I. Zapata, R. Bartussek, F. Sols, and P. Hänggi, Phys. Rev. Lett. **77**, 2292 (1996).
[12] P. Reimann, M. Grifoni, and P. Hänggi, Phys. Rev. Lett. **79**, 10 (1997).
[13] I. Zapata, J. Łuczka, F. Sols, and P. Hänggi, Phys. Rev. Lett. **80**, 829 (1998).
[14] G. Burmeister and K. Maschke, Phys. Rev. B **57**, 13050 (1998).
[15] W. Ruppel, R. von Baltz, and P. Würfel, Ferroelectrics **43**, 109 (1982).
[16] M. Wagner, Phys. Rev. B **49**, 16544 (1994).
[17] M. Wagner, Phys. Rev. A **51**, 798 (1995).
[18] M. Büttiker, Phys. Rev. Lett. **57**, 1761 (1986).
[19] F. Sols, Ann. Phys. (NY) **214**, 386 (1992).
[20] L. Lewin, *Polylogarithms and Associated Functions* (North Holland, New York, 1981).
[21] M. H. Pedersen and M. Büttiker, Phys. Rev. B **58**, 12993 (1998).

Part V

Transport in Semiconductor Superlattices

Transport in Semiconductor Superlattices: from Quantum Kinetics to Terahertz-Photon Detectors

A. P. Jauho[1], A. Wacker[2], and A. A. Ignatov[1,3]

[1] Mikroelektronik Centret, Technical University of Denmark, Bldg. 345east, DK-2800 Lyngby, Denmark
[2] Institute für Theoretische Physik, Technische Universität Berlin, Hardenbergstr. 36, 10623 Berlin, Germany
[3] Institute for Physics of Microstructures, Russian Academy of Science, 603600 Nizhny Novgorod, Russia

Abstract. Semiconductor superlattices are interesting for two distinct reasons: the possibility to design their structure (band-width(s), doping, etc.) gives access to a large parameter space where different physical phenomena can be explored. Secondly, many important device applications have been proposed, and then subsequently successfully fabricated. A number of theoretical approaches has been used to describe their current-voltage characteristics, such as miniband conduction, Wannier-Stark hopping, and sequential tunneling. The choice of a transport model has often been dictated by pragmatic considerations without paying much attention to the strict domains of validity of the chosen model. In the first part of this paper we review recent efforts to map out these boundaries, using a first-principles quantum transport theory, which encompasses the standard models as special cases. In the second part, focusing in the mini-band regime, we analyze a superlattice device as an element in an electric circuit, and show that its performance as a THz-photon detector allows significant optimization, with respect to geometric and parasitic effects, and detection frequency. The key physical mechanism enhancing the responsivity is the excitation of hybrid Bloch-plasma oscillations.

1 Introduction

Ever since the pioneering work of Esaki and Tsu [1], which drew attention to the rich physics and potential device applications of semiconductor superlattices, these man-made structures have remained a topic of intense research. Semiconductor superlattices have proven to be a fruitful platform for studying a wide range of transport phenomena, such as their intrinsic negative differential conductivity [2], the formation of electric field domains [3], Bloch oscillations [4], as well as dynamical localization [5] and absolute negative conductance [6] under external irradiation, just to mention a few.

These phenomena depend crucially on the relations of the energy scales involved, namely the zero-field miniband width (which is four times the interwell coupling T_1), the scattering rate $\Gamma/\hbar$, and the potential drop per period

($\equiv eFd$, where F is the applied static field and d is the superlattice period). Three distinct approaches have been used to describe transport in the parameter space spanned by (T_1, eFd, Γ): miniband conduction (MBC)[1], [7], Wannier-Stark hopping (WSH)[8], and sequential tunneling (ST)[9], [10]. Until recently, however, the precise range of validity of these various approaches had not been addressed quantitatively. To achieve this, one much use a first-principles approach, which reduces to the standard theories in the appropriate limits. After a brief review of the standard approaches in Sect. II, we introduce in Sect. III a nonequilibrium Green function formalism, which we have used to delineate the boundaries of the different domains of validity [12], [13]. We also present comparisons with Monte Carlo simulations. As we shall see, under favorable conditions it is quite possible to obtain a very accurate description with the standard methods, which is a great advantage because the first-principle Green function calculations are quite complicated and have so far successfully implemented only for rather simple scattering mechanisms (the scattering matrix elements are assumed to be independent of momentum transfer). Thus, in Sect. IV we adopt one of the standard approaches, i.e., the miniband approach, to model a superlattice THz-photon detector, taking into account the effects due to the external circuitry. We conclude that by detailed modeling substantial device performance optimization can be achieved, e.g. the detector sensivity may be improved by almost 50 % by a judicious choice of its parameters.

2 Standard transport models

Here we review the standard models used to describe transport in semiconductor superlattices. For simplicity, the results quoted in the next three subsections are written in a relaxation time approximation, but a generalization to more realistic scattering processes is possible. As an underlying Hamiltonian we use a tight-binding model:

$$\hat{H}^{\rm SL}_{n,m} = (\delta_{n,m-1} + \delta_{n,m+1})\, T_1 + \delta_{n,m}(E_k - neFd)\,. \tag{1}$$

Here T_1 is the overlap matrix element, $E_k = \hbar^2 k^2/(2m)$ is the kinetic energy perpendicular to the growth direction, and the electric field is taken into account by a shift in the site energies.

2.1 Miniband conduction (MBC)

For zero electric field Eq. (1) is diagonalized by a set of Bloch functions $\varphi_q(z) = \sum_n e^{inqd}\Psi_n(z)$ (here the wave function $\Psi_n(z)$ is localized in quantum well n, e.g., a Wannier function) and the dispersion relation is given by the miniband $E(q) = 2T_1\cos(qd)$. The stationary Boltzmann equation for the distribution function $f(q, \mathbf{k})$ is then

$$\frac{eF}{\hbar}\frac{\partial f(q,\mathbf{k})}{\partial q} = \frac{n_F(E(q)+E_k) - f(q,\mathbf{k})}{\tau(E(q)+E_k)} \tag{2}$$

where the relaxation-time is allowed to depend on energy, see, e.g., Ref.[12] for a suitable model. Once the solution to Eq. (2) is found, the current is calculated from

$$J(F) = \frac{e}{4\pi^3\hbar}\int \mathrm{d}^2k \int_{-\pi/d}^{\pi/d} \mathrm{d}q f(q,\mathbf{k}) \frac{\mathrm{d}E(q)}{\mathrm{d}q}\,. \tag{3}$$

The electron density per period is given by

$$N_{2D} = \frac{d}{4\pi^3}\int \mathrm{d}^2k \int_{-\pi/d}^{\pi/d} \mathrm{d}q f(q,\mathbf{k}) \tag{4}$$

and is used to determine the chemical potential for a given electron density. This approach can be extended beyond the relaxation time approximation[14], [15], but the generic features remain unchanged.

2.2 Wannier-Stark hopping (WSH)

In the presence of an electric field, the eigenstates of the Hamiltonian become the localized Wannier-Stark states,

$$\phi_\nu(z) = \sum_n J_{n-\nu}\left(\frac{2T_1}{eFd}\right)\Psi_n(z) \tag{5}$$

with energy $E_\nu = -\nu eFd$, and $J_n(z)$ is the Bessel function of the first kind. Scattering causes hopping between the different states. Within Fermi's golden rule, the current is given by

$$\begin{aligned} J(F) = & \sum_{l>0} l\frac{e}{\tau_0}\sum_n \left[J_n\left(\frac{2T_1}{eFd}\right)J_{n-l}\left(\frac{2T_1}{eFd}\right)\right]^2 \\ & \times\frac{1}{2\pi^2}\int \mathrm{d}^2k\,[n_F(E_k) - n_F(E_k + leFd)]\,. \end{aligned} \tag{6}$$

Here the term $\sum[J_nJ_{n-l}]^2$ arises due to the spatial overlap of the Wannier-Stark functions. The electron density per period is given by:

$$N_{2D} = \rho_0 k_B T_e \log\left[1+\exp\left(\frac{\mu}{k_BT_e}\right)\right] \tag{7}$$

which relates μ to N_{2D}. Again, it is possible to generalize this approach to more realistic scattering mechanisms [16], [17].

	coupling T_1	field drop eFd	scattering Γ
Miniband conduction 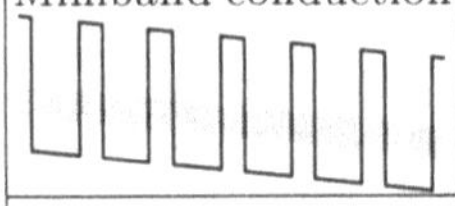	exact miniband	acceleration	golden rule
Wannier Stark hopping 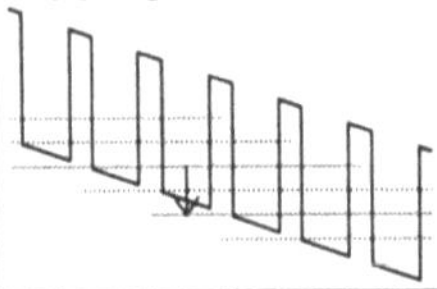	exact: Wannier Stark states		golden rule
Sequential tunneling 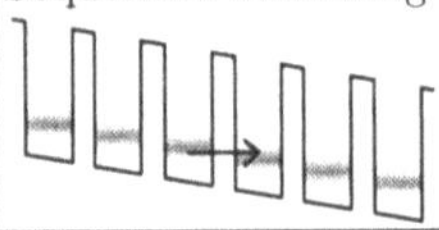	lowest order	energy mismatch	"exact" spectral function

Fig. 1. Overview of standard theoretical models for superlattice transport

2.3 Sequential tunneling (ST)

In this approximation the phase information is lost after each tunneling event between adjacent wells. The scattering within a well is treated self-consistently by solving for the spectral functions $A(\mathcal{E}, \mathbf{k})$; in this work we use the self-consistent Born-approximation[18] for the self-energy. The transitions to neighboring wells are calculated in lowest order of the coupling yielding [20], [18], [19]:

$$J(F) = \frac{e}{2\pi^2} \int \mathrm{d}^2k \int \frac{\mathrm{d}\mathcal{E}}{2\pi\hbar} T_1^2 A(\mathcal{E}, \mathbf{k}) A(\mathcal{E} + eFd, \mathbf{k}) \left[n_F(\mathcal{E}) - n_F(\mathcal{E} + eFd)\right] . \tag{8}$$

The carrier density is given by:

$$N_{2D} = \frac{1}{2\pi^2} \int \mathrm{d}^2k \int \frac{\mathrm{d}\mathcal{E}}{2\pi} n_F(\mathcal{E}) A(\mathcal{E}, \mathbf{k}) . \tag{9}$$

This approach gives quantitative agreement with experiments in weakly coupled structures when realistic models for impurity and interface scattering are employed [10], [20].

2.4 Summary

The important issue to recognize is that these three approaches treat scattering, external field, and coupling within different approximations. MBC does not properly include field-induced localization because of its inherent assumption of extended states, WSH treats scattering in lowest order perturbation theory (in particular, there is no broadening of the states), and ST is explicitly lowest order in the interwell coupling. The basic features of these models are summarized in Figure 1.

2.5 Nonequilibrium Green functions (NGF)

The basic building blocks are the correlation and retarded Green functions:

$$G^{<}_{m,n}(t,t',\mathbf{k}) = i\langle a^{\dagger}_{n}(t',\mathbf{k})a_{m}(t,\mathbf{k})\rangle \tag{10}$$

$$G^{\mathrm{ret}}_{m,n}(t,t',\mathbf{k}) = -i\Theta(t-t')\langle\{a_{m}(t,\mathbf{k}),a^{\dagger}_{n}(t',\mathbf{k})\}\rangle, \tag{11}$$

where and $a^{\dagger}_{n}(t,\mathbf{k})$ and $a_{n}(t,\mathbf{k})$ are the creation and annihilation operators for the state $\Psi_n(z)e^{i(\mathbf{k}\cdot\mathbf{r})}/A$ in well n. These functions obey the Dyson and Keldysh equations, respectively, given below for the superlattice case. Ref.[21] may be consulted for a text-book discussion. The observables, such as the momentum distribution, current, and electron density are computed from $G^{<}_{m,n}(t,t',\mathbf{k})$:

$$f_{\mathrm{QM}}(q,\mathbf{k}) = \int \frac{d\mathcal{E}}{2\pi i}\sum_{n} e^{-ihqd}G^{<}_{n,0}(\mathcal{E},\mathbf{k}) \tag{12}$$

$$J(F) = \frac{e}{2\pi^2}\int\frac{d\mathcal{E}}{2\pi}\int \mathrm{d}^2k\frac{2}{\hbar}\,\mathrm{Re}\left\{T_1G^{<}_{n+1,n}(\mathcal{E},\mathbf{k})\right\} \tag{13}$$

$$N_{2D} = \frac{1}{2\pi^2}\int\frac{d\mathcal{E}}{2\pi i}\int \mathrm{d}^2k\, G^{<}_{n,n}(\mathcal{E},\mathbf{k}) \tag{14}$$

These expressions exploit the fact that in the stationary state the Green functions only depend on the time difference $\tau = t - t'$, and one can define a Fourier transformation via [21]:

$$G_{m,n}(\mathcal{E},\mathbf{k}) = \int \mathrm{d}\tau e^{i\left(\mathcal{E}-eFd\frac{n+m}{2}\right)\frac{\tau}{\hbar}}G_{m,n}(t,t-\tau,\mathbf{k})\,. \tag{15}$$

Without scattering between the $\mathbf{k}$-states and at $T_1 = 0$ the Green-functions are diagonal in the well index: $G^{\mathrm{ret}}_{m,n}(\mathcal{E},\mathbf{k}) = \delta_{m,n}g^{\mathrm{ret}}_{n}(\mathcal{E},\mathbf{k})$ with the free particle Green-function $g^{\mathrm{ret}}_{n}(\mathcal{E},\mathbf{k}) = 1/(\mathcal{E}-E_k+i0^+)$. The full Green function is then determined by the Dyson equation:

$$\begin{aligned} G^{\mathrm{ret}}_{m,n}(\mathcal{E},\mathbf{k}) &= g^{\mathrm{ret}}_{m}(\mathcal{E},\mathbf{k}) + g^{\mathrm{ret}}_{m}\left(\mathcal{E}+eFd\frac{m-n}{2},\mathbf{k}\right) \\ &\times\sum_{l}\Sigma^{\mathrm{ret}}_{m,l}\left(\mathcal{E}+eFd\frac{l-n}{2},\mathbf{k}\right)G^{\mathrm{ret}}_{l,n}\left(\mathcal{E}+eFd\frac{l-m}{2},\mathbf{k}\right). \end{aligned} \tag{16}$$

The self-energy will be written as

$$\Sigma^{\rm ret}_{m,n}(\mathcal{E},\mathbf{k}) = \delta_{m,n}\tilde{\Sigma}^{\rm ret}_{n}(\mathcal{E},\mathbf{k}) + T_1\delta_{m+1,n} + T_1\delta_{m-1,n} \tag{17}$$

where $\tilde{\Sigma}$ contains contributions both from impurity and phonon scattering. The relevant expressions are

$$\tilde{\Sigma}^{\rm ret/<}_{n,\rm imp}(\mathcal{E}) = \frac{N_d}{A}\sum_{\mathbf{k}'} V^2_{\rm imp} G^{\rm ret/<}_{n,n}(\mathcal{E},\mathbf{k}') \tag{18}$$

for impurities, while for optical phonon scattering we take (see, e.g., Ch. 4.3 of Ref. [21] for the derivation):

$$\begin{aligned}\tilde{\Sigma}^{<}_{n,\rm o}(\mathcal{E}) = \frac{|M_{\rm o}|^2}{A}\sum_{\mathbf{k}'}\Big\{&N_{\rm o} G^{<}_{n,n}(\mathcal{E}-\hbar\omega_{\rm o},\mathbf{k}') \\ &+(N_{\rm o}+1)G^{<}_{n,n}(\mathcal{E}+\hbar\omega_{\rm o},\mathbf{k}')\Big\}\end{aligned} \tag{19}$$

$$\begin{aligned}\tilde{\Sigma}^{\rm ret}_{n,\rm o}(\mathcal{E}) = \frac{|M_{\rm o}|^2}{A}\sum_{\mathbf{k}'}\Big\{&(N_{\rm o}+1)G^{\rm ret}_{n,n}(\mathcal{E}-\hbar\omega_{\rm o},\mathbf{k}') + N_{\rm o}G^{\rm ret}_{n,n}(\mathcal{E}+\hbar\omega_{\rm o},\mathbf{k}') \\ &+i\int\frac{d\mathcal{E}'}{2\pi}G^{<}_{n,n}(\mathcal{E}-\mathcal{E}',\mathbf{k}')\left[\frac{1}{\mathcal{E}'-\hbar\omega_{\rm o}+i0^+} - \frac{1}{\mathcal{E}'+\hbar\omega_{\rm o}+i0^+}\right]\Big\}\end{aligned} \tag{20}$$

It is possible to simulate acoustic phonons with a similar expression, using a small fictitious discrete energy $\hbar\omega_{ac}$ [13]. In the above, we also gave the self-energy expressions needed in the Keldysh equation:

$$\begin{aligned}G^{<}_{m,n}(\mathcal{E},\mathbf{k}) = \sum_{m_1} &G^{\rm ret}_{m,m_1}\left(\mathcal{E}+eFd\frac{m_1-n}{2},\mathbf{k}\right)G^{\rm adv}_{m_1,n}\left(\mathcal{E}+eFd\frac{m_1-m}{2},\mathbf{k}\right) \\ &\times\tilde{\Sigma}^{<}_{m_1}\left(\mathcal{E}+eFd\left(m_1-\frac{m+n}{2}\right),\mathbf{k}\right).\end{aligned} \tag{21}$$

The numerical evaluation of these equations has been discussed in two recent publications [12], [13], and here we just summarize our basic strategy, and give a few representative results. The first step in the analysis consists of evaluating the current-voltage characteristics for the different models, and an example of such a calculation is given in Fig. 2. We note that the Boltzmann equation gives an excellent agreement with the full quantum result, in particular for low electric fields. The additional structure in the quantum mechanical curve is a phonon-replica: the Boltzmann equation cannot capture features like this because the electric field does not enter as an energy-scale in the collision integral. It is also of interest to note that the use of a realistic electron-phonon scattering model, instead of a simple relaxation time, leads to a current-voltage characteristic which deviates significantly from the simple Esaki-Tsu shape, $v_{\rm drift} \propto F/(F^2+F^2_{\rm crit})$. By performing a large number of calculations like the one depicted in Figure 2 it is possible to develop visual criteria as to the validity of the various simplified approaches discussed in

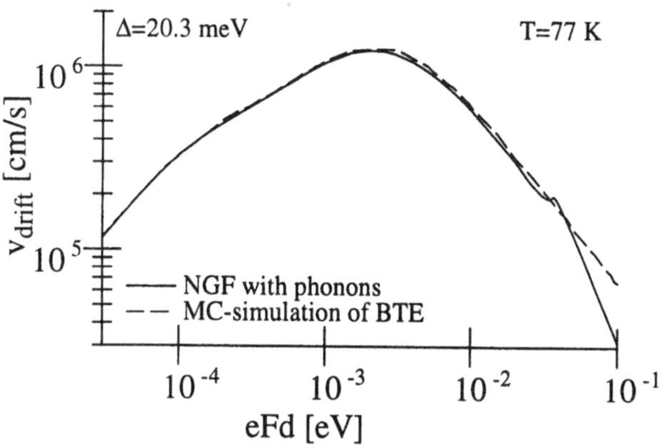

Fig. 2. Drift velocity for a wide-band superlattice. Full line: Calculation by nonequilibrium Green functions. Dashed line: Monte Carlo simulation of Boltzmann's transport equation (from Ref.[12]).

Section 2. These phenomenological considerations can be made more rigorous by examining in detail the retarded Green function, which determines the quantum mechanical correlation between quantum wells n and m [12]. For example, if the terms $G^{\mathrm{ret}}_{n\pm1,n}$ are of the order of $G^{\mathrm{ret}}_{n,n}$, the wave-function is delocalized. By using the expressions given in Ref.[12], one finds that the states are delocalized if $2|T_1| \gg \Gamma$ and $2|T_1| \gg eFd$, i.e., the Boltzmann miniband picture is a useful starting points. Similarly, by demanding that the $G^{\mathrm{ret}}_{m,n}$ vanish if $m \neq n$, one can find the regime where sequential tunneling dominates. The respective ranges of validity can be presented as a "phase digram", Fig. 3.

Additional insight to the differences between quantum kinetics and Boltzmann kinetics can be obtained by examining the momentum distribution functions. Figure 4 presents such a comparison. We direct attention to the following features. At low fields, Fig. 4(a), the distribution function can be viewed as a distorted thermal equilibrium function, and linear response theory holds, corresponding to the linear part of the current-voltage characteristics of Fig. 2 (however, at very low temperatures one should consider weak localization effects, not included in the present choice of the impurity self-energy, and the agreement between quantum and Boltzmann calculations may be weakened). If the electric field increases, electron heating becomes important. For moderate fields the distribution function resembles a distorted equilibrium function, but with an elevated electron temperature, Fig. 4(b). At even higher fields, close to the maximum in the IV-curve, the distribution function strongly deviates from any kind of equilibrium in q space, Fig. 2(c). The results for quantum and semiclassical calculations look similar in

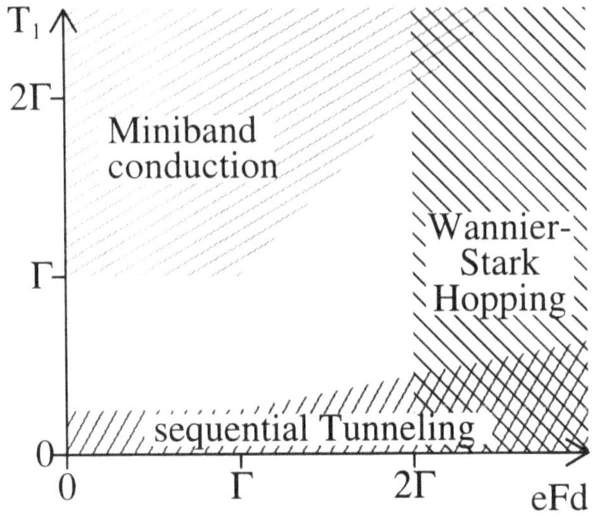

Fig. 3. Ranges of validity for different transport models

the negative differential conductance (NDC) regime (not shown in Fig. 4.). The situation changes dramatically when the field energy is larger than the mini-band width, $edF \geq \Delta$, Fig. 4(d). As the electrons can perform several Bloch oscillations in the semiclassical picture, the distribution function is almost flat withing the Brillouin zone of the miniband. The latter holds for the NGF result as well. However, the absolute values of the distribution functions differ significantly. The reason is the modification in scattering processes due to the presence of the electric field, leading to significant deviations in the distribution function. We noticed this phenomenon already in our discussion of the current-voltage characteristics of Fig. 2. The strong changes in the quantum momentum distribution are further illustrated in Fig. 5.

We can summarize the results given in the first part of this paper as follows. In wide-band superlattices the Boltzmann equation gives reliable results concerning linear response at low fields, electron heating at moderate fields, and the onset of negative differential conductivity. In contrast, for high electric fields or weakly coupled SLs significant differences appear. In this case the quantum nature of transport is important and a semiclassical calculation may be seriously in error.

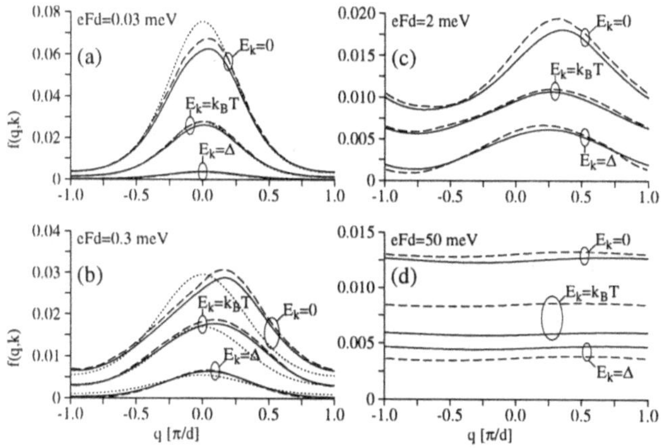

Fig. 4. Electron distribution versus quasimomentum in the Brillouin zone of the miniband for different values of k (from Ref.[12]). The parameters as Fig. 2. Full line: NGF calculation. Dashed line: MC-simulation of BTE. The dotted line shows the thermal distribution $\propto \exp[-E(q,\mathbf{k})/k_B T_e]$ with $T_e = T$ in (a) and $T_e = 140$ K in (b) for comparison.

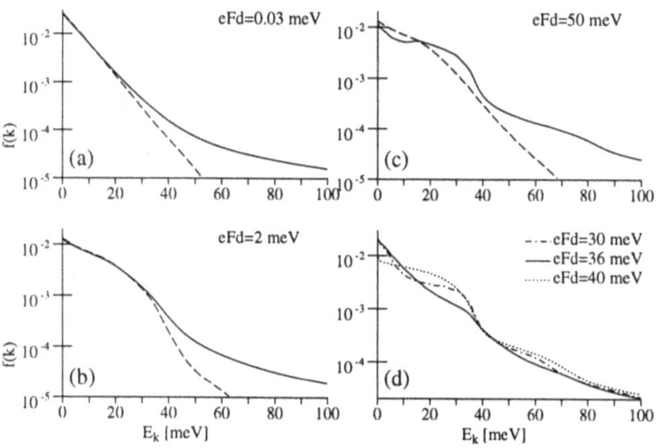

Fig. 5. Electron distribution versus quasimomentum k (from Ref.[12]). (a,b,c): Comparison between NGF calculation (full line) and BTE (dashed line); (d): Results from NGF for different fields. The BTE result (not shown) resembles the result from (c) for all three fields.

3 Superlattice as a THz-photon detector

3.1 Introduction

It has recently been estimated [22] that the room temperature current responsivity of a superlattice detector ideally coupled to the THz-photons can nearly reach the quantum efficiency $e/\hbar\omega$ in the limit of high frequencies $\omega \gg \nu$ (here ω is the incident radiation frequency and ν is a characteristic scattering frequency). This value of the responsivity is being normally considered as a quantum limit for detectors based on superconducting tunnel junctions operating at low temperatures [23]. For high frequencies the mechanism of the THz-photons detection in superlattices was described [22] as a bulk superlattice effect caused by dynamical localisation of electrons.

Here we describe a recent theory of the superlattice current responsivity [24]. We focus on relative broad-band superlattices ($\Delta \simeq 20$ meV), and field strengths smaller than the onset of negative differential conductivity. Thus, according to the analysis given above, a Boltzmann equation based description should suffice. One should note, however, that here we investigate a time-dependent situation, and that a microscopic analysis in the spirit of the first part of this paper has not, to our knowledge, been carried out. We use an equivalent circuit for the superlattice coupled to a broadband antenna (see Fig. 6), which is similar to the equivalent circuit used in resonant tunneling [25] and Schottky diode [26] simulations. The suggested equivalent circuit of the device allows one to treat microscopically the high-frequency response of the miniband electrons and, simultaneously, take into account a finite matching efficiency between the detector antenna and the superlattice in the presence of parasitic losses. Our analytic results lead to the identification of an important physical concept: the excitation of hybrid plasma-Bloch [27] oscillations *in the region of positive differential conductance of the superlattice*. Numerical computations [24], performed for room temperature behavior of currently available superlattice diodes, show that both the magnitudes and the roll-off frequencies of the responsivity are strongly influenced by this effect. The excitation of the plasma-Bloch oscillations gives rise to a resonant-like dependence of the responsivity on the incident radiation frequency, improving essentially the coupling of the superlattice to the detector antenna. We will also show that peak current densities in the device and its geometrical dimensions should be properly optimized in order to get maximum responsivity for each frequency of the incident photons. Finally, we will present numerical estimates of the responsivity for the 1-4 THz frequency band and compare its value with the quantum efficiency $e/\hbar\omega$ of an ideal detector.

3.2 Theoretical formalism

The exact solution of the time-dependent Boltzmann equation in the relaxation time approximation for an arbitrary time-dependent electric field can

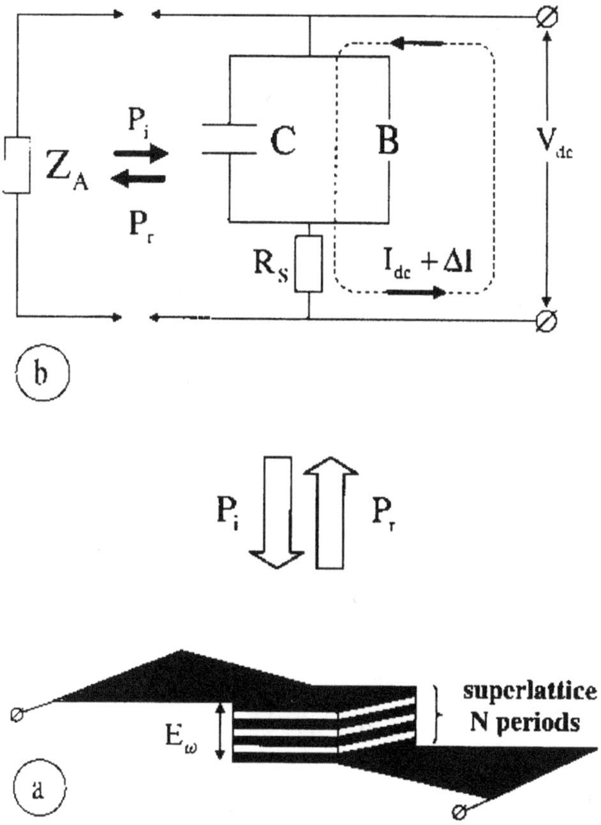

Fig. 6. (a) THz-radiation coupled to a N-period semiconductor superlattice by a co-planar broad band bow- tie antenna, P_i and P_r are the incident and reflected powers respectively. (b) Equivalent circuit for a THz-photon detector with a dc voltage bias source: B–miniband electrons capable to perform Bloch oscillations, C–superlattice capacitance, R_S–parasitic series resistance, Z_A–bow-tie antenna impedance, V_{dc}–dc bias voltage.

be presented in the form of a path integral [28]:

$$f(q, \mathbf{k}, t) = \int_{-\infty}^{t} \nu dt_1 \exp\left[-\nu(t - t_1)\right] f_0(q - \int_{t_1}^{t} e/\hbar F(t_2) dt_2, \mathbf{k}) \, . \quad (22)$$

Using Eqs.(3) and (22) we find the time-dependent current $I(t)$ describing ac transport in a superlattice, with electron performing ballistic motion in a mini-band according to the acceleration theorem and suffering scattering [29], [30]:

$$I(t) = 2I_p \int_{-\infty}^{t} \nu dt_1 \exp\left[-\nu(t-t_1)\right] \sin\left[\frac{e}{N\hbar}\int_{t_1}^{t} V(t_2)dt_2\right]; , \tag{23}$$

where $V(t) = LF(t)$ is the voltage across the superlattice perpendicular to the layers, $L = Nd$ is the superlattice length, N is the number of periods in the superlattice sample, $I_p = Sj_p$, $S = \pi a^2$ is the area of the superlattice, a is the superlattice mesa radius, and

$$j_p = e\frac{v_0}{2}\int \frac{2d\mathbf{k}dq}{(2\pi\hbar)^3}\cos(qd)\, f_0(q,\mathbf{k}) \tag{24}$$

is the characteristic current density. The integration over q in Eq.(24) must be carried out over the Brillouin zone $-\pi/d \le q \le \pi/d$.

The peak current density j_P and the scattering frequency ν can be considered as the main parameters of the employed model. They can readily be estimated from experimentally measured or numerically simulated values of I_P and V_P. For both degenerate and non-degenerate electron gas one gets [29], [30]

$$j_P = en\frac{v_0}{2} \tag{25}$$

if $\Delta \gg kT, \epsilon_F$, where kT is the equilibrium thermal excitation energy, $\epsilon_F = \hbar^2(3\pi^2 N_{2D})^{2/3}/(2m_{\text{eff}})$ is the Fermi energy of degenerate electrons, $m_{\text{eff}} = m_{zz}^{1/3}m^{2/3}$ is the density of states effective mass near the miniband bottom, and $m_{zz} = 2\hbar^2/\Delta d^2$ is the effective mass of electrons along the superlattice axis. In the particular case of the Boltzmann equilibrium distribution function Eq.(24) yields [30] $j_P = (env_0/2)[I_1(\Delta/2kT)/I_0(\Delta/2kT)]$, where $I_{0,1}$ are the modified Bessel functions.

We now suppose that in addition to the dc voltage V_{SL}, an alternating sinusoidal voltage with a complex amplitude V_ω is applied to the superlattice:

$$V(t) = V_{SL} + \frac{1}{2}\left[V_\omega \exp(i\omega t) + V_\omega^* \exp(-i\omega t)\right] . \tag{26}$$

Generally, V_{SL}, V_ω can be found from an analysis of the equivalent circuit given in Fig. 6. We write the ac voltage amplitude as $V_\omega = |V_\omega|e^{i\psi}$; both $|V_\omega|$ and ψ can be obtained self-consistently taking account of reflection of the THz photons from the superlattice and their absorption in the series resistor R_S.

Making use Eq.(23) we obtain [30]:

$$I(t) = 2I_P \int_0^{\infty} \nu dt_1 \exp(-\nu t_1) \sin\left[\frac{eV_{SL}}{N\hbar}t_1 + \Phi(t,t_1)\right]; , \tag{27}$$

where

$$\Phi(t,t_1) = \frac{e}{N\hbar\omega} \times \frac{1}{2}\left\{iV_\omega \exp(i\omega t)[\exp(-i\omega t_1) - 1] + c.c.\right\} . \tag{28}$$

According to Eq.(27), electrons in a superlattice miniband perform damped Bloch oscillations with the frequency $\Omega_B = eV_{SL}/N\hbar = eE_{SL}d/\hbar$, and the phase $\Phi(t, t_1)$ modulated by the external ac voltage.

Equation (27) contains, as special cases the following results: (i) a harmonic voltage $V(t)$ ($V_{SL} = 0$) leads to dynamical localisation, and current harmonics generation with oscillating power dependence [30]; (ii) a dc current- voltage characteristics of the irradiated superlattice $I_{DC}(V_{SL}, V_\omega) = (\omega/2\pi)\int I(t)dt$ shows resonance features ('Shapiro steps') leading to absolute negative conductance [30], [31], [32], [33]; (iii) and to generation of dc voltages (per one superlattice period) that are multiples of $\hbar\omega/e$[34].

3.3 Current responsivity

We define the current responsivity[23] of the superlattice detector as the current change ΔI induced in the external dc circuit per incoming ac signal power P_i:

$$R_i(\omega, V_{SL}) = \frac{\Delta I}{P_i} \,. \tag{29}$$

This definition takes into account both the parasitic losses in the detector and the finite efficiency for impedance-matching of the incoming signal into the superlattice diode.

It can be shown [24], that in the small-signal approximation both the dc current change ΔI_{DC}^{SL} and the power P_{abs}^{SL} absorbed in the superlattice are proportional to the square modulus of the complex voltage $|V_\omega|^2$. This circumstance permits us to calculate $|V_\omega|^2$ self-consistently for given values of the incoming power, making use a linear ac equivalent circuit analysis and, then, find the current responsivity $R_i(\omega, V_{SL})$.

The results of the calculation of the superlattice current responsivity $R_i(\omega, V_{SL})$ are presented in the following form:

$$R_i(\omega, V_{SL}) = \frac{R_i^{(0)}(\omega, V_{SL})A(\omega, V_{SL})}{1 + R_S(dI_{DC}^{SL}(V_{SL})/dV_{SL})} \,, \tag{30}$$

where

$$R_i^{(0)}(\omega, V_{SL}) = -\frac{e}{N\hbar\nu}\frac{(V_{SL}/V_P)[3 + (\omega\tau)^2 - (V_{SL}/V_P)^2]}{[1 + (V_{SL}/V_P)^2][1 + (\omega\tau)^2 - (V_{SL}/V_P)^2]} \tag{31}$$

is the superlattice current responsivity under conditions of a perfect matching and neglecting parasitic losses, $R_S \to 0$. [22]

The factor $A(\omega, V_{SL})$ in Eq. (30) describes the effect of the electrodynamical mismatch between the antenna and the superlattice and the signal absorption in the series resistance

$$A(\omega, V_{SL}) = \left[1 - \left|\frac{Z_A - (Z_{AC}^{SL}(\omega, V_{SL}) + R_S)}{Z_A + (Z_{AC}^{SL}(\omega, V_{SL}) + R_S)}\right|^2\right] \times \frac{\mathrm{Re}Z_{AC}^{SL}(\omega, V_{SL})}{\mathrm{Re}Z_{AC}^{SL}(\omega, V_{SL}) + R_S} \,. \tag{32}$$

The first factor in Eq. (32) describes the reflection of the THz-photons due to mismatch of the antenna impedance Z_A and the total impedance of the device $Z_{AC}^{SL}(\omega, V_{SL}) + R_S$, with the second one being responsible for sharing of the absorbed power between the active part of the device described by the impedance $Z_{AC}^{SL}(\omega, V_{SL})$ and the series resistance R_S.

The superlattice impedance is defined as

$$Z_{AC}^{SL}(\omega, V_{SL}) = 1/\left[G_{AC}^{SL}(\omega, V_{SL}) + i\omega C\right] , \tag{33}$$

where $G_{AC}^{SL}(\omega, V_{SL})$ is the superlattice conductance, $C = \epsilon_0 S/4\pi L$ is the capacitance of the superlattice, and ϵ_0 is the average dielectric lattice constant.

Finally, the last factor in the denominator of Eq. (30) describes the redistribution of the external bias voltage V_{DC} between the dc differential resistance of the superlattice $(dI_{DC}^{SL}(V_{SL})/dV_{SL})^{-1}$ and the series resistance R_S, with the dc voltage drop on the superlattice V_{SL} being determined by the solution of the well-known load equation [23]

$$V_{DC} = V_{SL} + I_{DC}^{SL}(V_{SL})R_S \, . \tag{34}$$

3.4 Superlattice dielectric function. Hybridisation of Bloch and plasma oscillations

We next analyze the condition of optimized matching of the superlattice to the incident radiation. Assuming the limit of negligible series resistance $R_S \to 0$ this condition can be obtained from the solution of the equation

$$Z_{AC}^{SL}(\omega, V_{SL}) = Z_A \tag{35}$$

for the complex frequency $\omega(V_{SL})$. This solution determines the resonant line position and the line width at which the absorption in the superlattice tends to its maximum value.

One can transform Eq. (35) to the following form:

$$\epsilon(\omega, E_{SL}) = \frac{\epsilon_0}{i\omega C Z_A} , \tag{36}$$

where

$$\epsilon(\omega, E_{SL}) = \epsilon_0 + \frac{4\pi\sigma_0}{i\omega} F_1(\omega, E_{SL}) \tag{37}$$

is the dielectric function of the superlattice, with the dc field E_{SL} being applied to the device [26], and $F_1(\omega, E_{SL})$ is defined by

$$F_1(\omega, V_{SL}) = \frac{1 + i\omega\tau - (V_{SL}/V_P)^2}{[1 + (V_{SL}/V_P)^2]\,[(1 + i\omega\tau)^2 + (V_{SL}/V_P)^2]} \, . \tag{38}$$

In the high-frequency limit $\epsilon_0/CZ_A\omega \to 0$ the solution of Eq. (35) coincides with the solution of the equation

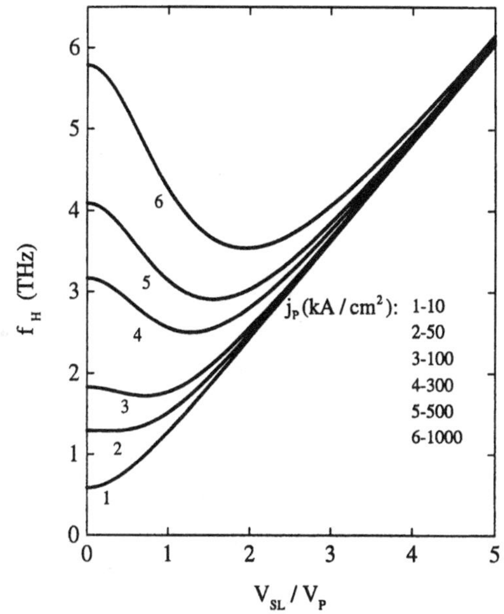

Fig. 7. The calculated hybrid plasma-Bloch oscillation frequency f_H as a function of the normalized superlattice voltage drop V_{SL}/V_P for different values of the peak current densities j_P = 10, 50, 100, 300, 500, and 1000 kA/cm^2 (from Ref.[24]). Typical values of the superlattice parameters (d = 50Å, E_P=10 kV/cm, ϵ_0 =13) were used for the calculations.

$$\epsilon(\omega, E_{SL}) = 0 \tag{39}$$

describing the eigenfrequencies $\omega_{\pm}^{H}$ of the hybrid plasma-Bloch oscillations in a superlattice, [26]

$$\omega_{\pm}^{H}(E_{SL}) = \pm\omega_P \left[\frac{1}{1+(E_{SL}/E_P)^2} + \left(\frac{\nu}{\omega_P}\right)^2 (E_{SL}/E_P)^2\right]^{1/2} + i\nu \tag{40}$$

where ω_P is the plasma frequency of electrons in a superlattice. The plasma frequency ω_P can be given in terms of the small-field dc conductivity σ_0 or, equivalently, in terms of the peak current density j_P

$$\omega_P = \left(\frac{4\pi\sigma_0\nu}{\epsilon_0}\right)^{1/2} = \left(\frac{8\pi j_P e d}{\epsilon_0 \hbar}\right)^{1/2} /;. \tag{41}$$

Equation (41) reduces in the particular case of wide-miniband superlattices ($\Delta \gg kT, \epsilon_F$) to the standard formula $\omega_P = (4\pi e^2 n/\epsilon_0 m_{zz})^{1/2}$.

In the limiting case of small applied dc electric fields $E_{SL}/E_P \to 0$ one finds from Eq. (40) the plasma frequency $\omega_\pm^H \to \pm\omega_P$, while in the opposite case $E_{SL}/E_P \to \infty$, the Bloch frequency $\omega_\pm^H \to \pm\Omega_B = \pm eE_{SL}d/\hbar$ is recovered. The scattering frequency ν in Eq. (40) is responsible for the line width of the plasma-Bloch resonance.

We have calculated the hybrid plasma-Bloch oscillation frequency $f_H = \omega_+^H/2\pi$, using Eqs. (40) and (41), for the typical values of the superlattice parameters $\epsilon_0 \simeq 13$, $d \simeq 50$ Å, $E_P \simeq 10$ kV/cm, $f_\nu = \nu/2\pi = 1.2$ THz for different values of the current densities j_p (see Fig. 7). For small values of the current densities $j_P \simeq 10$ kA/cm^2 the frequency of the hybrid oscillation increases with applied voltage in all range of the parameter V_{SL}/V_P . On the other hand, for higher values of the current densities $j_P \simeq (50 - 1000)$ kA/cm^2 the hybrid oscillation's frequency starts to decrease with increasing bias voltage in the sub-threshold voltage range $V_{SL} \le V_P$. Then, at super-threshold voltages $V_{SL} \ge V_P$, ω_H starts to increase again tending to the Bloch frequency. It is important to note that at high values of the dc current densities j_P the hybrid plasma-Bloch oscillations become well defined eigenmodes of the system ($f_H \ge f_\nu$). Therefore, an essential improvement of the matching efficiency between antenna and the superlattice can be expected in the high- frequency range due to a resonant excitation of this eigenmode in the device.

3.5 High-frequency limit

Let us compare the high-frequency limit of the responsivity of the superlattice with the quantum efficiency $R_{\max} = e/\hbar\omega$ which is believed to be a fundamental restriction for the responsivity of superconductor tunnel junctions[23]. This quantum efficiency (or quantum limit) corresponds to the tunnelling of one electron across the junction for each signal photon absorbed[23], with a positive sign of the responsivity.

In our case the mechanism of the photon detection is different (see Fig. 8). Electrons move against the applied dc electric force due to absorption of photons. At $V_{SL} = V_P$ the responsivity is negative, indicating that one electron is subtracted from the dc current flowing through the superlattice when the energy $2eV_P$ is absorbed from the external ac field. One half of this energy is needed for the electron to overcome the potential barrier which is formed by the dc force, with another half being delivered to the lattice due to energy dissipation. If the applied dc voltage is strong enough, i.e. $V_{SL} \gg V_P$, dissipation plays no essential role in the superlattice responsivity. In this case the energy eV_{SL} should be absorbed from the ac field in order to subtract one electron from the dc current simply due to the energy conservation law.

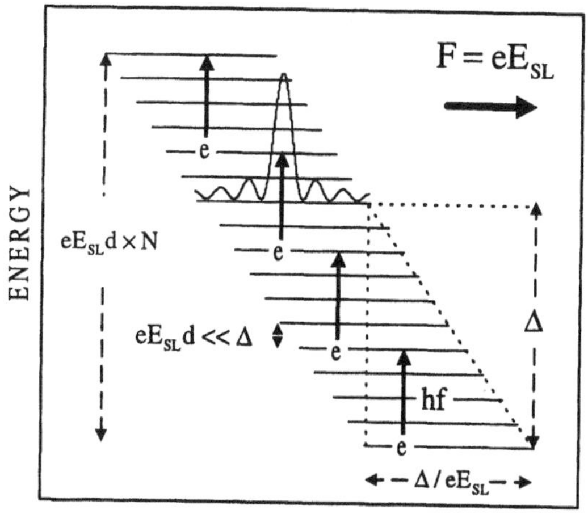

Fig. 8. Real space energy diagram illustrating THz- photon ($f \gg \nu/2\pi$) detection in the superlattice: DC electric field E_{SL} is applied to the N-period semiconductor superlattice with the miniband width Δ . Under the action of the dc field electrons perform Bloch oscillations with the spatial amplitude Δ/eE_{SL} . At critical dc electric voltage (field) $V_{SL} = V_P = N\hbar\nu/e$ ($eE_{SL}d = \hbar\nu$) electrons move against the dc electric force due to absorption of photons climbing up the Wannier- Stark ladder. The energy $2eV_P$ should be absorbed from external ac field in order to subtract one electron from the external circuit. One half of this energy is needed for the electron to overcome the potential barrier which is formed by the dc force, with the other half being delivered to the lattice due to energy dissipation. A quasi-classical description of the process is valid if $f \ll \Delta/\hbar$ when allowed transitions between different Wannier-Stark state exist.

3.6 Excitation of the plasma-Bloch oscillations

For demonstration of the frequency dependence of the superlattice current responsivity in the THz-frequency band we will focus on the $GaAs/Ga_{0.5}Al_{0.5}As$ superlattices specially designed to operate as millimeter wave oscillators at room temperature. In Ref. [35] wide-miniband superlattice samples with $d = 50$ Å, $\Delta \simeq 113$ meV, $n \simeq 10^{17}$ cm^{-3}, were investigated experimentally. They demonstrated a well-pronounced Esaki- Tsu negative differential conductance for $E_{SL} \geq E_P \simeq 4$ kV/cm with the high peak current of the order of $j_P \simeq 130$ kA/cm^2. The measured value of the peak current is in a good agreement with the estimate $j_P \simeq (80 - 160)$ kA/cm^2 for $n \simeq (1-2) \times 10^{17}$, $T = 300$ K based on Eq. (24), if one assumes an equilibrium Boltzmann distribution for the charge carriers. From the peak electric field and current we find the scattering and plasma frequencies $f_\nu \simeq 0.5$ THz, $f_P = 2$ THz,

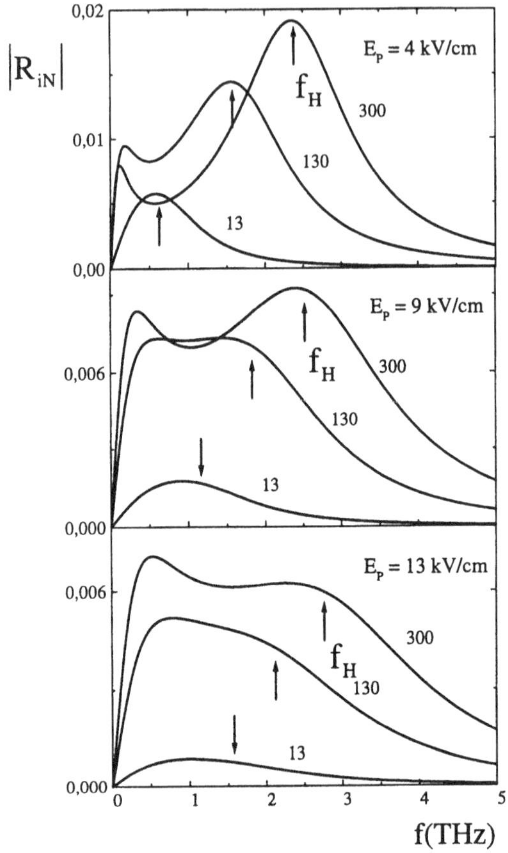

Fig. 9. The frequency dependence of the normalised current responsivity $|R_{iN}| = |R_i/(e/\hbar\omega)|$ of the superlattice THz-photon detector ($a = 2\mu$m, $L = 0.5\mu$m, $R_S = 10\Omega$, $V_{SL} = 0.95V_P$) for three values of the peak current density (j_P = 13, 30, and 300 kA / cm^2) and for three values of the peak electric field (E_P = 4, 9, and 13 kV/cm) (from Ref.[24]). The relevant positions of the hybrid plasma-Bloch frequencies f_H are indicated for each curve by arrows showing characteristic resonance (high peak current densities) and roll-off (low peak current densities) behavior.

respectively, assuming $\epsilon_0 = 13$ for the average dielectric lattice constant. The maximum frequency for the semiclassical approach to be valid for these samples is $f_\Delta \simeq 27$ THz.

Figure 9 shows the frequency dependence of the normalised current responsivity calculated for three values of the peak current density in the superlattice, i.e. $j_P = 13, 130$, and 300 kA/cm^2 and for three values of the peak electric field, $E_P = 4$, 9, and 13 kV/cm. We also use the typical values for the superlattice length $L = 0.5\mu$m (superlattice consists of 100 periods), and assume $a = 2\mu$m for the superlattice mesa radius [35], [36], [37]. We choose $R_S = 10\Omega$ for the series resistance of the device in the THz- frequency band, i.e. the same value as for resonant tunnelling diodes having the same radius of mesas [25]. The calculations are performed in the region of the positive differential conductance for dc bias voltage close to the peak voltage ($V_{SL} = 0.95$ V_P)

For $E_P = 4$ kV/cm ($f_\nu \simeq 0.5$ THz) Fig. 9 demonstrates well-pronounced resonant behavior of the normalised responsivity as a function of frequency. The resonance frequency and the maximum value of the responsivity rise if the peak current density increases. For $j_p = 300$ kA/cm^2 the normalised responsivity reaches its maximum value $-R_{iN} \simeq 0.02$ ($-R_i \simeq 2$ A/W) at frequency $f \simeq 2.5$ THz. For higher values of the peak electric fields $E_P = 9$ kV/cm ($f_\nu \simeq 1.08$ THz) and $E_P = 13$ kV/cm ($f_\nu \simeq 1.57$ THz) the resonance line-widths are broadened due to implicit increase of the scattering frequencies. In particular, for $E_P = 13$ kV/cm, $j_P = 300$ kA/cm^2 the normalised responsivity has an almost constant value $-R_{iN} \simeq 0.006$ ($-R_i \simeq 0.6$ A/W) up to $f \simeq 2.5$ THz and, then, rapidly decreases. The frequency behavior of the normalised responsivity originates from excitation of the plasma- Bloch oscillations in the superlattice. We indicate in Fig. 9 the positions of the hybrid frequencies $f_H = |\omega_\pm^H|/2\pi$ with arrows. For small peak electric fields (low values of the scattering frequencies) the hybrid frequency corresponds to the maximum of the normalised responsivity. For higher values of the peak field (higher values of the scattering frequencies) it corresponds to the roll-off frequency at which the responsivity starts to decline.

3.7 Optimized superlattice length

The enhancement of the normalised responsivity requires an optimum matching efficiency of the superlattice to the broad-band antenna and minimization of the parasitic losses in the series resistor. These requirements impose an optimum length of the superlattice for each chosen frequency of the incoming THz-photons and series resistance.

We show in Fig. 10 the dependence of the normalised responsivity on the number of the superlattice periods for $f = 2.5$ THz.We used for calculation $a = 2\mu$m, $j_P = 130$ kA/cm^2, $V_{SL} = 0.95$ V_P, and three values of the series resistance $R_S = 10$, 30 and 50 Ω. For all three values of the series resistance the responsivity displays a well pronounced maximum for the optimum number of the superlattice periods $N = N_{\max}$. The value of $N_{\max}$ increases with increasing of the series resistance ($N_{\max} \simeq 40$ for $R_S = 10$ Ω, $N_{\max} \simeq 60$ for $R_S = 30$ Ω and for $N_{\max} \simeq 90$ for $R_S = 50$ Ω). This result can be readily understood by recalling that a larger volume of the superlattice minimizes

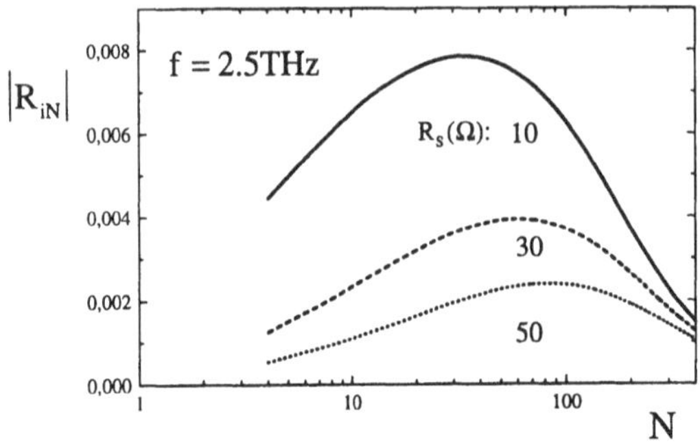

Fig. 10. The dependence of the normalised current responsivity $|R_{iN}| = |R_i/(e/\hbar\omega)|$ of the superlattice THz-photon detector ($a = 2\mu$m, $E_P = 4$kV/cm, $j_P = 130$ kA/cm^2) at $f = 2.5$ THz for three values of the series resistance (R_S=10, 30, and 50 Ω) as a function of number of the superlattice periods N (from Ref.[24]).

parasitic losses for higher values of the series resistance because of reduction of the sample's capacitance.

In addition to the examples discussed here, Ref. [24] reports several other aspects of the superlattice responsivity, such as bias voltage dependence, optimized peak current density etc.

3.8 Conclusions

We have illustrated here the steps required to perform a superlattice device optimization. These include: (i) Consideration of the frequency dependence of the superlattice. Here a truly microscopic analysis is not yet available, but a Boltzmann equation based theory should be applicable, with due caution. (ii) The impedance matching between the antenna and the superlattice must be optimized. (iii) The effect of parasitic losses must be included in the analysis. One of the main results emerging from our analysis is the importance of collective excitations: the hybridized plasma-Bloch oscillations both enhance the responsivity and increase the roll-off frequency. It is found that an optimized superlattice can have a responsivity which approaches 10 % of an ideal superconducting tunnel junction. The great advantage, of course, is that superlattice based detectors work even at room temperature, and this property should offer a wide range of applications.

References

1. L. Esaki and R. Tsu, IBM J. Res. Develop. **14**, 61 (1970).
2. A. Sibille, J. F. Palmier, H. Wang, and F. Mollot, Phys. Rev. Lett. **64**, 52 (1990).
3. H. T. Grahn, R. J. Haug, W. Müller, and K. Ploog, Phys. Rev. Lett. **67**, 1618 (1991).
4. C. Waschke *et al.*, Phys. Rev. Lett. **70**, 3319 (1993).
5. M. Holthaus, Phys. Rev. Lett. **69**, 351 (1992).
6. B. J. Keay *et al.*, Phys. Rev. Lett. **75**, 4102 (1995).
7. P. A. Lebwohl and R. Tsu, J. Appl. Phys. **41**, 2664 (1970).
8. R. Tsu and G. Döhler, Phys. Rev. B **12**, 680 (1975).
9. D. Miller and B. Laikhtman, Phys. Rev. B **50**, 18426 (1994).
10. A. Wacker and A. P. Jauho, Physica Scripta **T69**, 321 (1997).
11. R. Aguado, G. Platero, M. Moscoso, and L. L. Bonilla, Phys. Rev. B **55**, 16053 (1997)
12. A. Wacker and A. P. Jauho, Phys. Rev. Lett. **80**, 369 (1998).
13. A. Wacker, A. P. Jauho, S. Rott. A. Markus, P. Binder, and G. H. Döhler, Phys. Rev. Lett. **83**, 836 (1999).
14. A. A. Ignatov, E. P. Dodin, and V. I. Shashkin, Mod. Phys. Lett. B **5**, 1087 (1991).
15. X. L. Lei, N. J. M. Horing, and H. L. Cui, Phys. Rev. Lett. **66**, 3277 (1991).
16. S. Rott, N. Linder, and G. H. Döhler, Superlattices and Microstructures **21**, 569 (1997).
17. V. V. Bryksin and P. Kleinert, J. Phys.: Cond. Mat. **9**, 7403 (1997); S. Rott *et al.*, Physica E (Amsterdam) **2**, 511 (1998).
18. G. D. Mahan, *Many-Particle Physics* (Plenum, New York, 1990).
19. S. Q. Murphy, J. P. Eisenstein, L. N. Pfeiffer, and K. W. West, Phys. Rev. B **52**, 14825 (1995).
20. A. Wacker, in *Theory of transport properties of semiconductor nanostructures*, edited by E. Schöll (Chapman and Hall, London, 1998), Chap. 10.
21. H. Haug and A.-P. Jauho, *Quantum Kinetics in Transport and Optics of Semiconductors* (Springer, Berlin, 1996).
22. A. A. Ignatov, E. Schomburg, J. Grenzer, S. Winnerl, K. F. Renk and E. P. Dodin, Superlattices & Microstructures **22**, 15 (1997).
23. For a review, see J. R. Tucker and M. J. Feldman, Rev. Mod. Phys. **57**, 1055 (1985).
24. A. A. Ignatov and A. P. Jauho, J. Appl. Phys. **85**, 3643 (1999)
25. T. C. L. G. Sollner, W. D. Goodhue, P. E. Tannenwald, C. D. Parker, and D. D. Peck, Appl. Phys. Lett. **43**, 588 (1983).
26. H. C. Torrey and C. A. Whitmer, Crystal Rectifiers (McGraw-Hill, New York, 1948), p. 336.
27. A. A. Ignatov and V. I. Shashkin, Sov. Phys. JETP. **66**, 526 (1987).
28. R. G. Chambers, Proc. Phys. Soc. (London) A **65**, 458 (1952).
29. R. Tsu and L. Esaki, Appl. Phys. Lett. **19**, 246 (1971).
30. A. A. Ignatov and Yu. A. Romanov, Sov. Phys. Solid State **17**, 2216 (1975); Phys. Status Solidi B **73**, 327 (1976).
31. A. A. Ignatov and Yu. A. Romanov, Radiophysics and Quantum Electronics (Consultants Bureau, N.Y., 1978) Vol. **21**, p. 90.

32. M. Holthaus, Phys. Rev. Lett. **69**, 351 (1992).
33. A. A. Ignatov, K. F. Renk, and E. P. Dodin, Phys. Rev. Lett. **70**, 1996 (1993); J. B. Xia, Phys. Rev. B **58**, 3565 (1998).
34. A. A. Ignatov, E. Schomburg, J. Grenzer, K. F. Renk, and E. P. Dodin, Z. Phys. B **98**, 187 (1995).
35. E. Dutisseuil, A. Sibille, J. F. Palmier, F. Aristone, F. Mollot, and V. Thietty-Mieg, Phys. Rev. B **49**, 5093 (1994).
36. E. Schomburg, A. A. Ignatov, J. Grenser, K. F. Renk, D. G. Pavel'ev, Yu. Koschurinov, B. Ja. Melzer, S. Ivanov, S. Schaposchnikov, and P. S. Kop'ev, Appl. Phys. Lett. **68**, 1096 (1996).
37. S. Winnerl, E. Schomburg, J. Grenser, H.-J. Regl, A. A. Ignatov, A. D. Semenov, K. F. Renk, D G. Pavel'ev, Yu. Koschurinov, B. Ja. Melzer, V. Ustinov, S. Ivanov, S. Schaposchnikov, and P. S. Kop'ev, Phys. Rev. B **56**, 10 303 (1997).

Current Self-Oscillations and Chaos in Semiconductor Superlattices

H. T. Grahn

Paul-Drude-Institut für Festkörperelektronik, Hausvogteiplatz 5–7, 10117 Berlin, Germany

Abstract. Weakly coupled semiconductor superlattices represent a non-linear system, which exhibits tunable current self-oscillations and chaos. The non-linearity originates from resonant tunneling between two-dimensional subbands in adjacent wells. The current oscillations are due to a recycling motion of a charged monopole over several superlattice periods. The charged monopole appears, because the non-linearity of the system in connection with a large carrier density results in the formation of electric-field domains in these systems. The monopole separates the different field regions. Current self-oscillations have been observed in doped and undoped, photoexcited superlattices up to frequencies of several GHz. A single period of the current oscillations contains additional spikes with a frequency more than one order of magnitude above the fundamental oscillation frequency. These spikes are a signature of the well-to-well hopping of the monopole. The fundamental oscillation frequency can be varied over more than two orders of magnitude by changing the applied voltage within a single sample. For different samples, a variation of the barrier width by a factor of three has resulted in a change of the fundamental oscillation frequency by more than three orders of magnitude. The frequency scales with the resonant coupling of the subbands in adjacent wells. In several samples, current self-oscillations have been observed up to room temperature. Recently, undoped superlattices have been used to investigate the carrier density dependence of the boundary between static and dynamic domain formation by varying the photoexcitation intensity. With increasing carrier density, the current oscillations disappear via a supercritical Hopf bifurcation, a subcritical Hopf bifurcation, and a homoclinic connection. The chaotic behavior of such a system, which was predicted through calculations within a simple drift-diffusion model, has also been investigated. The bifurcation diagram of the power spectra under application of an external ac voltage shows the well-known route to chaos via alternating windows of frequency locking and quasi-periodicity. Real-time current traces have been used to construct Poincaré sections, which support this interpretation. However, for other dc voltages, the route to chaos can become much more complex. Recently, the multi-fractal dimension of the chaotic attractors has been determined as a function of the dc voltage using the experimentally derived Poincaré sections.

1 Introduction

Semiconductor superlattices were invented by Esaki and Tsu (1970). They consist of a periodic arrangement of two semiconductors A and B with different energy gaps producing a superperiodicity d, which is much larger than

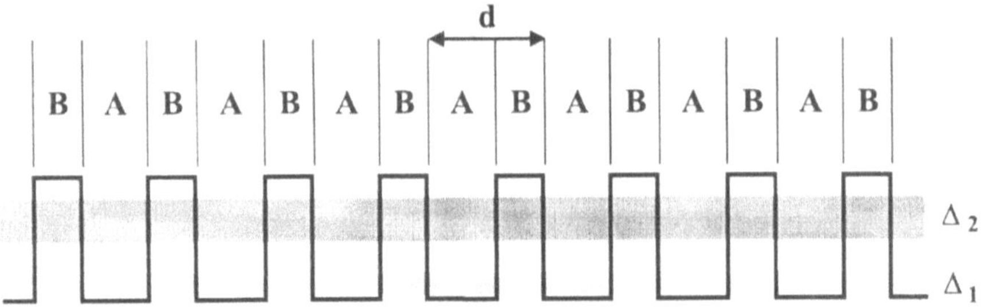

Fig. 1. Schematic diagram of the structural superlattice consisting of two semiconductors A and B with different energy gaps of period d (top) and the resulting modulation of the conduction band edge (bottom). The first two minibands are shown as shaded areas with energy widths Δ_1 and Δ_2.

the lattice constant a of the constituent materials (Grahn 1995). The periodic structural arrangement results in a periodic modulation of the valence and the conduction band edge as shown in Fig. 1. If the modulation of the valence and the conduction band edge occurs in the opposite (same) direction, the superlattice is called type I (type II). We will consider only one type of carrier, i.e., electrons in the conduction band, so that for the remainder of this article the distinction between type I and type II is not important. The coupling between different wells results in the formation of minibands as shown in the lower part of Fig. 1. The width of the miniband Δ is determined by the strength of the coupling between adjacent wells, i.e., the smaller or thinner the barrier, the larger the width of the miniband. Applying an electric field parallel to the superlattice direction results in carrier transport along the superlattice direction. In contrast to electrons in the conduction band of bulk semiconductors, where the drift velocity v_d is proportional to the electric field F, miniband transport results in a nonlinear drift-velocity versus electric-field characteristic. Esaki and Tsu (1970) determined the drift velocity of miniband transport, when scattering of carriers with a scattering time τ is taken into account in the simplest possible way. The drift velocity is given as a function of the applied electric field by

$$\mathrm{v}_d = \mathrm{v}_0 \frac{F/F_0}{1+(F/F_0)^2}, \tag{1}$$

where $\mathrm{v}_0 = \frac{\Delta\, d}{2\, \hbar}$ and $F_0 = \frac{\hbar}{e\, \tau\, d}$. The field dependence of the drift velocity for miniband transport is shown in Fig. 2. After an initial linear increase of the drift velocity with increasing electric field, v_d reaches a maximum and then decreases. The field region beyond the maximum of the drift velocity is called negative differential velocity (NDV) region. If the superlattice is weakly coupled, no minibands are formed. However, the transport characteristic remains qualitatively the same as shown in Fig. 2 due to a detuning of the resonance of the ground state subband E_1 in adjacent wells (Wacker 1998, Wacker

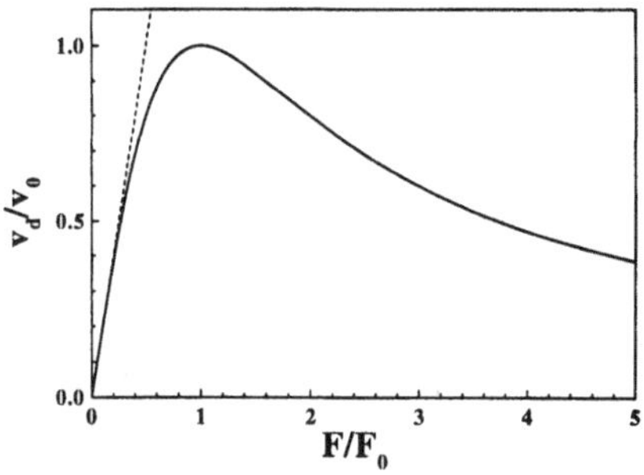

Fig. 2. Drift velocity vs electric field for miniband transport according to the Esaki-Tsu model.

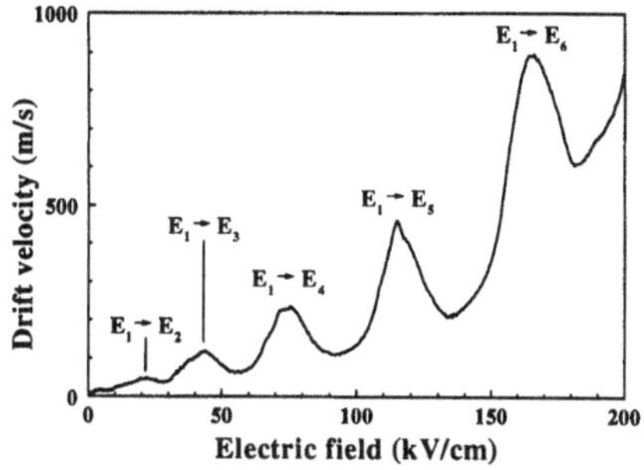

Fig. 3. Drift velocity vs electric field for a superlattice with 40 periods, 21 nm GaAs wells, and 2.5 nm AlAs barriers measured at 7.5 K.

and Jauho 1998). With a further increase of the field, the second E_2 or a higher subband E_n can be brought into resonance with the ground state subband in adjacent wells resulting again in an enhanced transport probability due to sequential resonant tunneling (Grahn 1995). Therefore, semiconductor superlattices can exhibit several regions of NDV as shown in Fig. 3 for a superlattice with 40 periods, 21 nm GaAs wells, and 2.5 nm AlAs barriers (Grahn 1998). For this particular system, resonant tunneling up to $E_1 \rightarrow E_6$ is observed. For the remainder of this article, we will assume that weakly coupled superlattices have at least two subbands E_1 and E_2. If the applied electric field is below the resonance condition for tunneling from $E_1 \rightarrow E_2$, we will refer to this field range as the *first plateau* of the I-V characteristics, since the current is limited by the first maximum in the drift velocity curve (cf. Fig. 4), while for a field above $(E_2 - E_1)/(e\,d)$ the term *second plateau* will be used to label the corresponding field range.

2 Theoretical Description

Transport in weakly coupled superlattices can be described by a combination of Poisson's equation and Ampere's law (Bonilla et al. 1994, Bonilla 1995). Using a discrete version of Poisson's equation, the electric field in the jth period is connected with the carrier density n_j and the doping density N_D by

$$F_j - F_{j-1} = \frac{e\,d}{\epsilon}\,(n_j - N_D)\,, \tag{2}$$

where ϵ denotes the dielectric constant of the respective material. Ampere's law relates the current density J with the time derivative of the electric field F_j, the drift velocity $\mathrm{v}(F_j)$, and the carrier density n_j

$$J = e\,\mathrm{v}(F_j)\,n_j + \epsilon\,\frac{dF_j}{dt}\,. \tag{3}$$

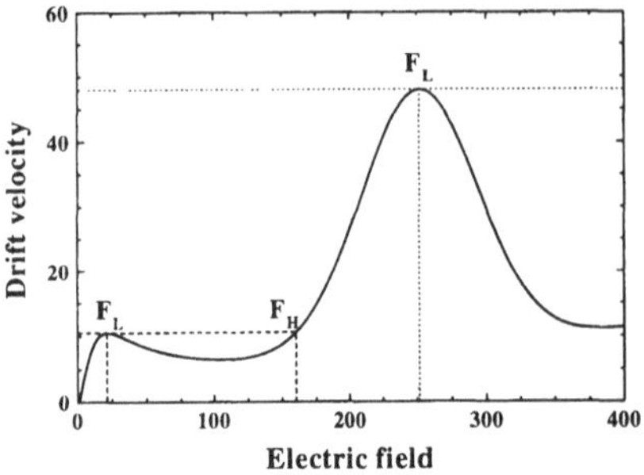

Fig. 4. Schematic drift velocity vs electric field characteristic for a weakly coupled superlattice with at least two subbands. F_L and F_H denote the field strength of the low- and high-field domains, respectively.

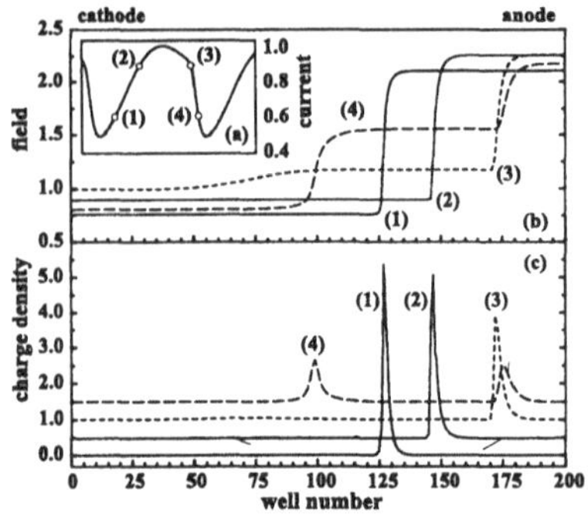

Fig. 5.
Calculated field and charge-density distribution within one oscillation period (see inset) for a doped superlattice with 200 periods showing the monopole recycling motion.

This set of equations is completed by boundary conditions. The first one connects the total field $\sum_{j=1}^{N} F_j = \frac{V}{d}$ to the applied voltage V, the second one determines the field change at the contacts. Assuming a typical drift-velocity vs electric-field characteristic as shown in Fig. 4, the set of nonlinear equations (2) and (3) results in an inhomogeneous field distribution throughout the superlattice, which for intermediate carrier densities can lead to spontaneous current oscillations as shown in Fig. 5. The oscillations of the current are due to a recycling motion of a charge monopole inside the superlattice (Kastrup et al. 1997). The electric field splits up into two domains, a low-field F_L and a high-field region F_H separated by a charge monopole called the domain boundary. In the following, we will discuss the experimental observations of these spontaneous current oscillations and the behavior under the influence of an external ac driving voltage.

3 Sample Configuration and Experimental Setup

The investigated samples consist of weakly coupled superlattices with different well and barrier widths. Sample 1 contains 40 periods with 9.0 nm GaAs wells and 4.0 nm AlAs barriers, while sample 2 has 50 periods with 13.3 nm GaAs wells and 2.7 nm AlAs barriers. They were produced by molecular beam epitaxy on (100) n^+-type GaAs substrates. The GaAs wells are Si-doped with a two-dimensional carrier density of 1.5×10^{15} m^{-2} in sample 1 and 1.0×10^{14} m^{-2} in sample 2. The n-type superlattice structures are embedded in n^+-n-n^+ diodes. Sample 3 contains an undoped superlattice with 100 periods of 6.2 nm GaAs wells and 3.4 nm AlAs barriers in a p^+-i-n^+ configuration, which acts in reverse bias as a photodetector. The samples are etched to yield mesas with a diameter of 120 μm and supplied with Ohmic contacts on top with a diameter of 70 μm.

The samples are mounted in a He-flow cryostat, which is equipped with high-frequency coaxial cables having a bandwidth of 20 GHz. The dc bias is applied perpendicular to the superlattice layers using a Keithley SMU 236. In addition, for the measurements of chaotic oscillations, an ac pulse/function generator (Wavetek 81) with a maximum frequency of 50 MHz is employed. For measuring the oscillatory component of the current, a 50-GHz sampling oscilloscope (Tektronix CSA 803), a real-time digital oscilloscope (Hewlett-Packard 54720A) with 1 GHz bandwidth or a 2.6-GHz spectrum analyzer (Advantest R3361) are used. The experimental setup is shown in Fig. 6. For the p^+-i-n^+ structure, a cw He-Ne laser (632.8 nm) with a maximum power of 21.3 mW is focused at normal incidence onto the p^+ cap layer to excite carriers in the superlattice region. Negative voltages refer to a negative bias applied to the p^+ top layer, i.e., reverse bias for the p^+-i-n^+ diode.

4 Spontaneous Current Oscillations

Typical time traces of spontaneous current oscillations are shown in Fig. 7 for samples 1 and 2 (Kastrup et al. 1995, Grahn et al. 1995, Kastrup et al. 1997). Both traces are recorded for electric fields above the resonance $E_1 \rightarrow E_2$, i.e., in the second plateau of the corresponding I-V characteristic. For the sample with a barrier width of 4.0 nm, the fundamental oscillation frequency is about 10 MHz. For sample 2, which has a thinner barrier of 2.7 nm, the oscillation frequency has increased to 95 MHz. We conclude that the frequency of the oscillations can be varied by changing the barrier width. Note that in the frequency spectra on the right side in Fig. 7 sample 1 contains a number of higher harmonics, while the oscillation in sample 2 is much more sinusoidal.

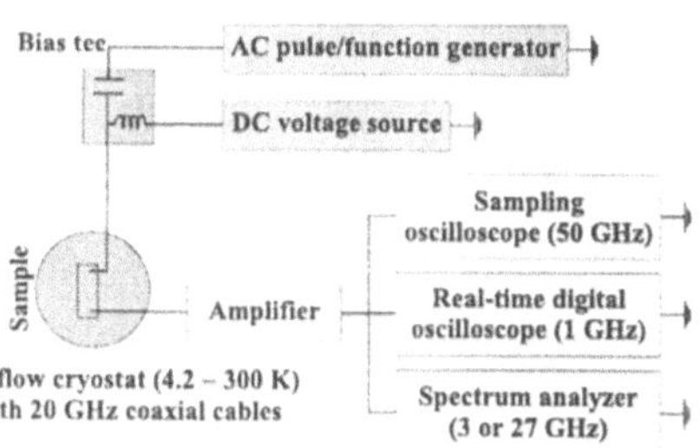

Fig. 6. Experimental configuration for measuring spontaneous current oscillations and driven chaotic oscillations.

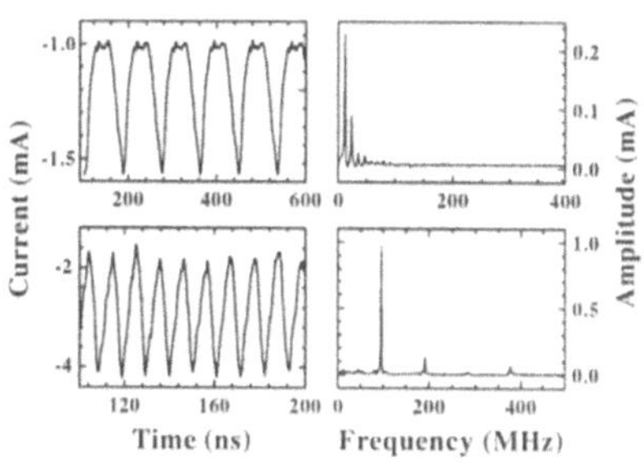

Fig. 7. Time traces (left) and frequency spectra (right) of spontaneous current oscillations in sample 1 (top) for −8 V and sample 2 (bottom) for −7.76 V recorded at 5–6 K.

We have investigated a number of samples with different barrier widths (Grahn et al. 1996, Kastrup et al. 1997). The measured frequency f_s of the spontaneous oscillations for different samples and for different current plateaus within a single sample are shown in Fig. 8 on a logarithmic scale as a function of the exponent of resonant coupling C in the WKB approximation, i.e.,

$$C = \frac{2\,d_B}{\hbar}\,\sqrt{2\,m^*(V-E)}\,, \tag{4}$$

where d_B denotes the barrier width, m^* the electron effective mass, V the height of the barrier, and E the energy of the subband, which the electrons tunnel into in the low-field domain. We clearly observe the expected exponential dependence of the WKB approximation. The dashed line shows a least square fit to the data points, resulting in a slope of -1.3 so that empirically $f_s = A\,\exp(-1.3\,C)$. We conclude that we can tune the frequency over a wide range by changing the barrier width. However, even within a single sample, we can vary the frequency over more than an order of magnitude by changing the applied field from the first to a higher plateau, which reduces the effective barrier height.

In addition to the frequency of the recycling motion of the domain boundary over many superlattice periods, there is another response present with a much higher frequency (Zhang et al. 1996a, Kastrup et al. 1997, Kantelhardt et al. 1997). In Fig. 9, a current trace recorded in the first plateau of sample 1 at 6 K is shown, which contains a number of spikes with a frequency of 10 MHz in addition to the fundamental oscillation frequency of 0.65 MHz. These spikes are due to the relocation of the domain boundary over a single period. There are 16 spikes between two maxima of the 0.65 MHz oscillation

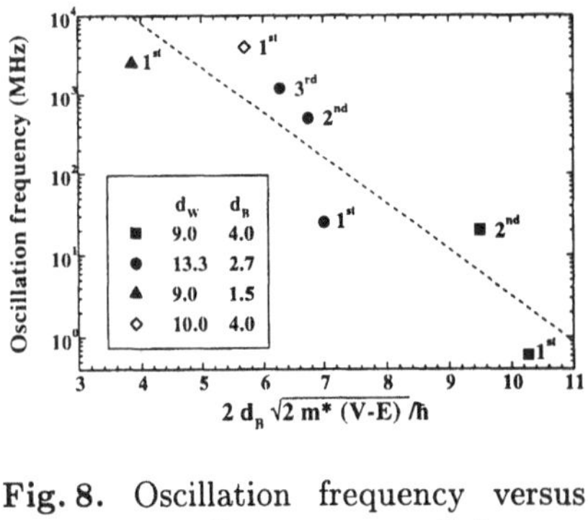

Fig. 8. Oscillation frequency versus resonant coupling exponent C for several GaAs/AlAs superlattices (solid symbols) and one $In_{0.53}Ga_{0.47}As$/$In_{0.52}Al_{0.48}As$ superlattice (open symbol) as well as for the indicated current plateaus.

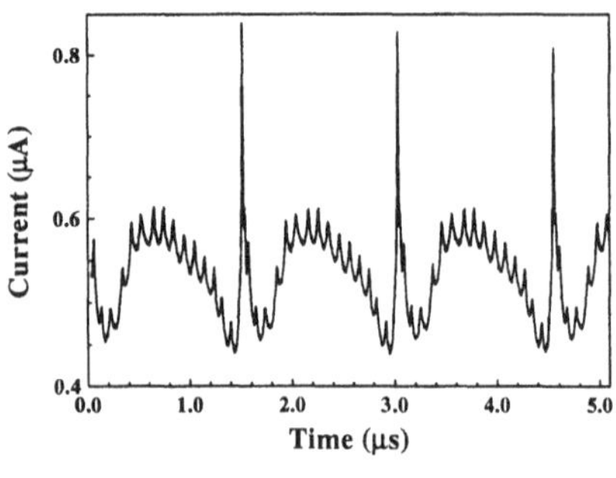

Fig. 9. Spontaneous current oscillations in sample 1 for an applied voltage of 2.77 V and a temperature of 6 K. There are 16 spikes in a single period of the 0.65 MHz oscillations.

indicating that the recycling motion of the charge accumulation layer covers about 40% of the total superlattice thickness. By introducing a delay time in the drift velocity in Ampere's law, spikes also appear in the calculated current versus time traces. The calculated time-dependent field distribution shows that this high-frequency component is due to the relocation of the domain boundary by a single period. Other samples also show this effect. The fraction of the superlattice involved in the recycling motion varies between 30 and 40%.

5 Bifurcation Scenarios under Photoexcitation

In order to experimentally investigate any carrier density of the current oscillations in the case of doped samples, it is necessary to grow a whole series of samples with different doping levels. This problem can be circumvented by using an undoped sample under photoexcitation. If a p^+-i-n^+ is biased in the reverse direction, no carriers are injected from the contacts so that only photoexcited carriers are produced in the intrinsic region, which contains the superlattice. By varying the laser power of the photoexcitation, the carrier density in the superlattice can be easily varied by several orders of magnitude within a single sample. Spontaneous current oscillations have been observed in undoped, photoexcited indirect-gap (Hosoda et al. 1996), i.e., type II, and direct-gap, i.e., type-I, superlattices (Ohtani et al. 1997, Ohtani et al. 1998a, Ohtani et al. 1998b, Ohtani et al. 1998c)

Figure 10 shows the time-averaged photocurrent (PC) versus applied voltage for several laser intensities for sample 3 (Luo et al. 1999). At 21.3 mW, a plateau is observed from about −1.45 to −8.88 V. This plateau exists for all laser powers shown in Fig. 10, but its width depends on the laser power. For laser powers below 1 mW, the plateau is replaced by a smooth NDC behavior as shown in the inset to Fig. 10. At 1 mW, the excited carrier density in the superlattice is still too low for the formation of electric-field domains. In this case, no plateau in the PC-V characteristics and no spontaneous current oscillations are observed. However, for larger laser powers, domain formation should occur as indicated by the presence of the photocurrent plateau shown in Fig. 10. At the same time, self-sustained photocurrent oscillations appear in the time domain (Ohtani et al. 1997). With increasing laser power, the carrier density becomes larger, and the electric-field domains exist over a wider bias voltage range so that the plateau covers a larger voltage region as shown in Fig. 10. For the laser powers used in these experiments, we do not observe the formation of stationary domains, where branch-like structures should appear in the plateaus (Grahn 1998). However, within a certain range of laser powers, a bistability region of the photocurrent is observed near the edge of the plateau as shown for 6.5, 8.6, and 15.9 mW in Fig. 10. When the reverse bias is increased, the photocurrent jumps down at a critical voltage of V_c^u. At the same time, the oscillations disappear. When the reverse bias

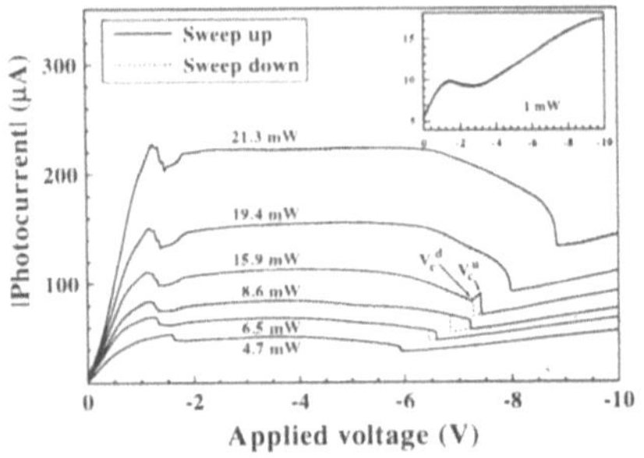

Fig. 10. Time-averaged PC-V characteristics of sample 3 recorded at 6.4 K for several laser powers as indicated. The inset shows the PC-V characteristics for 1 mW. The sweep directions are indicated by different lines.

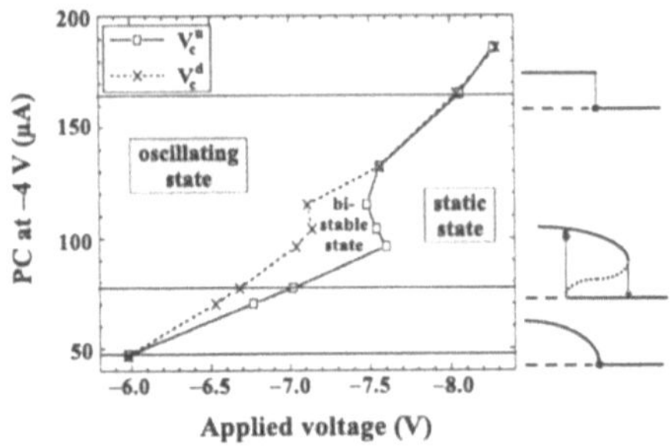

Fig. 11. Relationship between critical voltages V_c^u and V_c^d (cf. Fig. 10) and laser power (PC at −4 V) at 5 K. The lines between the data points are a guide to the eye. On the right hand side, the different bifurcation scenarios are indicated.

is decreased, the photocurrent jumps up at a critical voltage of V_c^d, and the oscillations re-appear.

The width of the bistable region can be controlled by changing the laser power. Figure 11 shows the relationship between the PC recorded at −4 V and the critical voltages V_c^u as well as V_c^d. At PC values below 47 μA, there is no bistability present. With increasing laser power, the bistable region appears, and its width gradually increases. At 132 μA, the width of the bistable region collapses, and the bistability disappears for all higher laser powers. Figure 11 represents a phase diagram, separating the oscillating from the static state. The region enclosed by the dashed and solid lines indicates the bistable region, where the oscillating and the static states coexist.

By tuning the laser power, not only the bistability, but also the bifurcation type from the oscillating to the static state can be controlled. At 6.4 K for a laser power of 4.7 mW [PC(−4 V)=51 μA] below the bistable regime, the frequency remains at a finite value, when the reverse bias voltage approaches V_c^u. At the same time, the amplitude of the oscillations decreases continuously to zero. This behavior shows that the oscillations disappear via a supercritical Hopf bifurcation in this laser power regime. For a supercritical Hopf bifurcation, there is only one critical point, where an unstable focus and a stable limit cycle change into a stable focus (cf. bottom schematic diagram in Fig. 11). No hysteresis should be present.

The behavior is quite different at high laser power outside the bistable regime. The frequency spectra, real-time current traces, and amplitude vs reduced bias of the PC oscillations are shown in Fig. 12 for a laser power of 21.3 mW [PC(−4 V)=222 μA]. The frequency of the PC oscillations tends gradually to zero with increasing reverse bias (the spectrum becomes broad-

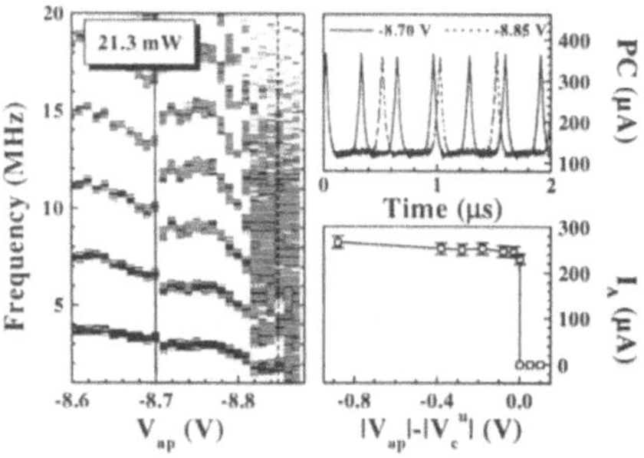

Fig. 12. Power spectra (left), real-time current traces (upper right), and amplitude I_A vs reduced bias (lower right) of the PC oscillations in sample 3 at 6.4 K for a laser power of 21.3 mW. The darker areas in the power spectra correspond to larger amplitudes.

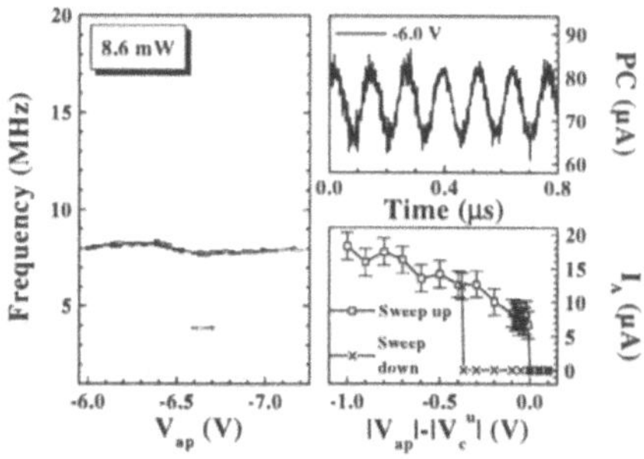

Fig. 13. Power spectra (left), real-time current traces (upper right), and amplitude I_A vs reduced bias (lower right) of the PC oscillations in sample 3 at 6.4 K for a laser power of 8.6 mW. The darker areas in the power spectra correspond to larger amplitudes.

band just before the oscillations completely disappear). At the same time, the amplitude of the oscillation hardly changes as shown for the real-time traces in Fig. 12. Such a behavior indicates that the oscillations disappear via a homoclinic connection (cf. top schematic diagram in Fig. 11), i.e., the oscillations vanish with a fixed amplitude and zero frequency (Patra et al. 1998, Moscoso et al. 1999). The broadband spectrum observed very close to the transition point is generally expected and results from the extreme sensitivity of the system near the saddle point of a homoclinic orbit.

The bifurcation scenario changes again for the bistable regime. As shown in Fig. 13 for a laser power of 8.6 mW [PC(-4 V)=84 μA], the frequency as well as the oscillation amplitude remain at finite values until the oscillations disappear. The static state is reached at -7.21 V. For the sweep-down condition, the oscillations suddenly re-appear also with a finite frequency and amplitude. The type of bifurcation is usually referred to as a subcritical Hopf bifurcation (Zhang et al. 1997b, Moscoso et al. 1999). There are two critical points for a subcritical Hopf bifurcation. At one critical point, the bifurcation takes place from an unstable focus (i.e., static state) to a stable focus and an unstable limit cycle (i.e., periodic oscillations), while at the other critical point the unstable limit cycle merges into a stable limit cycle (cf. central schematic diagram in Fig. 11). Bistability should be present between these two critical points.

Bistability and different bifurcation scenarios have already been predicted in theoretical work on doped as well as undoped, photoexcited superlattices (Schöll et al. 1996, Kastrup et al. 1997, Prengel et al. 1997, Moscoso et al. 1999). By changing both the dc bias and the carrier density, the theoretical investigations have revealed very complex dynamical phenomena. In these experiments, we have demonstrated the existence of bistability and three different bifurcation scenarios.

6 Chaotic Current Oscillations

If in addition to a dc voltage a modulating ac voltage is applied to this nonlinear system, different routes to chaos have been calculated (Bulashenko and Bonilla 1995, Bulashenko et al. 1996) and experimentally observed (Zhang et al. 1996b, Luo et al. 1998a). We will focus in terms of the dc voltage on the second plateau of the I-V characteristics of sample 1, which is shown in Fig. 14. There are two dc voltages marked in the second plateau, one near the center of the plateau V_{dc}^1 and one near the edge V_{dc}^2. Note that, in order to clearly show both plateaus in one plot, the current scale is logarithmic.

We then applied an ac driving voltage at a frequency f_d of the golden mean $[(1+\sqrt{5})/2 = 1.618]$ times the fundamental frequency f_s of the spontaneous current oscillations. For V_{dc}^1, f_d was set to 18.4 MHz, while for V_{dc}^2 we used 49.4 MHz. We varied the amplitude of the ac modulation voltage V_{ac} and recorded time traces as well as frequency spectra. In Fig. 15, several time traces are shown at V_{dc}^1 for different values of V_{ac}. At $V_{ac} = 75$ mV, the trace clearly exhibits frequency locking with a winding number as defined below of 5/8, while at $V_{ac} = 180$ mV a 2/3 frequency-locked state has been reached. At $V_{ac} = 175$ mV, the time trace shows some signature of chaos. In Fig. 16, selected time traces at V_{dc}^2 for different values of $V_{ac} = 34$, 40, and 63 mV exhibit a quasi-periodic, 2/3 frequency-locked, and chaotic state, respectively. All these time traces show only a small fraction of the recorded total time scale. We typically measured up to 1600 periods with a resolution of 20 points per period.

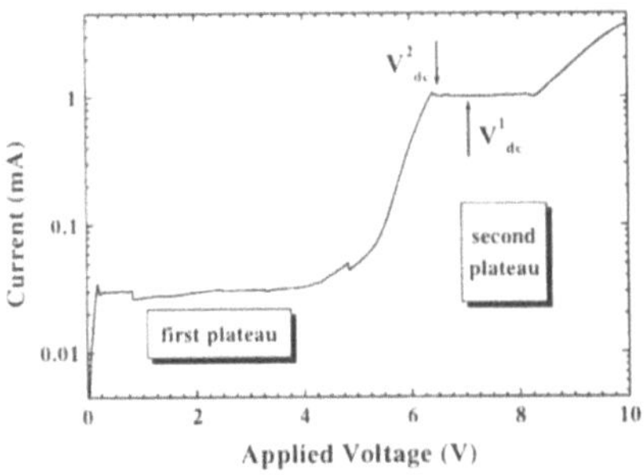

Fig. 14. I-V characteristics of sample 1 at 5 K. Two voltages are marked $V_{dc}^1 = 7.08$ and $V_{dc}^2 = 6.574$ V, the first one near the center and the second one near the edge of the second plateau.

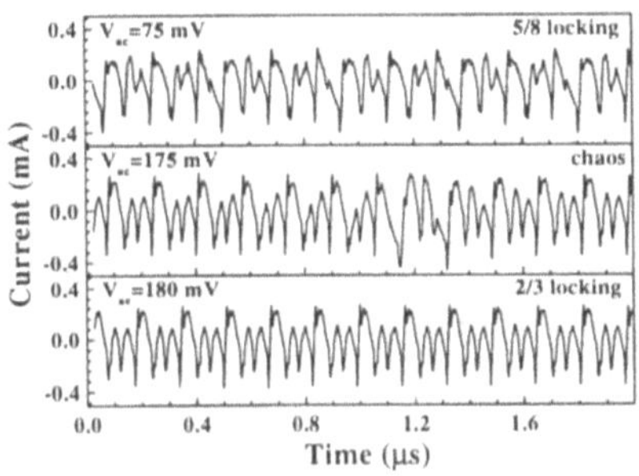

Fig. 15. Driven current oscillations for an applied voltage of $V_{dc}^1 = 7.08$ V in sample 1 recorded at 5 K for several amplitudes of the ac driving voltage as indicated.

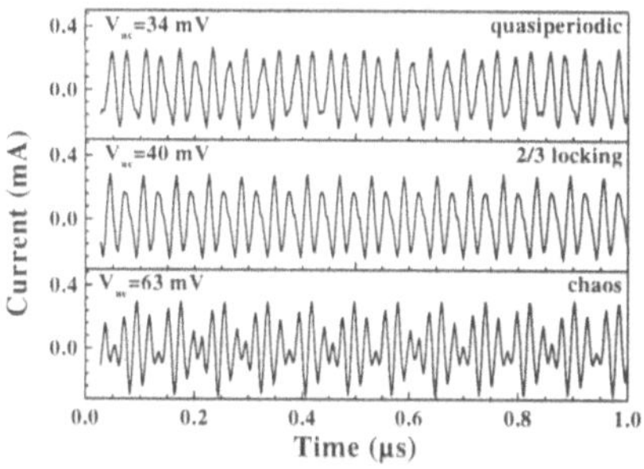

Fig. 16. Driven current oscillations for an applied voltage of $V_{dc}^2 = 6.574$ V in sample 1 recorded at 5 K for several amplitudes of the ac driving voltage as indicated.

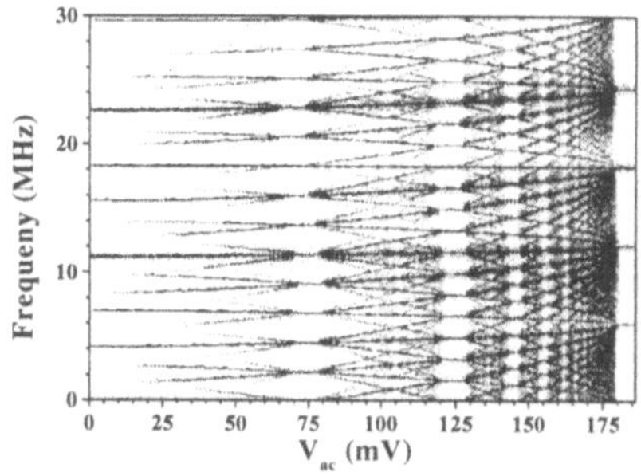

Fig. 17. Power spectra of the driven current oscillations as a function of the driving voltage amplitude V_{ac} for $V_{dc}^1 = 7.08$ V. The darker the area in the plot, the larger the absolute value of the amplitude of the oscillations.

The power spectra of the driven oscillations are shown in Figs. 17 and 18 for the dc voltages V_{dc}^1 and V_{dc}^2, respectively. For V_{dc}^1, there are alternating windows of quasi-periodicity and frequency locking. The frequency-locked windows starting at $V_{ac} = 75$ mV exhibit the following sequence of winding numbers as defined by the ratio of the number of frequencies up to the intrinsic frequency and the total number of frequencies up to the driving frequency, 5/8, 7/11, 9/14, (2n+1)/(3n+2) reaching 2/3 at very high values of V_{ac}. This number also reflects the ratio of the driving frequency to the intrinsic frequency at that particular value of V_{ac}. Note that the intrinsic frequency increases slightly with increasing V_{ac}. The quasi-periodic windows are characterized by a much richer frequency spectrum. For quasi-periodicity, the ratio of the driving to the intrinsic frequency equals an irrational number. The larger number of frequencies appears due to linear combinations of the these two frequencies. However, in the range of $V_{ac} = 170$ to 180 mV, the frequency spectra become rather smeared out so that the oscillations do not correspond to a quasi-periodic or frequency-locked state. The time trace at $V_{ac} = 175$ mV looks at first glance very similar to the one at $V_{ac} = 180$ mV. A closer look reveals a clear difference between the two traces (cf. Fig. 15). The relative amplitude of the two smaller peaks remains constant for $V_{ac} = 180$ mV, while it varies continuously for $V_{ac} = 175$ mV. Furthermore, around 1.2 μs, there is a strong change in the amplitude of the current oscillations. Both observations indicate that in this range of the driving voltage amplitude chaotic oscillations are present.

The frequency power spectra at $V_{dc}^2 = 6.574$ V shown in Fig. 18 exhibit a very different dependence on V_{ac}. At low values of V_{ac}, a quasi-periodic regime is observed, which ends at 37 mV. At this point, the frequency spectra begin to smear out over a limited spectral range. At $V_{ac} = 40$ mV, a

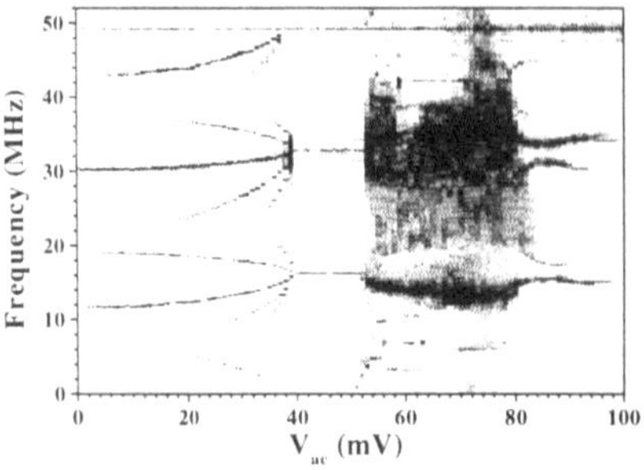

Fig. 18. Power spectra of the driven current oscillations as a function of the driving voltage amplitude V_{ac} for $V_{dc}^2 = 6.574$ V. The darker the area in the plot, the larger the absolute value of the amplitude of the oscillations.

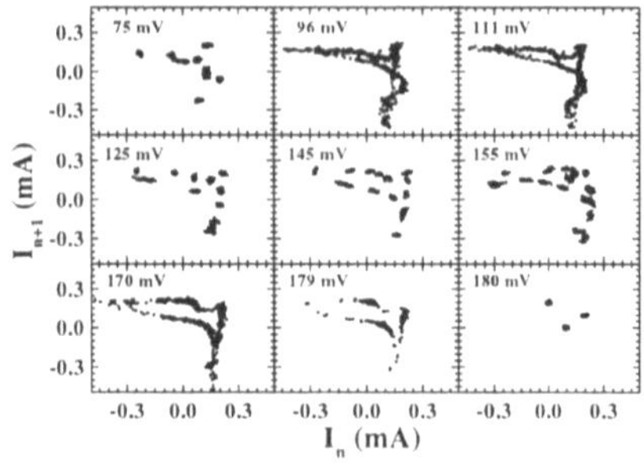

Fig. 19. Poincaré maps of the driven current oscillations for $V_{dc}^1 = 7.08$ V and different values of the driving voltage amplitude V_{ac} as indicated.

frequency-locked state with a winding number 2/3 appears, which extends up to 50 mV. For larger values of V_{ac}, the frequency spectra are more or less smeared out over the whole frequency range indicating the presence of chaotic oscillations. At $V_{ac} = 100$ mV, only a 1/1 frequency-locked state remains. The large difference of Figs. 17 and 18 are probably due to the fact that for a dc voltage near the edge of the plateau and above a certain driving voltage amplitude the transition from the oscillating to the static state strongly influences the response of the system to an ac modulation. For values of V_{ac} above 50 mV, the system may enter and leave the static field distribution periodically resulting in the observed complex behavior.

In order to obtain further insight in the chaotic behavior, we have analyzed time traces in terms of a return map, i.e., we plot the value of the current at the $(n+1)^{th}$ period as a function of the nth period using the driving frequency as the sampling frequency. These so-called Poincaré maps are shown for selected values of V_{ac} in Figs. 19 and 20 for the dc voltages V_{dc}^1 and V_{dc}^2, respectively. In the case of a dc voltage applied near the center of the plateau, we observe 8, 11, 14, 17, and 3 points at 75, 125, 145, 155, and 180 mV , respectively, as expected for frequency locking with the corresponding winding numbers 5/8, 7/11, 9/14, 11/17, and 2/3. For the quasi-periodic regime between 75 and 125 mV, two examples are shown for $V_{ac} = 96$ and 111 mV, which consist of a closed loop of points. A closed loop indicates the presence of quasi-periodicity. However, the loop is twisted in comparison to the Poincaré map of a single frequency oscillation driven by another frequency with an irrational frequency ratio. This twisting can be explained, when the higher harmonics of the intrinsic oscillations are taken into account (Luo et al. 1998b). It is also present for the frequency-locked state and can be explained in the same way (Luo et al. 1998b). At $V_{ac} = 170$ mV, the closed loop begins

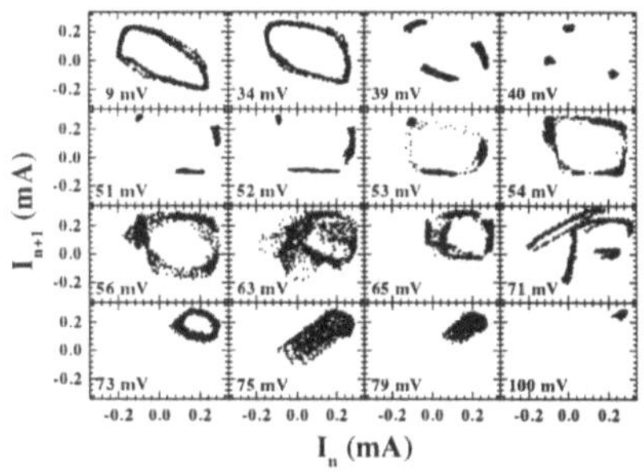

Fig. 20. Poincaré maps of the driven current oscillations for $V_{dc}^2 = 6.574$ V and different values of the driving voltage amplitude V_{ac} as indicated.

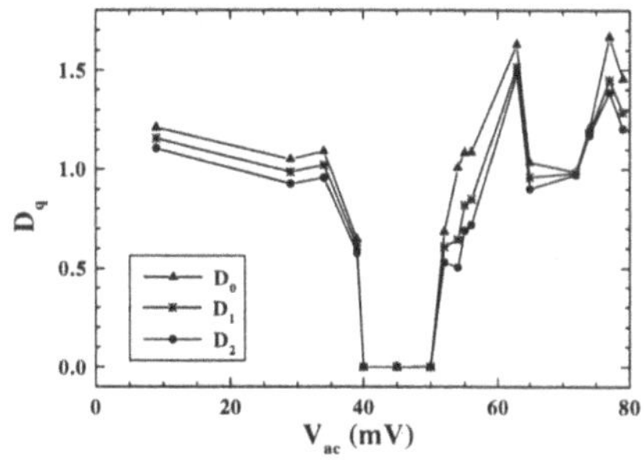

Fig. 21. Capacity D_0, information D_1, and correlation dimension D_2 as a function of the driving voltage amplitude V_{ac} for $V_{dc}^2 = 6.574$ V.

to break up. In the range between 170 and 180 mV, we do not observe a closed loop as for quasi-periodicity or isolated points as for frequency locking so that we conclude that chaotic oscillations are present in this regime. The route to chaos follows the well-known path of quasi-periodicity → frequency locking → quasi-periodicity → ... → frequency locking → chaos.

As expected from the frequency spectra under external driving voltage, also the Poincaré maps for a dc voltage near the edge of the plateau differ significantly from the ones for a dc voltage near the center of the plateau. Up to $V_{ac} = 37$ mV, typical Poincaré maps corresponding to quasi-periodic behavior are observed. Two examples are shown in Fig. 20 for 9 and 34 mV, which have the typical shape. Note that for this dc voltage, the intrinsic oscillations do not contain any higher harmonics. A very different type of Poincaré map is derived for 38 and 39 mV. This type of map shows some signature of synchronized chaos. Between 39 and 40 mV, there is a very abrupt change from a spatially extended Poincaré map to three isolated points indicating 2/3 frequency locking. This frequency-locked state remains up to $V_{ac} = 50$ mV. For V_{ac} larger than 50 mV, the Poincaré maps are spatially extended with various shapes, which are clearly not related to a quasi-periodic or frequency-locked state. The Poincaré maps support the interpretation of the frequency spectra that in this regime of driving voltage amplitudes the driven oscillations are chaotic. When V_{ac} is above 80 mV, the Poincaré map consists of an extended spot, which shrinks in size with increasing V_{ac}. At 100 mV, the 1/1 frequency-locked state is reached with a Poincaré map consisting of a single point. In this case, the route to chaos follows a different path, quasi-periodicity → synchronized chaos → frequency locking → chaos → more complex chaos.

In order to obtain information about the dimension of the attractor, we have derived the capacity D_0, information D_1, and correlation dimension D_2 from the recorded time traces (Bulashenko et al. 1999). Figure 21 shows the dependence of these three dimensions as a function of driving voltage amplitude for V_{dc}^2. In the quasi-periodic regime for values of V_{ac} below 38 mV, all three dimensions are about 1, which corresponds to the dimension of a closed loop with a constant point density. This value supports the interpretation of this regime as quasi-periodic. However, the values are actually somewhat larger than one. A closer look at the corresponding Poincaré maps reveals that there are probably two closed loops on top of each other. This regime may contain a more complicated dynamical behavior than plain quasi-periodicity. Between 40 and 50 mV, all three dimensions are zero indicating the presence of the already identified 2/3 frequency-locked state. For the region of synchronized chaos at 38 to 39 mV, the capacity dimension $D_0 = 0.64$ is significantly smaller than one, but still finite. For V_{ac} larger than 50 mV, all three dimensions strongly increase to values larger than one. A more detailed analysis, in particular of the chaotic regime, is presently performed, which will give further insight in the underlying chaos. Furthermore, a detailed theoretical investigation for dc biases near the center and edge of the plateau is necessary in order to completely understand the chaotic behavior of this system.

Finally, we should mention that we also observed undriven chaotic oscillations in doped (Zhang et al. 1996a, Zhang et al. 1997a) and undoped, photoexcited superlattices (Ohtani et al. 1997, Ohtani et al. 1998a, Ohtani et al. 1998b, Ohtani et al. 1998c). Presently, the appearance of undriven chaos is not understood. However, it may be related to the existence of another time scale such as the delay time, which was introduced to explain the current spikes within a single fundamental oscillation. We also studied the response of an undoped, photoexcited superlattice to an external ac driving voltage (Ohtani et al. 1998a). A very different bifurcation pattern is observed in comparison with the doped superlattices and the calculated results.

7 Summary and Conclusions

Weakly coupled superlattices represent a nonlinear system due to sequential resonant tunneling between different subbands in adjacent wells. Several regions of negative differential velocity exist in such systems resulting in the formation of static and dynamic electric-field domains. In the dynamic regime, spontaneous current oscillations appear with frequencies ranging from the sub-MHz regime up to several GHz. The frequencies are mainly determined by the resonant coupling between adjacent wells, which depends exponentially on the barrier width as well as the square root of the effective barrier height.

In an undoped, photoexcited superlattice, we have observed a tunable bistability and controllable bifurcation processes. When the bistability is absent at low (high) laser powers, the oscillations disappear via a supercritical Hopf bifurcation (homoclinic connection). However, in the bistable regime, the oscillations disappear via a subcritical Hopf bifurcation with increasing reverse bias.

When an ac modulation voltage is applied in addition to a dc voltage, frequency-locked, quasi-periodic and chaotic oscillations can be observed. Depending on the position of the applied dc bias with respect to the current plateau, different routes to chaos are observed, which differ in the complexity of the route. Finally, in the case of more complex chaos, the multifractal dimension of the chaotic attractors has been derived from the real-time current traces supporting the interpretation derived from the Poincaré maps.

Acknowledgments

The author would like to thank H. Asai, L. L. Bonilla, O. Bulashenko, A. Fischer, R. Hey, J. W. Kantelhardt, J. Kastrup, K. J. Luo, M. Rogozia, K. H. Ploog, A. Wacker, and Y. Zhang for their intensive collaboration and stimulating discussions as well as H. Kostial and E. Wiebicke for expert sample processing. Partial support of the Deutsche Forschungsgemeinschaft within the framework of Sfb 296 is gratefully acknowledged.

References

Bonilla L. L. (1995): in *Nonlinear Dynamics and Pattern Formation in Semiconductors and Devices*, edited by F.-J. Niedernostheide (Springer-Verlag, Berlin), Chap. 1

Bonilla L.L., Galán J., Cuesta J. A., Martínez F. C., & Molera J. M. (1994): Phys. Rev. B 50, 8644

Bulashenko O. M., and Bonilla L. L. (1995): Phys. Rev. B 52, 7849

Bulashenko O. M., García M. J., & Bonilla L. L. (1996): Phys. Rev. B 53, 10008

Bulashenko O. M., Luo K. J., Grahn H. T., Ploog K. H., & Bonilla L. L. (1999): Phys. Rev. B 60, 5694

Esaki L., and Tsu R. (1970): IBM J. Res. Develop. 14, 61

Grahn H. T. (1995): *Semiconductor Superlattices* (World Scientific, Singapore)

Grahn H. T. (1998): in *Hot Electrons in Semiconductors, Physics and Devices*, edited by N. Balkan (Clarendon Press, Oxford), pp. 357–381

Grahn H., Kastrup J., Ploog K., Bonilla L., Galán J., Kindelan M., & Moscoso M. (1995): Jpn. J. Appl. Phys. 34, 4526

Grahn H. T., Kastrup J., Klann R., Ploog K. H., & Asai H. (1996): in *Proceed. of the 23rd International Conference on the Physics of Semiconductors*, edited by M. Scheffler and R. Zimmermann (World Scientific, Singapore), p. 1671

Hosoda M., Mimura H., Ohtani N., Tominaga K., Watanabe T., Fujiwara K., & Grahn H. T. (1996): Appl. Phys. Lett. 69, 500

Kantelhardt J. W., Grahn H. T., Ploog K. H., Moscoso M., Perales A., & Bonilla L. L. (1997): Phys. Status Solidi B 204, 500

Kastrup J., Klann R., Grahn H. T., Ploog K., Bonilla L. L., Galán J., Kindelan M., Moscoso M., & Merlin R. (1995): Phys. Rev. B 52, 13761

Kastrup J., Hey R., Ploog K. H., Grahn H. T., Bonilla L. L., Kindelan M., Moscoso M., Wacker A., & Galán J. (1997): Phys. Rev. B 55, 2476
Luo et al. 1998a Luo K. J., Grahn H. T., Ploog K. H., & Bonilla L. L. (1998): Phys. Rev. Lett. 81, 1290
Luo et al. 1998b Luo K. J., Grahn H. T., Teitsworth S. W., & Ploog K.H. (1998): Phys. Rev. B 58, 12613
Luo K. J., Teitsworth S. W., Kostial H., Grahn H. T., & Ohtani N. (1999): Appl. Phys. Lett. 74, 3845
Moscoso M., Bonilla L. L., & Galán J. (1999): in *Proceed. the 24th International Conference on the Physics of Semiconductors* edited by D. Gershoni (World Scientific, Singapore), V-C-14 (0524.pdf)
Ohtani, N., Hosoda, M., & Grahn H. T. (1997): Appl. Phys. Lett. 70, 375
Ohtani et al. (1998a) Ohtani N., Egami N., Fujiwara K., & Grahn H. T. (1998): Solid-State Electron. 42, 1509
Ohtani et al. (1998b) Ohtani N., Egami N., Grahn H. T., Ploog K. H., & Bonilla L. L. (1998): Phys. Rev. B 58, R7528
Ohtani et al. (1998c) Ohtani N., Egami N., Grahn H. T., & Ploog K. H. (1998): Physica B 249–251, 878
Patra M., Schwarz G., & Schöll E. (1998): Phys. Rev. B 57, 1824
Prengel F., Patra M., Schwarz G., & Schöll E. (1997): in *Proceed. of the 23rd International Conference on the Physics of Semiconductors*, edited by M. Scheffler and R. Zimmermann (World Scientific, Singapore), p. 1667.
Schöll E., Schwarz G., Patra M., & Wacker A. (1996): in *Hot Carriers in Semiconductors*, edited by K. Hess, J. P. Leburton, and U. Ravaioli (Plenum Press, New York), p. 177
Wacker A. (1998): in *Theory of Transport Properties of Semiconductor Nanostructures*, edited by E. Schöll (Chapman and Hall, London), Chap. 10
Wacker A., and Jauho A. P. (1998): Phys. Rev. Lett. 80, 369
Zhang et al. (1996a) Zhang Y., Klann R., Ploog K. H., & Grahn H. T. (1996): Appl. Phys. Lett. 69, 1116
Zhang et al. (1996b) Zhang Y., Kastrup J., Klann R., Ploog K. H., & Grahn H. T. (1996): Phys. Rev. Lett. 77, 3001
Zhang et al. (1997a) Zhang Y., Klann R., Grahn H. T., & Ploog K. H. (1997): Superlattices Microstruct. 21, 565
Zhang et al. (1997b) Zhang Y., Klann R., Ploog K. H., & Grahn H. T. (1997): Appl. Phys. Lett. 70, 2825

Part VI

Spin Properties

Spintronic Spin Accumulation and Thermodynamics

A.H. MacDonald

Department of Physics, Indiana University, Bloomington, IN 47405

Abstract. The spin degree of freedom can play an essential role in determining the electrical transport properties of spin-polarized electron systems in metals or semiconductors. In this article, I address the dependence of spin-subsystem chemical potentials on accumulated spin-densities. I discuss both approaches which can be used to measure this fundamental thermodynamic quantity and the microscopic physics which determines its value in several different systems.

1 Introduction

The role of the electronic spin degree of freedom in theories of the electrical transport properties of paramagnetic metals is passive; usually it as appears only as an afterthought—a factor of two to account for spin degeneracy. All this changes profoundly in electronic systems with substantial spin-polarization, either spontaneous or induced, particularly so if the system is either electrically or magnetically inhomogeneous. Recently, interest in the role of the electronic spin has increased, in part because of the possibility of fabricating technologically useful magnetoresistive sensors and other devices based on spin-dependent transport effects, particularly giant magnetoresistance [1] and tunnel magnetoresistance [2]. *Spintronics* [3], the study of spin-dependent electronic transport effects in systems containing metallic ferromagnets, is now a large and active area of basic and applied physics. In this article, I discuss the possibility of using transport experiments not to make devices, but instead to measure a fundamental thermodynamic property of a spin-polarized electron system, the dependence of spin subsystem chemical potentials on accumulated spin-densities. This quantity can be important in modeling some spin-dependent transport effects. I will discuss several examples where it also provides a new and useful test of our understanding of the microscopic physics of a spin-polarized itinerant electron system.

In Section 2, I derive a formally exact expression for the dependence of chemical potentials on subsystem densities by defining a spin-dependent thermodynamic-density-of-states *matrix*. For bulk three-dimensional charged particle systems, the total density is fixed by electroneutrality requirements. In this case only the dependence of up and down spin chemical potentials on the difference between up and down spin densities is of interest. I show that these two quantities can be expressed in terms of the differential magnetic susceptibility, and a less familiar quantity, the derivative of chemical potential with respect to external field. The focus of the article is this latter quantity, which I will refer to as the ***inverse magnetic compressibility*** . In Section 3, I discuss two spintronics experiments which can be exploited to measure its value in particular systems. In Section 4, I discuss the microscopic physics which determines its value in three distinct spin-polarized electron systems. Section 5 contains a brief summary.

2 Thermodynamic Density-of-States Matrix

Non-equilibrium spin accumulation [4, 5] due to electronic transport occurs generically in inhomogeneous spin-polarized electron systems and is a ubiquitous feature of spintronics. Any theory of spin accumulation requires, explicitly or implicitly, a model for the relationship between the up and down spin densities and their chemical potentials. Linearizing around the equilibrium state, we can write

$$\begin{aligned} d\mu_\uparrow &= (D_\uparrow^{-1} + F_{\uparrow,\uparrow})\, dn_\uparrow + F_{\uparrow,\downarrow}\, dn_\downarrow \\ d\mu_\downarrow &= F_{\downarrow,\uparrow}\, dn_\uparrow + (D_\downarrow^{-1} + F_{\downarrow,\downarrow})\, dn_\downarrow . \end{aligned} \tag{1}$$

where μ_σ is the spin-σ chemical potential and n_σ is the density of spin-σ electrons. In these equations, I treat $n_\uparrow$ and $n_\downarrow$ as separate thermodynamic variables, something which is useful in discussing spin-accumulation since the processes which establish equilibrium between spin-$\uparrow$ and spin-$\downarrow$ subsystems are often slow. The spin-quantization axis has been chosen to lie along the direction of net spin-polarization [6].

It is normally convenient to measure the local chemical potential of a charged particle system from the local electrostatic potential and this common convention is implicit throughout these notes. Accordingly

$$\mu_\sigma \equiv \frac{1}{V}\frac{\partial F(T, n_\uparrow, n_\downarrow)}{\partial n_\sigma} \tag{2}$$

with the free-energy per volume, F/V, calculated excluding any electrostatic contributions. The matrix of coefficients in Eq. (1) is accordingly given by the matrix of second derivatives of F/V with respect to n_σ, with the electrostatic term (which would diverge because of the long-range of the electron-electron interaction) neglected. (Note that $F_{\uparrow,\downarrow} = F_{\downarrow,\uparrow}$.) To make contact with familiar descriptions of spin accumulation, I have introduced the band-theory spin-dependent density-of-states per volume, D_σ. If contributions due to correlation effects were neglected, the relationship between chemical potentials and densities would be diagonal in spin indices and only the density-of-states terms would appear on the right-hand-side of Eq. (1). The density-of-states contribution to the spin-σ chemical potential change is simply the result of changing the filling of the spin-σ energy band. I will refer to the additional correlation terms in Eq. (1), $F_{\sigma,\sigma'}$, as local-field corrections. The fact that the local-field corrections are in general off-diagonal in the spin indices can have a qualitative importance. Eq. (1) is formally exact, however, the values for the local field corrections, are not known in general, and their computation is a challenge to theory. In the following Section, I discuss two experiments which measure a particular combination of these coefficients.

Eq. (1) can be inverted to express spin-dependent density changes in terms of spin-dependent chemical potential changes

$$dn_\sigma = \sum_{\sigma'} \mathcal{D}_{\sigma,\sigma'} d\mu_{\sigma'} \tag{3}$$

where

$$\begin{aligned} \mathcal{D}_{\sigma,\sigma} &= \frac{D_\sigma(1 + D_{\bar\sigma} F_{\bar\sigma,\bar\sigma})}{1 + D_\sigma F_{\sigma,\sigma} + D_{\bar\sigma} F_{\bar\sigma,\bar\sigma} - D_\sigma D_{\bar\sigma} F^2_{\sigma,\bar\sigma}} \\ \mathcal{D}_{\sigma,\bar\sigma} &= -\frac{D_\sigma D_{\bar\sigma} F_{\bar\sigma,\sigma}}{1 + D_\sigma F_{\sigma,\sigma} + D_{\bar\sigma} F_{\bar\sigma,\bar\sigma} - D_\sigma D_{\bar\sigma} F^2_{\sigma,\bar\sigma}}, \end{aligned} \tag{4}$$

where $\bar\sigma =\downarrow$ if $\sigma =\uparrow$ and *vice-versa.* $\mathcal{D} \equiv \sum_{\sigma,\sigma'} \mathcal{D}_{\sigma,\sigma'}$, the rate of change of total density with chemical potential when the spin-subsystems are in equilibrium, is the total thermodynamic density-of-states of an electron system. If interactions are neglected $\mathcal{D} = D_\uparrow + D_\downarrow$. In spintronics it is useful to generalize this concept by defining a thermodynamic density-of-states *matrix* as in Eq. (4).

To evaluate the inverse magnetic compressibility, I add to the Hamiltonian a Zeeman coupling [7] term which contributes $-g^*\mu_B H(n_\uparrow - n_\downarrow)/2$ to the free energy per unit volume. Here g^* is the system's g-factor, μ_B is the electron Bohr magneton, and H is the field strength. It is convenient to use a notation where μ_σ is defined as the chemical potential without its Zeeman contribution, whereas μ is the full field-dependent chemical potential. Then the condition for equilibrium between up and down spins is

$$\mu = \mu_\uparrow - g^*\mu_B H/2 = \mu_\downarrow + g^*\mu_B H/2, \tag{5}$$

or differentiating with respect to field strength:

$$\frac{\partial\mu}{\partial g^*\mu_B H} = \frac{\partial\mu_\uparrow}{\partial g^*\mu_B H} - \frac{1}{2} = \frac{\partial\mu_\downarrow}{\partial g^*\mu_B H} + \frac{1}{2} \tag{6}$$

At fixed total electron density ($dn_\uparrow + dn_\downarrow = 0$), differentiating Eq. (3) with respect to $g^*\mu_B H$ gives

$$\frac{\partial\mu_\uparrow/\partial H}{\partial\mu_\downarrow/\partial H} = -\frac{\mathcal{D}_{\uparrow,\downarrow} + \mathcal{D}_{\downarrow,\downarrow}}{\mathcal{D}_{\uparrow,\uparrow} + \mathcal{D}_{\downarrow,\uparrow}} = -\frac{D_\uparrow^{-1} + F_{\uparrow,\uparrow} - F_{\uparrow,\downarrow}}{D_\downarrow^{-1} + F_{\downarrow,\downarrow} - F_{\uparrow,\downarrow}}. \tag{7}$$

Combining Eq. (7) and Eq. (6), I obtain that for fixed total density

$$\frac{\partial \mu}{\partial g^* \mu_B H} = \frac{D_\uparrow^{-1} - D_\downarrow^{-1} + F_{\uparrow,\uparrow} - F_{\downarrow,\downarrow}}{2[D_\uparrow^{-1} + D_\downarrow^{-1} + F_{\uparrow,\uparrow} + F_{\downarrow,\downarrow} - 2F_{\uparrow,\downarrow}]}. \tag{8}$$

Similar considerations lead to the following expression for the differential magnetic susceptibility,

$$\chi_S = \frac{g^* \mu_B}{2} \frac{\partial (n_\uparrow - n_\downarrow)}{\partial H} = \frac{(g^* \mu_B)^2}{D_\uparrow^{-1} + D_\downarrow^{-1} + F_{\uparrow,\uparrow} + F_{\downarrow,\downarrow} - 2F_{\uparrow,\downarrow}}. \tag{9}$$

Both of these expressions are formally exact. I show in the following paragraph that the dependence of the chemical potentials on spin accumulation is specified by the inverse magnetic compressibility and χ_S.

Theories of spin accumulation contain in general three elements: i) a theory of the spin and space dependent transport coefficients which lead to non-equilibrium spin-densities, ii) a theory for the disequilibration $(\mu_\uparrow - \mu_\downarrow)$ produced by these spin densities, and iii) a theory of the relaxation process which attempts to establish equilibrium between the spin subsystems. The thermodynamic property we are discussing is related to the second element. Assuming electroneutrality $(dn_\downarrow = -dn_\uparrow)$ it follows from Eq. (1) that

$$\begin{aligned} \frac{d\mu_\uparrow}{d(n_\uparrow - n_\downarrow)} &= \frac{D_\uparrow^{-1} + F_{\uparrow,\uparrow} - F_{\uparrow,\downarrow}}{2} \\ \frac{d\mu_\downarrow}{d(n_\uparrow - n_\downarrow)} &= -\frac{D_\downarrow^{-1} + F_{\downarrow,\downarrow} - F_{\uparrow,\downarrow}}{2}. \end{aligned} \tag{10}$$

Note that $d(\mu_\uparrow - \mu_\downarrow)/d(n_\uparrow - n_\downarrow) = (g^* \mu_B)^2/2\chi_S$. This relationship should not be a surprise since an external magnetic field maintains a chemical potential difference $\mu_\uparrow - \mu_\downarrow = g^* \mu_B H$ and induces a magnetization per volume $m = \chi_S H = g^* \mu_B (n_\uparrow - n_\downarrow)/2$. The linear relationship between $\mu_\uparrow - \mu_\downarrow$ and the non-equilibrium spin accumulations is thus completely characterized by χ_S. It is only if we want to know the chemical potential shifts of up-spin and down-spin subsystems individually that the inverse magnetic compressibility is required. The individual chemical potential shifts driven by a non-equilibrium spin accumulation are:

$$\frac{4\chi_S}{(g^* \mu_B)^2} \frac{d\mu_\uparrow}{d(n_\uparrow - n_\downarrow)} = 1 + 2\frac{d\mu}{d(g^* \mu_B H)} = \frac{D_\uparrow^{-1} + F_{\uparrow,\uparrow} - F_{\uparrow,\downarrow}}{D_\uparrow^{-1} + D_\downarrow^{-1} + F_{\uparrow,\uparrow} + F_{\downarrow,\downarrow} - 2F_{\uparrow,\downarrow}}$$

$$\frac{4\chi_S}{(g^* \mu_B)^2} \frac{d\mu_\downarrow}{d(n_\uparrow - n_\downarrow)} = -1 + 2\frac{d\mu}{d(g^* \mu_B H)} = -\frac{D_\downarrow^{-1} + F_{\downarrow,\downarrow} - F_{\uparrow,\downarrow}}{D_\uparrow^{-1} + D_\downarrow^{-1} + F_{\uparrow,\uparrow} + F_{\downarrow,\downarrow} - 2F_{\uparrow,\downarrow}}$$

3 Spintronic Thermodynamic Measurements

3.1 Field-dependent Coulomb Blockade Peaks

The first type of experiment I discuss was pioneered by Ono and co-workers [8] and takes advantage of the equally spaced conductance peaks which occur in Coulomb blockade devices. The experimental geometry is that of a single-electron-transistor (SET) in which current flow from source to drain through a small metallic particle is influenced by a gate voltage. In general source lead, drain lead, and metallic particle can be either paramagnetic or ferromagnetic. For definiteness, I assume that the only the small metallic particle is ferromagnetic and address a situation which is simpler than what has been encountered in experiments by also assuming a single domain. The following paragraph provides a simplified explanation the operation of a SET which is sufficient for present purposes.

The dependence of the ground-state energy of an isolated metallic grain on its net charge is dominated by an electrostatic contribution and has the form

$$E_0(N) = \frac{e^2(N-N_0)^2}{2C} + V\epsilon_0(N/V) \tag{11}$$

where $e(N-N_0)$ is the net charge on the grain and the effective capacitance of the grain $C \sim R$ where R is the grain diameter. Here $\epsilon_0(n)$ is the energy per unit volume V calculated for a macroscopic grain which is electrically neutral. The conductance for current flow through the grain between lead electrodes is sharply peaked when the addition energy of the island is equal to the chemical potential in the lead electrodes μ_L, *i.e.* when

$$\mu(N) \equiv E_0(N+1) - E_0(N) = \mu_L. \tag{12}$$

In a SET the chemical potential for electrons on the metallic grain is manipulated by a gate voltage U:

$$E_0(N) \to \frac{e^2(N-N_0)^2}{2C} + V\epsilon_0(N/V) - NeU. \tag{13}$$

As U is varied, the equilibrium number of particles on the grain changes. The number of particles in the grain's ground state changes between N and $N+1$ when Eq.(12) is satisfied:

$$eU_N^* = e^2(N+1/2)/C + \mu(N/V) - \mu_L. \tag{14}$$

A peak occurs in the source-drain conductance at this value of the gate voltage. In this equation $\mu(n)$ is the bulk chemical potential of an electrically neutral system, the quantity examined in Section 2.

As discussed in Section 2, $\mu(n)$ is field-dependent in general. It follows from Eq. (14) that provided the field-dependence of the lead chemical potentials can be ignored

$$\frac{d\mu(n)}{d(g^*\mu_B H)} = \frac{d(eU_N^*)}{d(g^*\mu_B H)}; \tag{15}$$

the quantity of interest can simply be read off the gate voltage dependence of the Coulomb blockade conductance peak. More realistic models are complicated by geometry dependent cross capacitances between different circuit elements which invalidate the simple relationship between the chemical potential on the ferromagnetic grain and the gate voltage assumed here. However these two quantities are still proportional and the proportionality constant can be sorted out experimentally by measuring the spacing between Coulomb blockade conductance peaks at fixed voltage. Estimates of $d(\mu_0(n)/d(g^*\mu_B H)$ have already been obtained [8] using this approach for one ferromagnetic transition metal.

3.2 Field-dependent double-layer compressibility measurements

The second potential experiment I discuss is a variant on one which has been used in the past [9, 10, 11, 12] to measure the compressibility of two-dimensional electron gas layers. It exploits techniques which have been developed [13] to make separate contact to nearby two-dimensional electron layers. The experimental set up can be thought of as a parallel plate capacitor, where one plate is a metal or heavily doped semiconductor and the second plate consists of two separately contacted two-dimensional electron layers, one on top of the other and closer to the metallic gate. A change in the charge density on the surface of the metal induces an opposing charge density distributed between the two two-dimensional electron layers. The equilibrium condition which determines the distribution of charge is:

$$\mu_T(n_T) = \mu_B(n_B) + 4\pi e^2 d(n - n_T - n_0/2). \tag{16}$$

Here n_T is the areal density in the top two-dimensional layer, n_B is the density in the bottom two-dimensional layer, d is the separation between the two layers, and $n = n_T + n_B$ is the total density in the two-layers. The second term on the right hand side of Eq. (16) is the electrostatic potential drop due to the electric field which exists between top and bottom two-dimensional layers. Note that the electric field between the gate and the top layer has magnitude $4\pi e(n - n_0/2)$, *i.e.* en equals the surface charge density on the metallic gate up to a constant. If the top two-dimensional layer is held at ground, the gate voltage V_G is therefore proportional to n. The magnitude of the electric field below the bottom two-dimensional layer, $4\pi e n_0/2$, does not change during the experiment.

To determine the compressibility [9], it is necessary only to measure the current which flows to the bottom layer when the gate voltage V_G (or equivalently n) is changed. Differentiating Eq. (16) with respect to n with $n_B + n_T = n$ leads to [9] a relationship between a fundamental thermodynamic quantities and a conveniently measurable dimensionless experimental quantity exploited in previous experimental work:

$$\frac{dn_B}{dn} = \frac{d\mu_T/dn_T}{4\pi e^2 d + d\mu_T/dn_T + d\mu_B/dn_B} \tag{17}$$

Since the first term in the denominator of Eq. (17) is normally dominant the experiment provides a direct measurement of the dependence of chemical potential on density, inversely proportional to the compressibility, for the top layer.

I propose using the same experimental setup to measure the dependence of the charge in the bottom layer on magnetic field at fixed gate voltage. Such a measurement can be related to the inverse magnetic compressibility; differentiating Eq. (16) with respect to field at fixed n I find that

$$\frac{dn_B}{d(g^*\mu_B H)} = \frac{d\mu_T/d(g^*\mu_B H) - d\mu_B/d(g^*\mu_B H)}{4\pi e^2 d + d\mu_T/dn_T + d\mu_B/dn_B}. \tag{18}$$

Since the denominator is dominated by the first term, the experiment provides a direct measurement of the difference between the rate of change chemical potential with field in top and bottom 2D layers.

4 Microscopic Theory of the Thermodynamic Density-of-States Matrix

4.1 Band Ferromagnets

In order to contextualize the issues raised by these relatively new measurement possibilities, I first discuss the estimate for $d\mu/d(g^*\mu_B H)$ which follows from the simplest possible mean-field theory of a ferromagnetic metal, Stoner-Wohlfarth [14] theory. In modern work this approach is wrapped in the cloak of spin-density-functional theory [15]. In the Stoner-Wohlfarth theory, quasiparticle spin-up and spin-down energies for each band and for each wavevector in the crystal's Brillouin-zone are split by an amount

$$\Delta = 2\mu_B H_{eff} = 2(Im + \mu_B H). \tag{19}$$

The effective magnetic field includes an exchange contribution which is proportional to the magnetic moment per Bohr magneton per volume of the system,

$$m = \frac{M}{V\mu_B} = n_\uparrow - n_\downarrow. \tag{20}$$

where $M = \mu_B(N_\uparrow - N_\downarrow)$ is the total magnetic moment. The exchange integral I is a phenomenological material property which is characteristic of a given system. In a given field, m and the chemical potential μ are determined by self-consistently occupying spin-split bands. At $T = 0$ the self-consistent mean-field equations are

$$m = \int_{\mu-\Delta/2}^{\mu+\Delta/2} d\epsilon\, D(\epsilon)$$
$$n = 2\int_{-\infty}^{\mu-\Delta/2} d\epsilon D(\epsilon) + \int_{\mu-\Delta/2}^{\mu+\Delta/2} d\epsilon D(\epsilon) \tag{21}$$

where $D(\epsilon)$ is the density-of-states per volume per spin at $\Delta = 0$. Let Δ_0 and I denote the self-consistent exchange splitting of the band and the self-consistent value of m in the absence of an external field respectively. Expanding μ and Δ to first order in H, I find that

$$\frac{d\mu}{d(g^* \mu_B H)} = \frac{D_\downarrow^{-1} - D_\uparrow^{-1}}{2[D_\downarrow^{-1} + D_\uparrow^{-1} - 4I]}. \tag{22}$$

Here $D_\uparrow = D(\mu_0 + \Delta_0/2)$ and $D_\downarrow = D(\mu_0 - \Delta_0/2)$ are the majority and minority spin densities of states in the absence of a field.

This expression can be recognized as a special case of the general result derived in Section 2 which is obtained by holding the bands rigid and taking $F_{\uparrow,\downarrow} = -F_{\uparrow,\uparrow} = -F_{\downarrow,\downarrow} = I$. In the formally exact theory, the three local-field-factors are independent. Density-functional theory [15] is a practical approach to many-particle physics which has been applied successfully to evaluate the thermodynamic properties of metals and underpins the modern theory [16] of ferromagnetic transition metals. In density-functional theory, the local field factors arise from an interplay between band-structure details and exchange-correlation single-particle potentials. Expectations based on the electron gas case, discussed for two-dimensions below, suggest that the diagonal local-field-factors should be negative and larger in magnitude than the positive off-diagonal local-field-factor. It is not at all obvious that these expectations apply to transition metals, especially because the majority spins at the Fermi energy tend to have predominantly itinerant s-electron character while the minority spins tend to have predominantly localized d-electron character. Comparison of theoretical and experimental values for both the differential magnetic susceptibility, and the magnetic compressibility, presents a serious and interesting challenge to the density-functional theory of metallic magnetism.

4.2 Zero Field Two-Dimensional Electron Gas

The ground state of the two-dimensional electron gas is not ferromagnetic except at extremely low densities [18, 19]. Spin-polarization can, however, be induced by application of an external magnetic field [20]. I now discuss a simple theory for the value of $d\mu/d(g^*\mu_B H)$ in a spin-polarized two-dimensional electron gas which is based on the Hartree-Fock approximation [21] for its energy: two-dimensional electron system:

$$\frac{E_{n_\uparrow,n_\downarrow}}{A} = \frac{n_\uparrow^2 + n_\downarrow^2}{2D_{2D}} - \frac{8e^2[n_\uparrow^{3/2} + n_\downarrow^{3/2}]}{3\pi^{1/2}} \tag{23}$$

Here $D_0 = m^*/(2\pi\hbar^2)$ is the density-of-states per area A in a 2D system. The first term on the right hand side of Eq.(23) is the band energy and the second term is the exchange energy. Differentiating twice with respect to density we obtain for the diagonal local-field-correction

$$F_{\sigma,\sigma} = -\frac{2e^2 n_\sigma^{-1/2}}{\pi^{1/2}}. \tag{24}$$

The off diagonal local-field-corrections vanishes in this approximation because of the neglect, in the Hartree-Fock approximation, of correlations between electrons of opposite spin. In the presence of an external magnetic field the partitioning of density between spin-subsystems is determined by setting $\mu_\uparrow - \mu_\downarrow = g^*\mu_B H$ where $\mu_\sigma = n_\sigma/D_{2D} - 4e^2 n_\sigma^{1/2}/\pi^{1/2}$. For electron-gas density[17] parameter $r_s = 2$, the Hartree-Fock ground state is paramagnetic and the spin-polarization is initially linear in field. At stronger fields, the exchange term causes the minority-spin chemical potential to increase with decreasing density. The polarization must then increases more rapidly with field in order to establish the required chemical potential difference. The polarization eventually jumps from a partial value to full polarization when $g^*\mu_B H \sim 0.034$Ry. (The atomic energy unit $1\text{Ry} = e^4m^*/2\epsilon^2\hbar^2$ is ≈ 5.5meV in GaAs.)

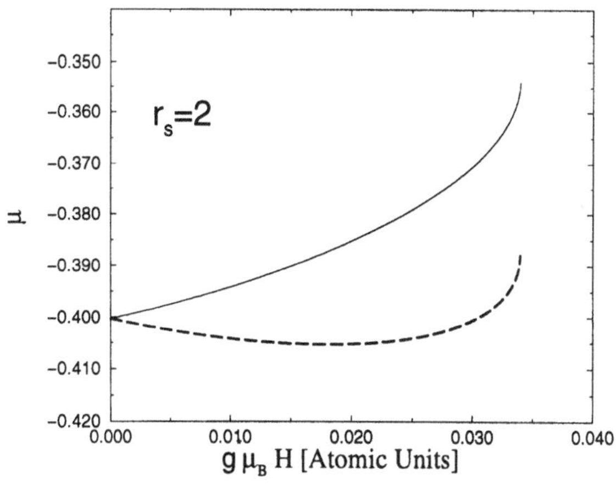

Fig. 1. Hartree-Fock approximation majority-spin (solid line) and minority-spin (dashed line) chemical potentials *vs.* Zeeman coupling strength for a two-dimensional electron gas with density parameter $r_s = 2$.

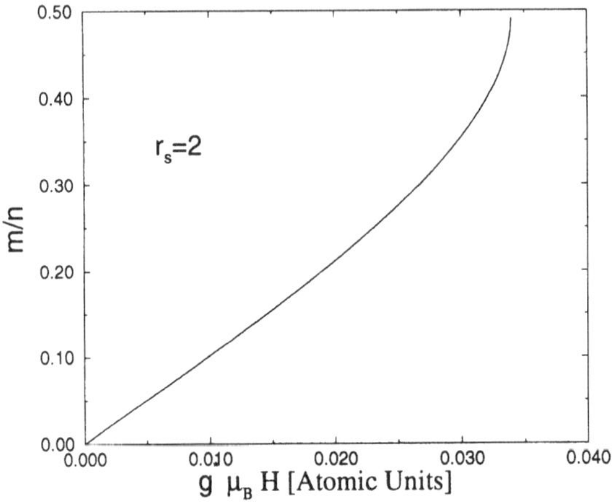

Fig. 2. Hartree-Fock approximation $m/n \equiv (n_\uparrow - n_\downarrow)/(n_\uparrow + n_\downarrow)$ for a two-dimensional electron gas with density parameter $r_s = 2$ *vs.* Zeeman coupling strength $g^* \mu_B H$ in atomic units $(e^4 m^*/2\hbar^2)$. For GaAs the Zeeman coupling strength in atomic (Ry) units is $0.0052H$[Tesla]. The system jumps to full polarization at $g^* \mu_B H \sim 0.034$Ry.

The results for the Zeeman coupling denpendence of subsystem chemical potentials and spin-polarization, shown in Fig.(1) and Fig.(2) respectively, can lead to substantial values of $d\mu/d(g^* \mu_B H)$ which can even exceeds one over a wide range of Zeeman coupling strength. This is despite the fact that the minority and majority spin density-of-states are identical in two-dimensions so that the inverse magnetic compressibility would be identically zero for a non-interacting electron system. Numerical results for the inverse magnetic compressibility for this illustrative example are shown in Fig.(3). Of course, we do not expect the Hartree-Fock approximation theory to be reliable in an electron gas for $r_s > 1$. The large spin-polarizations at relatively weak Zeeman couplings in Fig. (2), occur because the Hartree-Fock approximation ground state is ferromagnetic for $r_s > \pi/2^{1/2}$. We know from reliable quantum Monte Carlo calculations[18, 19] that, in reality, this instability does not occur until much lower densities are reached. Nevertheless illustrative calculation does suggest what should be expected in low-density two-dimensional electron gas layers. For modern samples, which have high mobility at densities a few $\times 10^{10}\text{cm}^{-2}$, it is clear that a substantial degree of spin-polarization (and possibly complete spin polarization) can be achieved in laboratory superconducting magnets. Measurement of the inverse magnetic compressibility, coupled with simultaneous measurements of the spin polarization, will provide a strict test of our understanding of magnetism in strongly correlated electron gas systems.

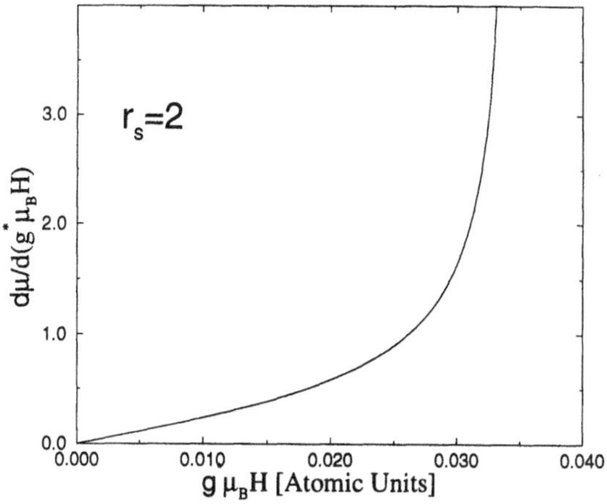

Fig. 3. Hartree-Fock approximation $d\mu/d(g^*\mu_B H)$ *vs.* Zeeman coupling strength, for a two-dimensional electron gas with density parameter $r_s = 2$. For $g^*\mu_B H > 0.034$, only the majority spins are occupied and $d\mu/d(g^*\mu_B H) \equiv -0.5$.

4.3 Quantum Hall Ferromagnets

In the quantum Hall regime, the 2D electron density is measured in units of the density which can be accommodated by a full Landau level; the Landau level filling factor $\nu \equiv (2\pi\ell^2)n$ where $\ell \equiv (\hbar c/eB)^{1/2}$ is the magnetic length. At Landau level filling factor $\nu = 1$, the 2D electron system has a strong ferromagnet ground state, *i.e. all* spins can be aligned with the field by arbitrarily weak Zeeman coupling. Quantum Hall ferromagnets have unusual charged excitations known as Skyrmions [22, 23]. Skrymions are topologically non-trivial configurations of the magnetization orientation distribution which occur in any 2D magnet, but carry an electrical charge e only in the case of quantum Hall ferromagnets. Skyrmions have an internal [24] integer quantum number K which specifies the number reversed spins in their interior. When the Landau level filling factor is close to $\nu = 1$, the low energy states of the system can be described in terms of Skyrmion degrees of freedom.

A simple model of a 2DES, valid at low temperature in this regime because Skyrmions are dilute, is obtained by ignoring Skyrmion-Skyrmion interactions. The following grand-canonical ensemble expressions specify the occupation probabilities of the $N_\phi = A/(2\pi\ell^2)$ Skyrmion quasielectron and quasihole states with K excess reversed spins:

$$
\begin{aligned}
n_{Ke} &= f(\epsilon_K + K\mu_\uparrow - (K+1)\mu_\downarrow) \\
n_{Kh} &= f(\epsilon_K + (K+1)\mu_\uparrow - K\mu_\downarrow).
\end{aligned} \tag{25}
$$

Here $f(\epsilon) = (\exp(\epsilon/k_B T) + 1)^{-1}$ is a Fermi factor [25], ϵ_K is the energy of a Skyrmion quasiparticle, $(2\pi\ell^2)^{-1}$ is the density of a full Landau level, and we have chosen the zero of energy so that quasielectron and quasihole skyrmion states have the same energy[26]. When the spin subsystems are in equilibrium, we can use Eqs. (25) to calculate the chemical potential, given the Landau level filling factor. The Landau level filling factor is increased by quasielectron excitations and decreased by quasihole excitations:

$$\nu = 1 + \sum_K (n_{Ke} - n_{Kh}). \tag{26}$$

Eq.(25) follows from the property that formation of the $K-th$ quasielectron Skyrmion requires the addition of $K+1$ spin-down electrons and the removal of K spin-up electrons from the $\nu = 1$ ground state, while formation of the $K - th$ quasihole Skyrmion requires the addition of K spin-down electrons and the removal of $K+1$ spin-up electrons. For non-interacting electrons only the $K = 0$ quasiparticles occur; for typical 2DES's, on the other hand, the lowest energy quasiparticles have $K = 3$; these quasiparticles dominate the low-temperature properties of the system for ν close to 1. From Eqs.(25) I obtain the following thermodynamic density-of-states matrix:,

$$\mathcal{D}_{\uparrow,\uparrow} = (2\pi\ell^2)^{-1} \sum_K \left[(K+1)^2 \Delta(\epsilon_K + \mu) + K^2 \Delta(\epsilon_K - \mu)\right] \tag{27}$$

$$\mathcal{D}_{\uparrow,\downarrow} = -(2\pi\ell^2)^{-1} \sum_K K(K+1) \left[\Delta(\epsilon_K + \mu) + \epsilon_K - \mu)\right] \tag{28}$$

where $\Delta(x) = \mathrm{sech}^2(x/2)/4k_B T$.

These expressions have been used previously to analyze [28] spin bottlenecks which have been observed [29] in the quantum Hall regime. Here I evaluate the dependence of the equilibrium chemical potential on Zeeman coupling strength. The possibility of measuring this quantity in two-dimensional electron systems was discussed in the previous section. Zeeman coupling adds [22, 26] to the Skyrmion quasiparticle energies, $\epsilon_K \to \epsilon_K + g^* \mu_B H(K + 1/2)$. In order for the filling factor to be held fixed as the Zeeman coupling strength varies, the chemical potential must change:

$$\frac{\partial \mu}{\partial(g^* \mu_B H)} = -\frac{\partial n/\partial(g^* \mu_B H)|_\mu}{\partial n/\partial \mu|_H} = \frac{\sum_K (K+1/2)[\Delta(\epsilon_K - \mu) - \Delta(\epsilon_K + \mu)]}{\sum_K [\Delta(\epsilon_K - \mu) + \Delta(\epsilon_K + \mu)]} \tag{29}$$

We see in Fig. (4) that for an ideal disorder-free two-dimensional electron system, $d\mu/d(g^* \mu_B H)$ is typically larger than one and changes sign when the filling factor crosses $\nu = 1$. These results were calculated using Skrymion state energies evaluated by Palacios *et. al.* [27]. The value of this quantity gives information about the Skyrmion system complementary to that available from other experiments[23]. Its value depends both on the average K value of the Skrymions present in the system, the quantity revealed by Knight

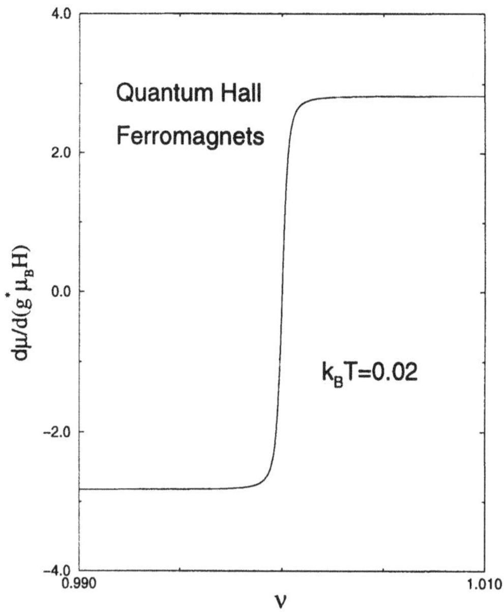

Fig. 4. $d\mu/d(g^*\mu_B H)$ *vs.* Landau level filling factor for a quantum Hall system near $\nu = 1.$ at temperature $k_B T = 0.02e^2/\ell$.

shift measurements, but also on the way in which the distribution of K values changes with field. Unlike transport experiments, this thermodynamics measurement would be sensitive to Skrymions localized by disorder in the system.

5 Summary

In this article, I have discussed the dependence of chemical potential on external magnetic field in a spin-polarized electron system. This quantity appears directly in theories of electron systems disturbed by non-equilibrium spin accumulations generated by spin and space-dependent electronic transport. I have discussed techniques which can be used to measure this fundamental thermodynamic quantity, both in two-dimensional electron systems and in ferromagnetic metals. I believe that its measurement can provide an important and interesting test of theory in several quite different spin-polarized electron systems.

References

1. T. Valet, and A. Fert, Phys. Rev. B **48**, 7099 (1993); M.A.M. Gijs and G.E.W. Bauer, Adv. Phys. **46**, 285 (1997); J-Ph Ansermet, J. Phys. C **10**, 6027 (1998).
2. J.S. Moodera, L.R. Kinder, T.M. Wong, and R. Meservey, Phys. Rev. Lett. **74**, 3273 (1995); Jagdeesh S. Moodera and Lisa R. Kinder, J. Appl. Phys. **79**, 4724 (1996); T. Miyazaki and N. Tezuka, J. Magn. Mater. **139**, L231 (1995). For a review see Jagadeesh S. Moodera, Joaquim Nassar, and George Mathon, Annu. Rev. Mater. Sci. **29**, 381 (1999).
3. See for example G. Prinz, Physics Today **48**, 58 (1995) and related articles in the same issue.
4. A. G. Aronov, Pis'ma Zh. Eksp. Teor. Fiz. **24**, 37 (1976) [JETP Lett. **24**, 32 (1976)]; R. Merservey and P.M. Tedrow, Phys. Rev. Lett. **70**, 2142 (1978); P.M. Tedrow and R. Merservey, Phys. Rep. **238**, 174 (1994).
5. M. Johnson and R.H. Silsbee, Phys. Rev. Lett. **55**, 1790 (1985); M. Johnson, Phys. Rev. Lett. **70**, 2142 (1993).
6. I will assume here that we are interested only in states with a collinear spin-polarization configuration.
7. In these notes I will always neglect the coupling of an external field to orbital degrees of freedom of the electrons. Although this coupling does not play an essential role in the physics discussed here, the role it plays in particular experiments should always be carefully assessed.
8. K. Ono, H. Shimada, S. Kobayashi, and Y. Ootuka, J. Phys. Soc. Jpn. **65**, 3449 (1996); K. Ono, H. Shimada, and Y. Ootuka, J. Phys. Soc. Jpn. **66**, 1261 (1997).
9. J.P. Eisenstein, L.N. Pfeiffer, and K.W. West, Phys. Rev. B50, 1760 (1994).
10. S.J. Papadakis, J.P. Lu, M. Shayegan, S. Parihar, and S. Lyon, Phys. Rev. B55, 9294 (1996).
11. N.K. Patel, I.S. Millard, E.H. Linfield, P.D. Rose, M.P. Grimshaw, D.A. Ritchie, G.A.C. Jones, and M. Pepper, Phys. Rev. B 53, 15443 (1996).
12. X. Ying, S.R. Parihar, H.C. Manoharan, and M. Shayegan, Phys. Rev. B 52, R11611 (1995).
13. J.P. Eisenstein, L.N. Pfeiffer, and K.W. West, Appl. Phys. Lett. **57**, 2324 (1991).
14. E.P. Wohlfarth, Rev. Mod. Phys. **25**, 211 (1953).
15. See for example, R.M. Dreizler and E.K.U. Gross, *Density Functional Theory: An Approach to the Quantum Many-Body Problem* (Springer, Berlin, 1990).
16. See for example, D.A. Papaconstantopoulos, *Handbook of the Band Structure of Elemental Solids* (Plenum Press, New York, 1986).
17. In 2D dimensions the electron gas density is related to the dimensionless parameter r_s by $n = 1/(\pi r_s^2 a_0^2)$ where $a_0 \equiv \hbar^2\epsilon/(m^* e^2)$ is the host system dielectric constant. In GaAs $a_0 \approx 10$nm.
18. B. Tanatar and D.M. Ceperley, Phys. Rev. B **39**, 5005 (1989).
19. Sergio Conti and Gaetano Senatore, Europhys. Lett. **36**, 695 (1996).
20. In two-dimensions it is possible to largely the coupling to orbital degrees-of-freedom of an external magnetic field oriented in the plane of the 2D system can usually be neglected.

21. This result is a simple generalization of literature results for unpolarized two-dimensional electron systems: F. Stern, Phys. Rev. Lett. **30**, 278 (1973); A. Isihara, Solid State Physics **42**, 271 (1989).
22. S.L. Sondhi, A. Karlhede, S.A. Kivelson, and E.H. Rezayi, Phys. Rev. B **47**, 16419 (1993); H.A. Fertig, L. Brey, R. Côté, and A.H. MacDonald, Phys. Rev. B **50**, 11018 (1994); H.A. Fertig, L. Brey, R. Côté, and A.H. MacDonald, Phys. Rev. Lett. **77**, 1572 (1996).
23. R. Tycko, S.E. Barrett, G. Dabbagh, L.N. Pfeiffer, and K.W. West Science **268**, 1460 (1995); N.N. Kuzma, P. Khandelwal, S.E. Barrett, L.N. Pfeiffer, and K.W. West, Science **281**, 686 (1998); E.H. Aifer, B.B. Goldberg, and D.A. Broido, Phys. Rev. Lett. **76**, 680 (1996); A. Schmeller, J.P. Eisenstein, L.N. Pfeiffer, and K.W. West, Phys. Rev. Lett. **75**, 4290 (1996).
24. A.H. MacDonald, H.A. Fertig, and Luis Brey, Phys. Rev. Lett. **76**, 2153 (1996).
25. Our primary interest is in the dilute high-temperature regime where the use of Fermi quantum statistics for skyrmions does not play any essential role.
26. Kun Yang and A.H. MacDonald, Phys. Rev. B **51**, 17247 (1995).
27. J.J. Palacios, D. Yoshioka, and A.H. MacDonald, Phys. Rev. B **54**, R2296 (1996).
28. A.H. MacDonald, Phys. Rev. Lett. **83**, 3262 (1999).
29. Ho-Bun Chan and Ray Ashoori, Phys. Rev. Lett. **83**, 3258 (1999).

Mesoscopic Spin Quantum Coherence

J.M. Hernandez[1], J. Tejada[1], E. del Barco[1], N. Vernier[2], G. Bellessa[2], and E. Chudnovsky[3]

[1] Dept. de Física Fonamental, Universitat de Barcelona, Diagonal 647, 08028 Barcelona, Spain.
[2] Laboratoire de Physique des Solides, Bâtiment 510, Université Paris-Sud, 91405 Orsay, France.
[3] Physics Department, CUNY Lehman College, Bronx, NY 10468-1589, USA.

Abstract. In this paper we present resonance experimental results performed on spin-10 Fe8 molecular clusters to elucidate the so called Schrödinger's cat paradox. We explain theoretically that the resonances detected in the millikelvin regime correspond to transitions between the symmetric and antisymmetric combinations of the degenerate ground state with $|S_Z| = 10$. The back and forth quantum tunnelling oscillations of the spin are superimposed during a time lower than 10^{-7} sec.

1 Introduction

Magnetic molecules with high magnetic anisotropy represent a very interesting family of materials which can be used to explore the border between quantum and classical mechanics (Chudnovsky 1996, Stamp 1996, Schwarzchild 1997). Molecules like Mn12 and Fe8 have spin 10 and at low temperature are equivalent to a single domain particle with constant modulus of its magnetic moment, $20\mu_B$, the orientation of which depends on the ratio between the temperature and the barrier height existing between the up and down orientations. A crystal of such molecules consist of molecules arranged in a tetragonal (Mn12, Lis 1980) or triclinic (Fe8, Wiedghardt *et al.* 1984) lattice, existing, in both cases, a very weak interaction between the molecules. The occurrence of thermally assisted resonant spin quantum transitions between the degenerate levels in the two wells of the magnetic anisotropy in such molecules was first clearly elucidated using Mn12-acetate clusters by performing dynamical hysteresis and magnetic relaxation experiments (Friedman *et al.* 1996, Hernandez *et al.* 1996, Thomas *et al.* 1996, Luis *et al.* 1997, Chudnovsky *et al.* 1998).

The spin Hamiltonian of the Fe8 molecular clusters may be written in a first approximation as

$$\mathcal{H} = -DS_z^2 + ES_x^2 - g\mu_B \mathbf{H} \cdot \mathbf{S} \,, \qquad (1)$$

where z and x denote the easy and hard axes respectively. The values of $D = 0.31$K and $E = 0.092$K are well known from EPR experiments (Sangregorio *et al.* 1997). The longitudinal component, H_Z, of the external magnetic field, H, modifies the barrier height between the two classical spin orientations while the transversal components, H_X and H_Y, affect the overlapping of the wave functions, the so called quantum splitting, of the spin levels existing in the two potential wells of the magnetic anisotropy barrier. That is, the quantum splitting, Δ, and consequently the rate of resonant tunnelling between the spin levels depend on both the magnitude of the applied, H, and its angle ϕ with the hard axis. The dependence of Δ on H_Z, for a fixed value of $H = 3$T, obtained by numerical diagonalization of the Hamiltonian (1) using the values of D and E above mentioned, is plotted in Fig. 1. The complex structure of the dependence of Δ on H is due to the non-Kramers topological quenching of tunneling first suggested by Garg (Garg 1993).

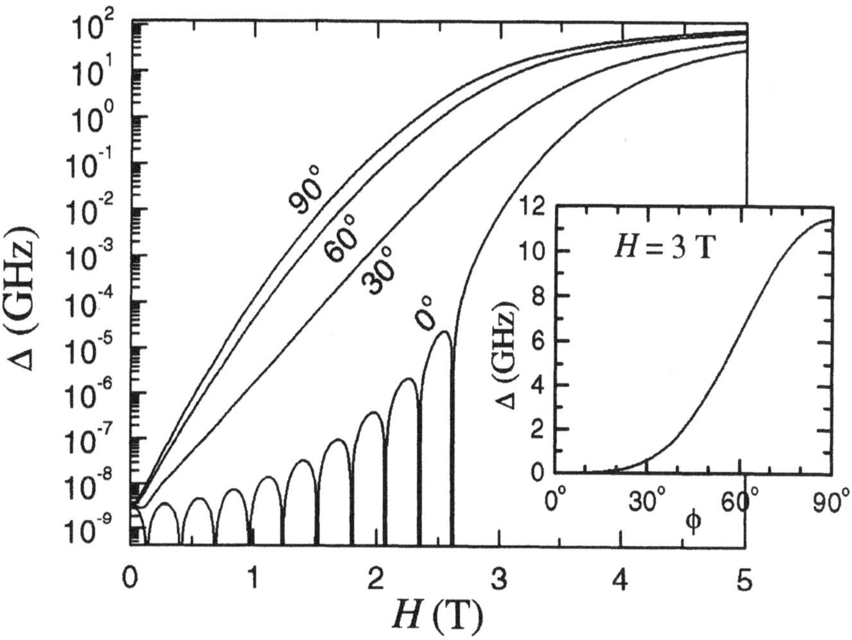

Fig. 1. Dependence of the ground-state splitting on the transverse field in Fe_8 for different orientations of the field in the plane perpendicular to the easy axis. The insert shows the angular dependence of the splitting at a fixed field.

The physics of the thermally assisted resonant tunnelling in Fe8 clusters is illustrated in Fig. 2. This system can be most easily interpreted in terms of the eigenstates of S_Z, ($|m\rangle$, $-10 < m < 10$), and in terms of the energy barrier caused by the magnetic anisotropy. The classical energy barrier between the states with $m = 10$ and $m = -10$, due to magnetic anisotropy, is about 35K.

If the external magnetic field is applied along the positive direction of the easy axis of the cluster, z-axis, that is the so called longitudinal component of magnetic field, the states with positive m values are more favourable than those for which m is negative. At field values $H_Z = nH_r$ with $H_r \sim 0.23$ T ($n = -10, -9, ..., -1, 0, 1, ..., 9, 10$) the levels m and $n-m$ match and resonant tunnelling occurs between these levels. The symmetry-violating terms in the Hamiltonian inducing tunnelling are those associated to the transverse component of both the magnetic field and magnetic anisotropy. At the resonance fields, the magnetisation relaxation can occur on measurement time scales and give rise to the step-like changes in the magnetisation. Otherwise the relaxation rate is lower, leading to plateaus in the magnetic hysteresis loops. The relaxation rate from any level is proportional to the product of the tunnelling rate, $\Delta/\hbar$, and the thermal activation rate, $\Gamma_m = 1/\tau_0 \exp[E_m - E_{10})/K_B T]$. The first level from the top of the barrier for which the tunnelling rate become small compared with the thermal activation rate, is called blocking level, m_B, and determines the effective height of the barrier at resonance. Since the tunnelling rate is intrinsically small for low-lying levels, levels for which $\Delta m = m - n$ is large, large transversal fields are necessary at low temperature to produce observable resonant tunnelling relaxation (Garanin 1991, Garanin *et al.* 1997, Fernandez *et al.* 1998).

In the experiments discussed in this paper, which have been performed at mK, the only populated spin levels are those at the bottom of the barrier. In zero longitudinal field these levels are the degenerate $|S_Z| = 10$ levels. The application of a large transversal field induces rapid back and forth tunnelling transitions between these states. Our aim was, therefore, the detection of the symmetric and antisymmetric combination of the $|S_Z| = 10$ states as a consequence of the coherent quantum oscillations of the spin similar to the textbook example of ammonia molecule.

2 Experimental Results

The magnetic characterisation of the Fe8 crystallites was performed by using a dc-SQUID inside a top loading dilution refrigerator. In Fig. 3 we show the low field characterisation corresponding to the zero field, ZFC, and field cooled, FC, magnetisation in the presence of a field $H = 70$Oe. Above $T = 0.8$K the susceptibility of the sample is described well by the Curie-Weiss law while below that temperature occurs the irreversibility between the ZFC and FC curves which may be indicative of the blocking of the spin 10 of the molecule. The existence of jumps, as a consequence of the occurrence of thermal resonant tunnelling is also clear from dc and ac measurements, see, for example, the inset of Fig. 3 in which we show the dependence of the real part of the ac-susceptibility at 1KHz and 2K as a function on the dc longitudinal field. The peaks appear at approximately equal spaced intervals

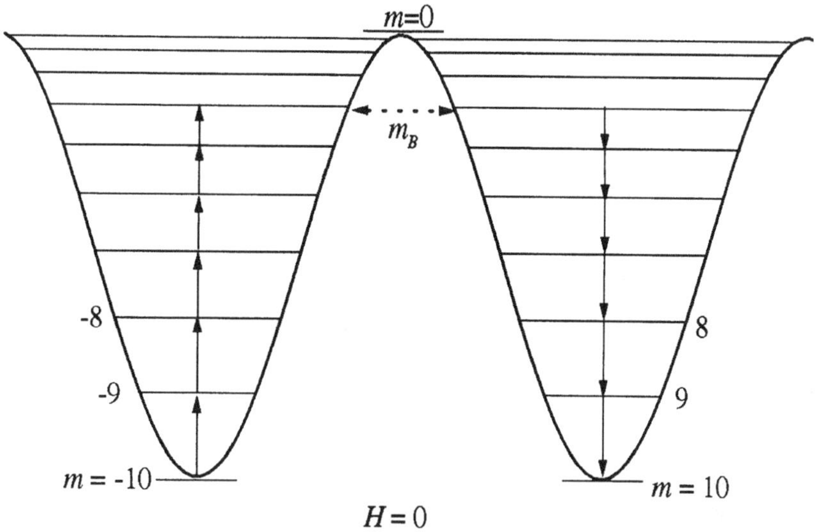

Fig. 2. Sketch showing the physical interpretation of the thermally asisted resonant tuneling effect.

of 0.23T, in agreement with the theoretical siggestion for the resonant field values $H_Z = nH_r$.

The resonance experiments were performed on an oriented powder sample having the easy axes of a large number of the crystallites predominantly in one direction. The dc magnetic field was applyed perpendicular to the easy of the crystallites while the resonant ac field of 680MHz was parallel to such direction. In Fig. 4 we show the absorption energy as a function of the values of the transversal field. The most remarkable experimental fact is the existence of two peaks in the absorption energy at $H = 2.25 \pm 0.05$T and $H = 3.60 \pm 0.05$T which position and intensity do not depend on temperature. The peaks, however, disappear at temperature higher than 200 mK.

3 Physical Interpretation

The intensity of the resonance in the absence of dissipation and for a given value of the applied magnetic field is proportional to

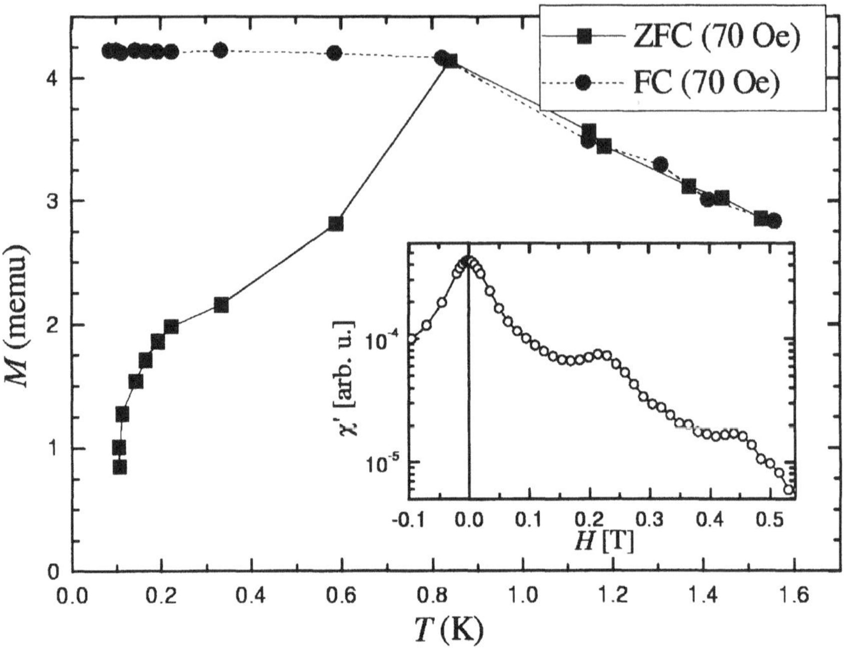

Fig. 3. Zero field cooled (ZFC) and field cooled (FC) magnetization data at low field. The inset shows the dependence of the real part of the ac-susceptibility on the dc-field.

$$\int_0^\pi g(\phi)L(\omega - \Delta[\phi, H])d\phi \; , \tag{2}$$

where $L(x)$ is a lorentzian, $g(\phi)$ is the distribution of Fe8 crystallites over the angle, ϕ, and $\omega = 2\pi f$, f being the frequency of the ac field which in our experiment is 680MHz. As we are mostly interested in the computation of the field position of the maximum of the resonance as a function of w and no preferred orientation on the angle ϕ is expected we may rewrite (2) as

$$\int_0^\pi \delta(\omega - \Delta[\phi, H])d\phi = \left| \frac{\partial \Delta[\phi, H]}{\partial \phi} \right|^{-1}_{\Delta[\phi,H]=\omega} \; . \tag{3}$$

It is clear, therefore, from both (3) and the dependence of the quantum splitting on the angle ϕ showed in the inset of Fig. 1 that two maxima in the resonance at the angles $\phi = 0$ and $\phi = \pi/2$ are expected. These two maxima of resonance absorption correspond, consequently, to two different field values which are solutions of the equations

$$\begin{array}{l} \Delta(\frac{\pi}{2}, H_1) = \omega \\ \Delta(0, H_2) = \omega \end{array} \; . \tag{4}$$

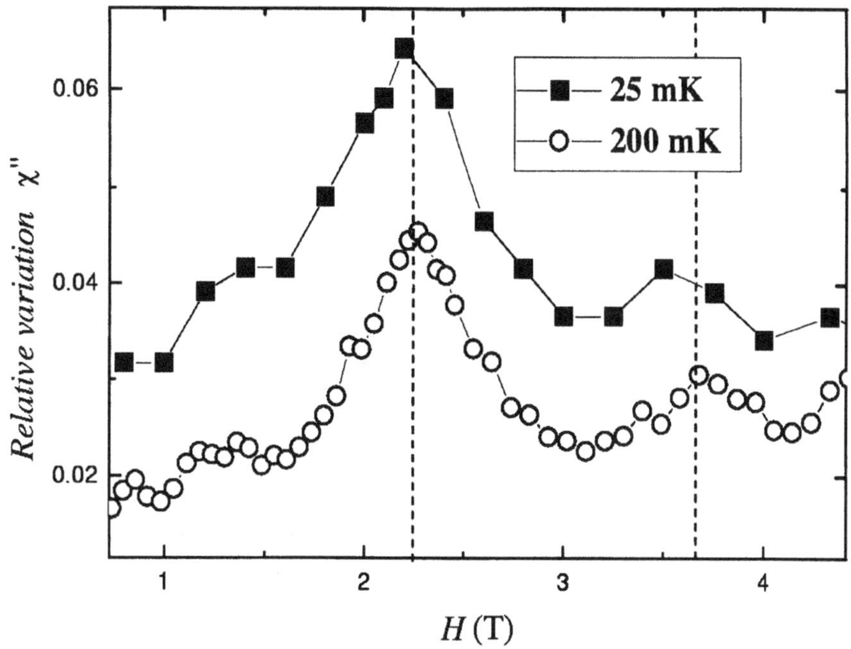

Fig. 4. The imaginary part of the high-frequency susceptibility of Fe8 as a function of transverse magnetic field. The dashed lines indicate the peak locations H_1 and H_2 obtained from the Hamiltonian using the parameters $D = 0.31$ K and $E = 0.092$ K. Since the absolute value of the power absorption is not known, the curves, for clarity, are arbitrary shifted in the vertical direction.

That is, the two resonance peaks we have experimentally observed correspond to the quantum splittings of the ground state for the cases when the magnetic field is applied perpendicular and parallel to the hard axis of the crystallites. From the position of these two peaks we have got the values $D = 0.275 \pm 0.005$K and $E = 0.092 \pm 0.005$K of (1). In Fig. 5 we show the results of our numerical computation of the wave function of the two resonant levels, $|g\rangle$, ground state, and $|e\rangle$, excited state, separated in energy by the quantum splitting, Δ, between which we have observed the two maxima for the resonance absorption.

In conclusion, we have presented experimental data for which the most plausible interpretation is the occurrence of the coherent quantum oscillations in spin-10 Fe8 molecules. Our explanation is granted by the remarkable agreement between the parameters D and E of the magnetic Hamiltonian deduced from our experimental results and those previously reported. In other words, the resonances experimentally detected may suppose a clear experimental elucidation of the Schrödinger's cat paradox for spin systems.

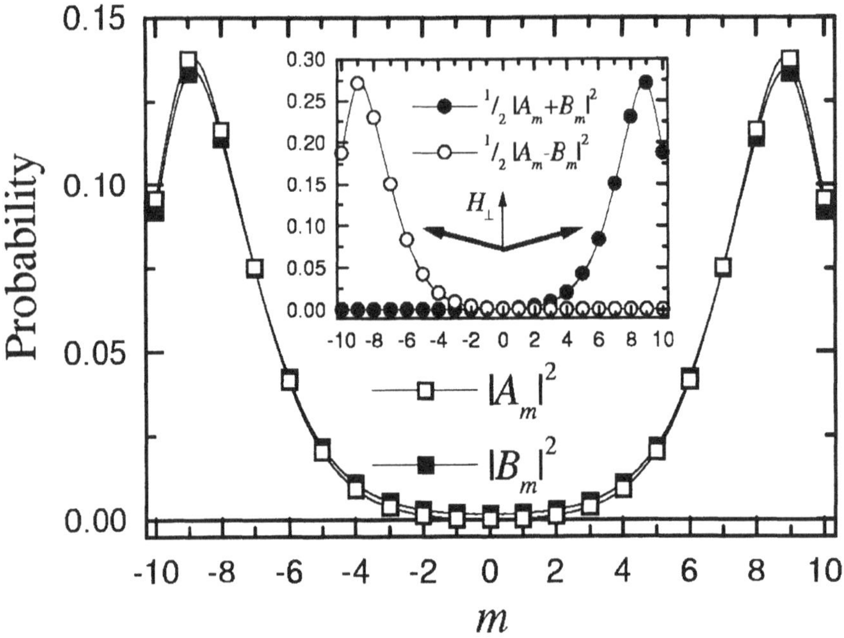

Fig. 5. Probability distribution over m in the ground state ($|g\rangle$) and the first excited state ($|e\rangle$), $|A_m|^2$ and $|B_m|^2$, respectively. The inset shows a double-degenerate classical ground state.

References

Chudnovsky E. M. *Science*, **274** (1996) 938.

Chudnovsky E. M. and Tejada J. *Macroscopic Quantum Tunneling of the Magnetic Moment* Cambridge University Press (1998).

del Barco E., Vernier N., Hernandez J.M., Tejada J., Chudnodvsky E.M., Molins E., and Bellessa G.*Europhys. Lett.*, **47** (1999) 722.

Fernandez J. F., Luis F., Bartolomé J. *Phys. Rev. Lett.* **80** (1998) 5659.

Friedman J. R., Sarachik M. P., Tejada J. and Ziolo R. *Phys. Rev. Lett.*, **76** (1996) 3830.

Garanin D.A. *J. Phys.*, **A24** (1991) L61.

Garanin D.A and Chudnovsky E.M. *Phys. Rev.*, **B56** (1997) 11102.

Garg A. *Europhys. Lett.*, **22** (1993) 205.

Hernandez J. M., Zhang X.X., Luis F., Bartolomé J., Tejada J. and Ziolo R. *Europhys. Lett.*, **35** , (1996) 301.

Lis T. *Acta Cryst.* **B36** (1980) 2042.

Luis F., Bartolomé J., Fernandez J. F., Tejada J., Hernandez J.M., Zhang X.X. and Ziolo R. *Phys. Rev. B* **55** (1997) 11448.

Sangregorio S., Ohm T., Paulsen C., Sessoli R. and Gatteschi D. *Phys. Rev. Lett.*, **78** (1997) 4645.

Schwarzchild B. *Phys. Today*, **50** (1997) 17.

Stamp P. C. E. *Nature*, **383** , (1996) 125.

Thomas L., Lionti F., Ballou R., Gatteshi D., Sessoli R. and Barbara B. *Nature*, **383** (1996) 145.

K. Wiedghardt, K. Pohl, I. Jibril and G. Huttner. *Angew. Chem. Int.Ed. Engl.* **23** (1984) 77.

Part VII

Random Systems and Localization

Numerical-Scaling Study of the Statistics of Energy Levels at the Anderson Transition

I.Kh. Zharekeshev and B. Kramer

Institut für Theoretische Physik, Universität Hamburg, Jungiusstraße 9, D-20355 Hamburg

Abstract. The statistical properties of the energy spectra of the Anderson Hamiltonian for random systems are numerically investigated near the disorder-induced metal-insulator transition. The level spacing distribution, the level density correlation function and the spectral form-factor are shown to be size-invariant at the critical point. They exhibit a crossover from the *critical orthogonal* to the *critical unitary* ensembles, which is controlled by the magnetic flux. One-parameter finite size scaling of the level statistics is used to detect with high precision the critical parameters: the critical exponent and the disorder dependence of the correlation length.

1 Introduction

The problem of localization of quantum states in disordered systems has been formulated more than three decades ago (Anderson 1958). The key result has been that depending on the randomness of the potential energy quantum objects can show a characteristic transition from extended to localized behavior corresponding to metallic and insulating behavior at zero temperature. This disorder-induced metal-insulator transition (MIT), conventionally called the Anderson transition, has been the subject of considerable experimental and theoretical work (Lee and Ramakrishnan 1985). During the past years, efforts have been concentrating on statistical and scaling properties in disordered mesoscopic electron systems (Efetov 1997), (Janssen 1998).

It is now commonly believed that the critical behavior at the MIT, described by a critical exponent ν of the localization length, is entirely determined by the spatial dimensionality d and by the fundamental symmetry. According to the single-parameter scaling theory developed by (Abrahams et al. 1979) an Anderson transition can occur only for $d > 2$. For $d \leq 2$, all quantum states are localized independent of energy and disorder. As for the symmetry, all systems are classified into three classes

1. the orthogonal class (time reversal and spin rotation symmetry),
2. the unitary class (broken time-reversal symmetry), and
3. the symplectic class (broken spin-rotational symmetry).

In random matrix theory (RMT) they correspond to the Gaussian orthogonal (GOE), unitary (GUE) and symplectic (GSE) ensembles, respectively

(Mehta 1991). Using the finite-size scaling hypothesis, considerable computational efforts have been performed in order to find numerical values for the critical exponents. The most recent estimates are known in three dimensions (3D) with and without magnetic field (Slevin and Ohtsuki 1997) and in two dimensions (2D) with a magnetic field (Weymer and Janssen 1998) and including spin-orbit coupling (Schweitzer and Zharekeshev 1997). Their results are consistent with the one-parameter scaling hypothesis.

Recently, it has been established that the statistical properties of energy levels in single-electron disordered systems are closely related to the localization properties (Altshuler and Shklovskii 1986), (Shklovskii et al. 1993). In the metallic region, they are described by the random-matrix theory (RMT) developed by (Wigner 1951) and (Dyson 1962). In the insulating regime they obey the Poissonian statistics of uncorrelated random variables. Close to the transition, the level statistics exhibit behaviors very similar to that of transport quantities, as the conductance obtained by the transfer-matrix method (Kramer and MacKinnon 1993). This implies that the knowledge of the spectral fluctuations is sufficient to extract the required critical parameters. Precisely at the transition, a *novel* universal statistics in 3D have been predicted (Shklovskii et al. 1993) and extensively studied (Kravtsov and Lerner 1995), (Hofstetter and Schreiber 1994), (Aronov et al. 1994), (Evangelou 1994), (Zharekeshev and Kramer 1995). The key feature of the statistics is size invariance at the transition. However their forms depend crucially on the presence or absence of time-reversal (Batsch et al. 1996) and spin-rotational symmetry (Kawarabayashi et al. 1996).

In 2D orthogonal case the level statistics always scale towards the Poisson distribution exhibiting no critical behavior (Zharekeshev et al. 1996). When applying the magnetic field or strong spin-orbit interactions, the corresponding *critical* statistics appear at non-vanishing disorder (Feingold et al. 1995), (Schweitzer and Zharekeshev 1995), in full agreement with the scaling theory. By using the finite–size scaling argument for the spacing distribution a new criterion for locating the metal–insulator transition has been established. This argument was cross–checked by numerically determining level statistics, and investigating the scaling properties of spectral correlations not only in 2D and 3D, but also in higher dimensions (Zharekeshev and Kramer 1998). One advantage of this method is that one needs only the energy spectrum without direct calculating eigenfunctions or the conductivity.

In this paper, we provide an overview of the numerical work during the past years. The spectral fluctuations in 3D one-electron disordered systems with and without time-reversal symmetry are examined. Several statistical quantities, as the level spacing distribution $P(s)$, the two-level correlation function $R(s)$ and the spectral structure factor $S(t)$ are considered in the vicinity of the Anderson transition. Using the finite-size scaling method we determine numerically the shape of *critical* statistics. We focus also on the critical behavior of the level statistics near the disorder–induced MIT and determine precisely the critical value of the disorder W_c and the correlation length exponent ν. We provide an independent determination of the correlation length as a function of the disorder, $\xi(W)$. Our results demonstrate unambiguously and strikingly, that the level spacing distribution plays not only the role of a scaling variable at the Anderson transition, but can also be used to define qualitatively and accurately the critical properties. The influence of the boundary conditions (BC) on the level statistics is briefly discussed.

2 The Anderson model

We consider the spectrum of disordered electrons using the Anderson model on a simple cubic lattice. The Hamiltonian is given by

$$\mathcal{H} = \sum_n \epsilon_n c_n^\dagger c_n + \sum_{\langle nm \rangle} t_\phi (c_n^\dagger c_m + c_n c_m^\dagger) \ , \tag{1}$$

where $\langle nm \rangle$ implies that n and m are nearest neighbors and $c_n^\dagger$ creates an electron on the nth site of a lattice. The random potential is described by the random site energies ϵ_n. They are uniformly and independently distributed in the interval from $-W/2$ to $W/2$. Thus, W parameterizes the strength of disorder. An Aharonov-Bohm flux ϕ is applied in all three directions, so that the transfer integral $t_\phi = t \exp(-2\pi \mathrm{i} \phi a/L)$, contains a phase factor which is given by the flux in units of the flux quantum $\phi_0 = h/e$. We assume t and the lattice spacing a to be the units of energy and spatial distance, respectively. The critical disorder at the band center of this model corresponds to $W = 16.5$ for $\phi = 0$.

The Hamiltonian Eq. (1) is numerically diagonalized for 3D lattices of different linear sizes L. The computational procedure based on the Lanczos algorithm is especially designed for solving eigenvalue problem of very large sparse matrices (Zharekeshev and Kramer 1999). Non-zero values of ϕ require diagonalization of Hermitian matrices. Discrete eigenenergies are calculated in an energy interval near the band center $\varepsilon = 0$ which contains up to 50% of the levels in the whole spectrum. On the one hand, due to the peculiarity of the trajectory of the mobility edge for the box distribution, the energy levels covered by this interval are statistically equivalent (Zharekeshev and Kramer 1997). On the other hand, such a wide interval is necessary for computing long-range spectral correlations. In order to take the energy dependence of the density of states $\rho(\varepsilon) := \sum_n \delta(\varepsilon - \varepsilon_n)$ properly into account, the spectrum has been unfolded by using parabolic splines.

We have computed sets of energy levels for fixed pairs of parameters L and W and many realizations of the randomness. The number of realizations depended on the size of the system. The total number of eigenvalues in the ensemble has been as large as 10^7 for the smallest size $L = 5$ and 10^5 for the largest size $L = 24$.

3 The level spacing distribution

The level spacing distributions for the GOE and GUE of random matrices are well approximated by

$$P_{\mathrm{W}}^{o}(s) = \frac{\pi}{2} s \exp\left(-\frac{\pi}{4}s^2\right), \quad P_{\mathrm{W}}^{u}(s) = \frac{32}{\pi} s^2 \exp\left(-\frac{4}{\pi}s^2\right) , \tag{2}$$

respectively (Wigner 1951). The level spacings s are measured in units of the mean separation Δ of adjacent levels $s \equiv (\varepsilon_{i+1} - \varepsilon_i)/\Delta$. The Wigner surmises, Eq. (2), deviate from the exact distributions $P_{\mathrm{GOE}}(s)$ and $P_{\mathrm{GUE}}(s)$ of the RMT, by less than 5% in the interval $0 < s < 2$. For example, the small-s behavior and the asymptotic form of the exact result for GOE are as follows

$$P_{\mathrm{GOE}}(s) = \begin{cases} \frac{\pi^2}{6} s + \mathcal{O}(s^2), & s \ll 1 \\ \exp\left[-\frac{\pi^2}{16}s^2 - \frac{\pi}{4}s + \mathcal{O}(\ln s)\right], & s \gg 1 \end{cases} \tag{3}$$

Indeed, our numerical results of $P(s)$ in the metallic regime for orthogonal symmetry are markedly different from the Wigner surmise Eq. (2), particularly in the tail. For $0 \leq s \leq 2$ it gets closer to the exact RMT-result from (Mehta 1991), so that the relative deviation from $P_{\mathrm{GOE}}(s)$ is much smaller than that of the Wigner surmise $P_{\mathrm{W}}^{o}(s)$. To achieve this precision one needs so many spacings that the error bar is less than 0.3-0.5%.

We have applied two different types of boundary conditions: periodic and the Dirichlet ("hardwall") BC. Figure 1 shows good coincidence of the numerical $P(s)$ for $W = 5$ (metallic regime) with the exact GOE result, unless the spacings s exceed the Thouless parameter E_{T}/Δ. The energy $E_{\mathrm{T}} = hD/L^2$ defines the border of the diffusive regime, where D is the diffusion constant (Thouless 1974). Deviations from $P_{\mathrm{GOE}}(s)$ occur for energies $s \geq E_{\mathrm{T}}$, where in this case $E_{\mathrm{T}} \approx 2$ or 2.5 depending on the type of the BC. Similar deviations from $P_{\mathrm{W}}^{u}(s)$ are found for the unitary case. The importance of the energy scale E_{T} in level statistics has first been understood by (Altshuler and Shklovskii 1986), who have calculated perturbative contributions beyond the validity of the RMT, using impurity diagram techniques. Non-perturbative corrections to the Wigner-Dyson statistics in the metallic regime have been analytically investigated (Kravtsov and Mirlin 1994) by using the combination of the nonlinear σ-models with the renorm-group transformation. Influence of the different types of the BC on the level statistics at the mobility edge has been numerically studied (Braun et al. 1998).

While explicit analytical results are known only in the asymptotic regimes $s \to 0$ and $s \to \infty$, the numerical results for $P(s)$ that are obtained for the random tight-binding Hamiltonian of condensed matter theory can be considered as standard data, since they are more accurate than those of non-linear dynamics. Usually, $P(s)$ for the ergodic regime is calculated directly for an ensemble of finite random matrices, whose *all* entries are randomly distributed with the same variance. It turned out that $P(s)$ obtained numerically from

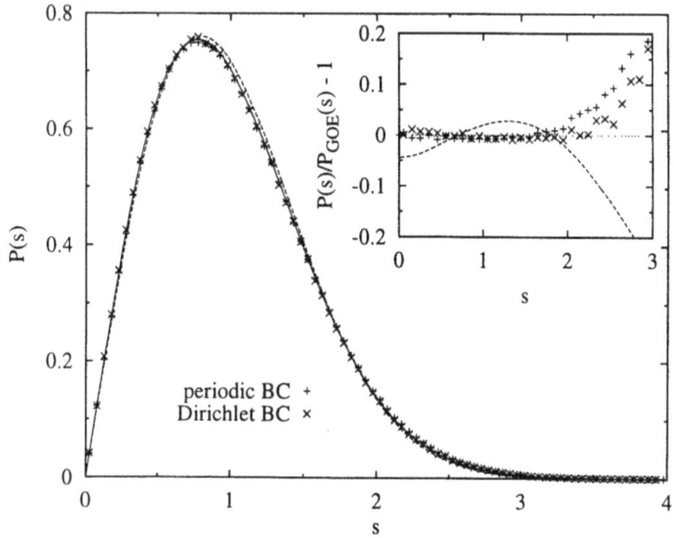

Fig. 1. Spacing distribution $P(s)$ of system size $L = 8$ and disorder $W = 5$ for periodic and Dirichlet boundary conditions (orthogonal case). Solid line: GOE data, which have been taken from (Mehta 1991). Dashed line: Wigner surmise $P(s)_W^o$ from Eq. (2). Inset: relative deviation of $P(s)$ from the GOE result. Number of spacings is about 10^7.

Eq. (1) for sufficiently large systems can be more precise than that from the table-given data in (Mehta 1991). For the GUE this can also be obtained from the zeros of the Riemann-function. In principle, one can now produce extremely precise data due to present day's immense computer power. Since the level statistics possess a generic universality, the knowledge of $P(s)$ provides a very useful and important information for purposes of metrology and standardization.

It has been shown earlier (Efetov 1983), (Altshuler et al. 1988) that the level statistics of disordered systems is governed by the Wigner-Dyson theory in the metallic regime, while it approaches the Poissonian distribution

$$P_{\mathrm{P}}(s) = e^{-s} \tag{4}$$

in the insulator. The reason is that the wavefunctions for small disorder $W \ll W_c$ overlap strongly, while for large disorder, $W \gg W_c$, they decay exponentially and their mutual overlap vanishes. Deep in the insulating region, the energy eigenvalues are completely uncorrelated variables. As an example we have calculated the level statistics of a sample with time-reversal symmetry ($\phi = 0$) for $L = 6$ for different disorders. Fig. 2 demonstrates how the distribution $P(s)$ changes from the GOE-result to the Poissonian, Eq. (4), when the disorder W increases.

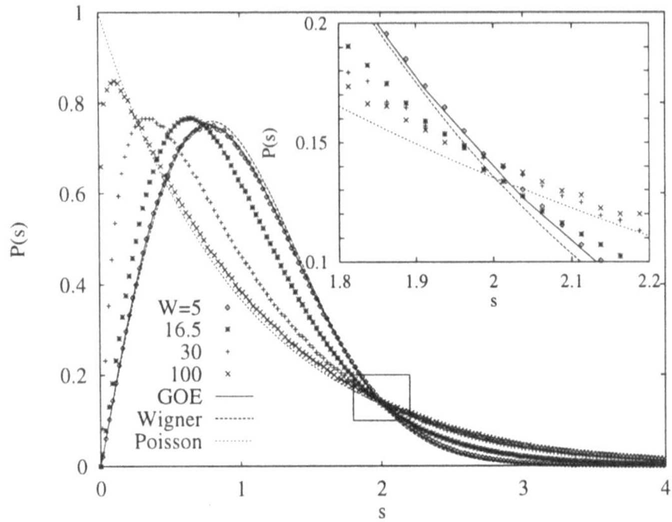

Fig. 2. Level spacing distribution $P(s)$ for an orthogonal system of linear size $L = 6$ for various disorder strengths W. Solid line: GOE data, which have been taken from (Mehta 1991). Dashed line: Wigner surmise, Eq. (2). Dotted line: Poisson distribution $P_P(s)$. Inset: $P(s)$ near $s \approx 2$, the crossing point of the Poisson and the Wigner distributions. Total number of spacings is about 10^7.

Of particular interest is the region of spacings around $s^\star = 2.002$, where the Wigner surmise $P_W^o(s)$ and the Poisson distribution $P_P(s)$ intersect. It has been suggested (Shklovskii et al. 1993) that independently of the disorder W, all $P(s)$ should also intersect at the same point $s^\star$, which would then play the role of a universal energy. We have performed detailed calculations with large number of realizations. Careful analysis of our data did not show any common crossing point (inset of Fig. 2). For instance, the data of $P(s)$ for $W = 16.5$ intersect with $P_{GOE}(s)$ at $s = 2.04 \pm 0.02$ and with $P_P(s)$ at $s = 2.00 \pm 0.02$, while $P(s)$ for $W = 100$ does with $P_{GOE}(s)$ at $s = 1.99 \pm 0.02$ and with $P_P(s)$ at $s = 2.72 \pm 0.06$, which is definitely far from $s^\star$. They cross each other at $s = 1.96 \pm 0.02$.

We consider now broken time reversal symmetry. This can be achieved by applying Aharonov-Bohm (AB) fluxes along all three perpendicular directions in a 3D lattice. Performing diagonalization for different magnitudes of the flux ranging from $\phi = 0$ to $\phi = 1/4$, we found the critical statistics to be sensitive to the flux (Batsch et al. 1996). As a function of ϕ, the distribution $P(s)$ at the transition changes smoothly from the critical orthogonal $P_c^o(s)$ to the critical unitary form $P_c^u(s)$ at $\phi = 1/4$ (see Fig. 3). This flux-controlled crossover of the critical level statistics repeats periodically, resuming the orthogonal form at $\phi = 0.5$, since a half of the flux quantum corresponds to the real Hamiltonian defined in Eq. (1) with the antiperiodic BC. For fixed flux,

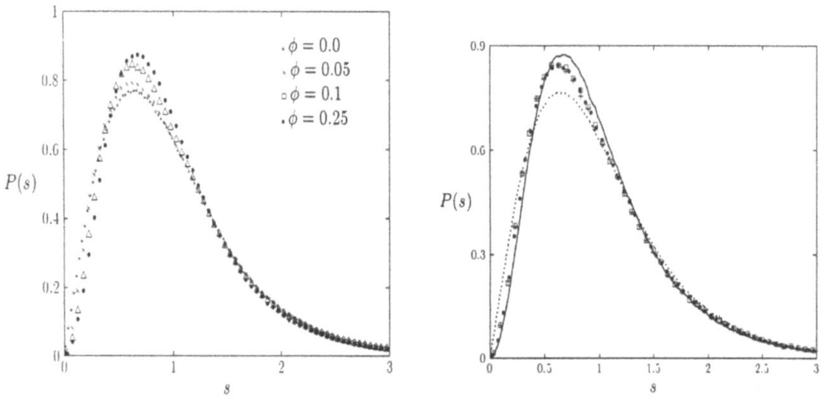

Fig. 3. The level spacing distribution $P(s)$ for the unitary case at the critical disorder $W = 16.5$. *Left*: at different values of the Aharonov-Bohm flux ϕ for linear size $L = 5$. *Right*: at the fixed flux $\phi = 0.2$ for different sizes $L = 5$ (+); 10 (□); 20 (•). Solid and dashed lines are the critical $P(s)$ for two limiting phases $\phi = 0.25$ and $\phi = 0$, respectively. Number of spacings $\sim 10^6$ for each ϕ.

all $P(s)$ at W_c proved to be insensitive to varying the size of the system $L = 5$, 10 and 20, as shown on the right hand side in Fig. 3 for $\phi = 0.2$. The same L-invariant behavior has been observed for various ϕ. We have checked that the critical disorder within our accuracy does not change with ϕ and equals $W_c \approx 16.5$. In other words, the mobility edge is basically flux-independent, in contrast to the behavior of the critical $P(s)$. It has recently been argued (Montambaux 1998) that the crossover between critical orthogonal and critical unitary shapes is governed by the product $ng\phi^2$, where n is the number of applied fluxes and g is the dimensionless conductance.

4 Two-level correlation function

One of the important statistical measures which characterizes the spectral correlations is the level density correlation function

$$R(s) = \frac{\langle \rho(\varepsilon + s\Delta)\rho(\varepsilon)\rangle}{\rho_0^2}, \qquad \rho_0 = \frac{1}{\Delta L^d}, \tag{5}$$

where $\langle ... \rangle$ denotes averaging over various realizations of the random potential and ρ_0 is the averaged density of states. The expressions for $R(s)$ for the three symmetry classes of the RMT are well known (Mehta 1991). As expected from the above, they can be applied for describing the spectral correlations in a disordered metal. For example, the asymptotic behavior for large s is $R(s) - 1 \propto s^{-2}$. In the insulating regime, no level correlations exist and

$R(s) = 1$, irrespective of energy. Similarly to the level spacing distribution, it is interesting to study $R(s)$ at the Anderson transition.

We have computed energy spectra for 3D systems with sizes ranging from $L = 6$ to 20 with large number of realizations at $W = 16.5$. Numerically, it is easier to calculate $R(s)$ by using

$$R(s) = \sum_{n=0}^{\infty} p(n,s), \quad p(0,s) \equiv P(s) \ , \tag{6}$$

where $p(n,s)$ is the distribution function of the separation s between the levels ε_{i+n+1} and ε_i. Figure 4 shows our results for orthogonal symmetry, where the upper limit $n_{\max} \approx 100$ in Eq. (6) is assumed. As expected, $R(s)$ at the transition is different from both, the GOE and the Poissonian limits.

It has been suggested (Kravtsov et al. 1994) that the correlation function at the metal-insulator transition depends only on d and the symmetry class, and in the asymptotic region $s \gg 1$ has the form

$$R(s) - 1 \propto \frac{1}{s^{\alpha}}, \qquad \text{with} \quad \alpha \equiv 1 + \frac{1}{\nu d} \ . \tag{7}$$

In 3D from the available scaling result $\nu \approx 1.5$ one obtains $\alpha \approx 1.2$. While previous numerical calculations (Braun and Montambaux 1995) were in agreement with the proposed power law yielding $\alpha = 1.17$, our present high-precision results do not confirm the above analytical expression Eq. (7).

Despite that the system size $L = 12$ and number of realizations are larger than in previous calculations, thus giving a better accuracy, our results for the tail of $R(s)$ have not been found to be reliably consistent with the power law (inset of Fig. 4). The data fluctuate very strongly when s becomes large ($s \geq 10$), so that no definite conclusion about the asymptotic behavior could be drawn. We note here in passing that the statistical uncertainties of our data are controlled by the central limit theorem. This means that the accuracy of the calculated $R(s)$ for large s is restricted by the total number N of the computed eigenvalues: $\delta R(s)/R(s) \propto 1/\sqrt{N}$. There is a certain maximal value of $s_{\max}$ up to which a fitting procedure for the asymptotic form is meaningful. A simple estimate yields $s_{\max} \approx 10 \div 12$ for the number of eigenvalues $N \sim 10^7$. Any interpolation beyond this value does not make any sense. Even though one can interpolate the decay of $1 - R(s)$ only in the region $s < 10$ by the power law, the fitted value of $\alpha \approx 2.3$ deviates strongly from the expected theoretical estimate $\alpha = 1 + 1/(3\nu)$. In this view the agreement between analytical expression Eq. (7) and previous numerical results seem to be a bit optimistic. Therefore, the exact asymptotic behavior of the correlation function at the Anderson transition has to be considered as an open problem.

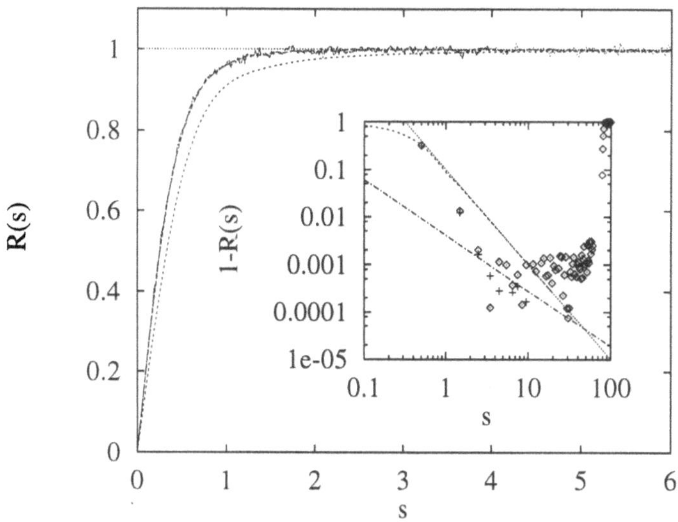

Fig. 4. Two-level correlation function $R(s)$ at the critical disorder $W = 16.5$ for linear sizes $L = 8$ and 12. Dotted line: Poisson limit, $R(s) = 1$. Dashed line: GOE result. Inset: large-s behavior of $R(s) - 1$ for $L = 8$ (+) and 12 (◇); dashed line: GOE result; dotted line: asymptotic of GOE ($\alpha = 2$); dashed-dotted line: power-law Eq. (7) with $\alpha = 1.2$. Total number of spacings 10^7.

5 Spectral form-factor

By definition, the *spectral two-level form factor* is the Fourier transform of $R(s)$,

$$S(t) \equiv \int_{-\infty}^{\infty} R(s) \exp(2i\pi st) ds \ , \qquad (8)$$

The exact expressions of the form-factor in RMT are known for all three universality classes (Mehta 1991). For instance, for the GUE

$$S_{\mathrm{GUE}}(t) = \begin{cases} 1 - |t|, & |t| \leq 1 \\ 0, & |t| \geq 1 \end{cases} \qquad (9)$$

The point $t = 1$ corresponds to the Heisenberg time $\tau_{\mathrm{H}} = h/\Delta$, where the form factor has a singularity (the divergence starts from the second derivative).

Here we focus the attention only on the unitary symmetry. In practice, it is more accurate to determine $S(t)$ directly from the computed eigenvalues ε_i, rather than using the Eq. (8),

$$S(t) = 2 \sum_{i>j} \cos(\pi t s_{ij}), \quad \text{with} \quad s_{ij} := \frac{\varepsilon_i - \varepsilon_j}{\Delta} \ . \qquad (10)$$

Figure 5 shows the results of numerical calculations of the form-factor $S(t)$ for the linear size $L = 20$ at fixed three-component AB-flux $\phi = 0.25$. One observes that the "knee" of $S(t)$ at the Heisenberg time $t = 1$ is washed out at finite disorder. The whole function changes monotonically from the GOE limit Eq. (9) to the Poissonian $S(t) = 0$. For weak disorder W, the Thouless

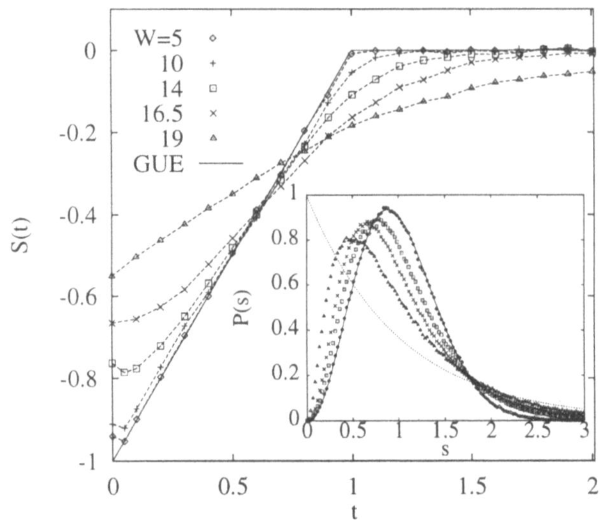

Fig. 5. The form-factor $S(t)$ of the unitary case, $\phi = 0.25$, for system size $L = 20$ and different disorders. The GUE-result shown by the solid line corresponds to Eq. (9), the Poissonian case to $S(t) = 0$. Inset: level spacing distribution $P(s)$ for different disorders. Solid line: exact $P_{\text{GUE}}(s)$; dashed line: Poissonian distribution. Number of spacings $\sim 10^5$ for each W.

time $\tau_T = L^2/D$ separates the ergodic and the diffusive regimes, with D being the diffusion constant. For small times, $t \ll \tau_T/\tau_H$, we observe for metallic systems ($W \leq 14$) a pronounced decrease of $S(t)$ in the diffusive regime. For $t \geq \tau_T/\tau_H$, one finds negligible weak localization corrections to $S_{\text{GOE}}(t)$. The fact that the character of the deviation of the form-factor from the ergodic limit in the mesoscopic conductor is governed only by the large, but finite conductance $g = G/(e^2/h) \gg 1$ has been established earlier (Andreev and Altshuler 1995), (Agam et al. 1995).

When approaching the critical region ($\xi(W) \geq L$), both the RMT and the perturbative approximations are no longer valid, because the Thouless and the Heisenberg times become of the same order, $\tau_T \sim \tau_H$, i. e. $g \simeq 1$. Precisely at $W = W_c$ where ξ diverges, the data for $S(t)$ do not depend on L. Obviously, this corresponds to the *the critical form-factor* and establishes the third universal level statistics in the thermodynamic limit. For comparison, the level spacing distribution $P(s)$ for the same values of disorder is shown in the inset of Fig. 5. The crossover of $P(s)$ from $P_{\rm GUE}(s)$ to $P_{\rm P}(s)$ Eq.(4) is similar as for the orthogonal symmetry (cp. Fig. 2).

6 The correlation length exponent

Since $P(s)$ scales differently with L for various regions of s, it is more convenient to consider the integrated probability function $I(s) = \int_s^\infty P(s')\,ds'$, which equals a fraction of those spacings which are larger than s. Because all spacings are positive, $I(0) = 1$. In addition, due to normalization to the total number of spacings $\int_o^\infty I(s)\,ds = 1$. For the two limiting regimes of metal and insulator Eqs. (2) and (4) yield

$$I_{\rm W}^o(s) = \exp(-\pi s^2/4), \qquad I_{\rm P}(s) = \exp(-s), \tag{11}$$

respectively. For numerical purposes we introduce the function

$$A(s) := \frac{I(s) - I_{\rm W}^o(s)}{I_{\rm P}(s) - I_{\rm W}^o(s)} \; . \tag{12}$$

It describes the relative deviation of $I(s)$ from $I_{\rm W}^o(s)$. In order to extract the dependence only on W and L and to minimize numerical errors we define a certain value of the spacing s_0 for $A(s)$. The choice of the average spacing, $s_0 = 1$, is not favorable because $A(\Delta)$ exhibits a very small size effect. One can see in Fig. 2 that $P_{\rm W}^o(s)$ and $P_{\rm P}(s)$ cross at two points $s_1 = 0.473$ and $s^\star = 2.002$. Here we studied $\alpha(W,L) \equiv A(s_1)$ representative for the small-s part of $P(s)$, where $I_{\rm W}^o(s)$ and $I_{\rm P}(s)$ are significantly different. This variable reflects the "strength" of the repulsion of two consecutive levels when the separation between them is less than s_1. For $L \to \infty$ the value of α is equal to zero for $W < W_c$ and to unity for $W > W_c$.

We have computed many data of the function $A(s)$ for system sizes $L = 6$, 8, 12, 16, 20, 24, 28 and various strength of disorder ranging from $W = 12$ to 20. Figure 6 shows the disorder dependence of α. The curves for different sizes cross at one point, W_c, which corresponds to the disorder–induced MIT. Near the critical point, as long as $|W - W_c| < 2$, the data of $\alpha(W)$ depend stronger on W for larger cubes. In the thermodynamic limit, α should eventually experience a discontinuous jump from zero to unity exactly at W_c. A different variable, $\int_{s^\star}^\infty I(s)\,ds + s^\star - 1$, was studied in (Hofstetter and Schreiber 1994), which corresponds to using $P(s)$ in the range of large spacings. In fact, the critical disorder W_c and the scaling properties of the integrated level spacing distribution do not depend on the choice of s_o. We have checked this s_o–independence by repeating the calculations for $s_o = 0.05$ and 4. However, the results for s_1 yields the highest accuracy. Using the data for $\alpha(W,L)$ one can determine very accurately the critical value of the disorder, $W_c = 16.4 \pm 0.1$.

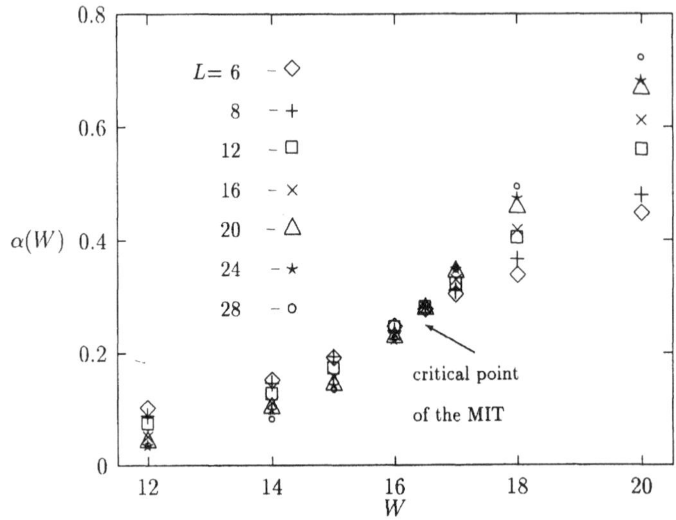

Fig. 6. Scaling variable $\alpha(W)$ near W_c for various system sizes L. Number of level spacings is about 10^5 for each pair of parameters $\{L, W\}$.

In what follows we discuss the critical behavior of α. It characterizes the slope of the linearly growing part of $P(s)$. Assuming the validity of the one–parameter scaling $\alpha(W, L) = f(L/\xi(W))$, one can replot the data of Fig. 6 by rescaling L. We numerically chose values of $\xi(W)$ by shifting the logarithm of the length L separately for each W. An overlap between data for adjacent values of W allows to fit most of the points onto a common curve with two branches, growing one ($W > W_c$) and decaying one ($W < W_c$) (Fig. 7).

Thus, we find a function $\xi(W)$ up to an arbitrary factor ξ_o. This arbitrariness can be avoided. The points for the maximum disorder $W = 20$ in Fig. 7 except of the sizes $L = 6$ and 8 correspond to the localized regime $\xi < L$. Supposing that the statistics of the levels of localized states is not sensitive to the dimensionality of system, one can match ξ for the above disorder to the relevant values of the localization length calculated by the transfer-matrix method in the quasi–1D case (Kramer and MacKinnon 1993). The function $\xi(W)$, as found by this procedure, is plotted in the inset of Fig. 7. Our values of ξ coincide with those found by the transfer-matrix method. The agreement is slightly better for the insulating than for the metallic side.

The next step is to extract the critical exponent of the localization length. Using the singularity of the correlation length near W_c

$$\xi(W) = l \left| \frac{W - W_c}{W_c} \right|^{-\nu}, \tag{13}$$

where the scale l is of order of the lattice constant, one can expand the scaling function $\alpha(L/\xi)$ into a power series

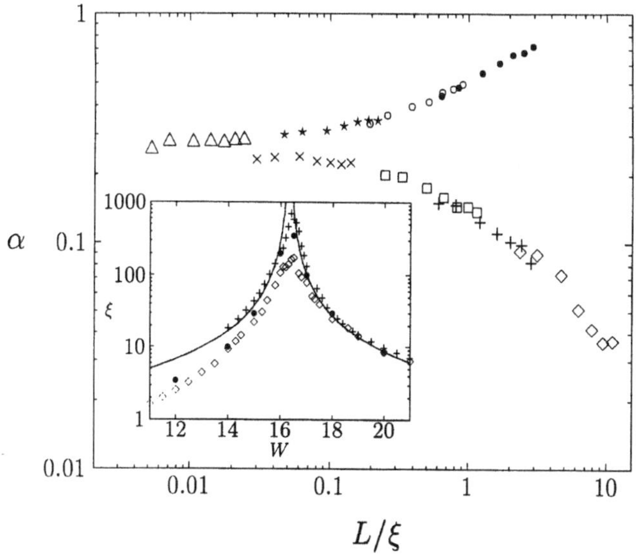

Fig. 7. Scaling variable α as a function of $L/\xi(W)$, showing one–parameter scaling. $W = 12(\diamond)$, $14(+)$, $15(\square)$, $16(\times)$, $16.5(\triangle)$, $17(\star)$, $18(\circ)$, and $20(\bullet)$. Inset: correlation length ξ as function of the disorder W. Present method ($\bullet$), Transfer-matrix method ($\diamond$) (Kramer and MacKinnon 1993); ($+$) (Schreiber and Ottomeier 1992). Continuous curve: plot of (13) with $W_c = 16.4$, $\nu = 1.45$, and $l \approx 1.1$.

$$\alpha(W, L) = \alpha_c(L) + A(W - W_c)L^{1/\nu} \ . \tag{14}$$

By applying a standard χ^2–criterion for the verification of the linearized scaling hypothesis with several parameters to fit the data, we found $\nu = 1.45 \pm 0.08$. It is well consistent with the critical exponent $\nu \simeq 1.5$ obtained recently by the transfer-matrix method (Slevin and Ohtsuki 1997). A slightly smaller value of $\nu \simeq 1.34$ for the large-s region of $P(s)$ has been computed in (Hofstetter and Schreiber 1994) by using the level–statistics method similar to ours. The dependence $\xi(W)$ from Eq. (13) with $\nu = 1.45$ is shown in the inset of Fig. 7 for comparison. We note that the critical exponent could be also estimated by fitting the data for $\xi^{-1}(W)$ to the power–law function without linearizing of $\alpha(W)$ near W_c. However the relative numerical errors become markedly larger similar as in the earlier approach (Kramer and MacKinnon 1993). Certainly the accuracy of ν can then be improved by considering a narrower interval δW near the critical point W_c or by computing larger systems.

7 Conclusions

We have numerically studied statistical properties of fluctuations in electron spectra of 3D disordered systems. The nearest-neighbor level spacing distribution, the two-point correlation function and the form-factor have been shown with high precision to exhibit size-independence at the critical point of the Anderson transition. With increasing of the size of the system they scale towards the corresponding results of the RMT below the critical disorder ($W < W_c$), and to the Poissonian distribution above it ($W > W_c$). As a result, in the thermodynamic limit only three universal statistics exist, including *the critical statistics* precisely at the transition. In the metallic phase, the range of validity of the Wigner-Dyson statistics extends to the infinite energy, because the Thouless conductance $g \propto E_{\mathrm{T}}/\Delta$ diverges for $L \to \infty$. In the insulating phase, the Poissonian limit is expected to apply since the statistics is a consequence of the superposition of the contributions from many independent localization volumes $V_\xi = \xi^d(W)$.

In the presence of time-reversal symmetry we have compared $P(s)$ in the metallic regime for periodic and Dirichlet boundary conditions. We have found that it is important to distinct between the Wigner surmises and the results of the RMT, in order to extract the weak-localization corrections. It is consistent with our results to assume that the level spacing distribution depends on the boundary conditions, not only in the metallic regime, but also at the critical point, since at $W = W_c$ the localization length $\xi(W)$ diverges. We have also studied the influence of the broken time-reversal symmetry on the critical level statistics. In the Aharonov-Bohm geometry we found a family of scale-invariant level statistics, which are controlled by the AB-flux. By increasing the flux, the level spacing distribution shows a crossover from the *critical orthogonal* to the *critical unitary* form.

Near the critical point the finite–size scaling properties of the integrated probability $A(s)$ of neighboring spacings were examined. For the small–s part of $P(s)$ this probability exhibits the transition between the Wigner and the Poisson distributions which corresponds to the delocalization–localization transition. The value of the critical disorder was found, $W_c = 16.4$, for the 'box distribution' of the site energies. Finally, we determined the disorder dependence of the correlation length $\xi(W)$ and the critical exponent $\nu = 1.45 \pm 0.08$ which are in good agreement with previous results obtained by the transfer–matrix method. In general, numerical calculations confirm the one–parameter scaling hypothesis. However analytically, in spite of considerable progress, the problem of the eigenvalue statistics at the Anderson transition remains still unsolved.

The authors are indebted to A. D. Mirlin and L. Schweitzer for discussions. This work was supported by the Deutscheforschungsgemeinschaft via Sonderforschungsbereich 508 "Quantenmaterialien" of the University of Hamburg.

References

E. Abrahams, P. W. Anderson, D. C. Licciardello, and T. V. Ramakrishnan, Phys. Rev. Lett. **42**, 673 (1979).

O. Agam, B. L. Altshuler, and A. V. Andreev, Phys. Rev. Lett. **75**, 4389 (1995).

B. L. Altshuler and B. I. Shklovskii, Zh. Eksp. Teor. Fiz. **91**, 220 (1986).

B. L. Altshuler, I. Kh. Zharekeshev, S. A. Kotochigova, and B. I. Shklovskii, Zh. Eksp. Teor. Fiz. **94**, 343 (1988).

A. V. Andreev and B. L. Altshuler, Phys. Rev. Lett. **75**, 902 (1995).

P. W. Anderson, Phys. Rev. **109**, 1492 (1958).

A. G. Aronov, V. E. Kravstov, and I. V. Lerner, Phys. Rev. Lett. **74**, 1174 (1995).

M. Batsch, L. Schweitzer, I. Kh. Zharekeshev, and B. Kramer, Phys. Rev. Lett. **77**, 1552 (1996).

D. Braun and G. Montambaux, Phys. Rev. B **52**, 13903 (1995).

D. Braun, G. Montambaux, and M. Pascaud, Phys. Rev. Lett. **81**, 1062 (1998).

F. J. Dyson, J. Math. Phys. **3**, 140 (1962); **3**, 1199 (1962).

K. B. Efetov, Adv. Phys. **32**, 53 (1983).

K. B. Efetov, *Sypersimmetry in disorder and chaos*, (Cambridge University Press, 1997).

S. N. Evangelou, Phys. Rev. B **49**, 16805 (1994).

M. Feingold, Y. Avishai and R. Berkovitz, Phys. Rev. B **52**, 8400 (1995).

T. Guhr, A. Müller-Groeling, and H. A. Weidenmüller, Phys. Rep. **299**, 189 (1998).

E. Hofstetter and M. Schreiber, Phys. Rev. B **49**, 14726 (1994).

M. Janssen, Phys. Rep. **295**, 1 (1998).

T. Kawarabayashi, T. Ohtsuki, K. Slevin, Y. Ono, Phys. Rev. Lett. **77**, 3593 (1996).

B. Kramer and A. MacKinnon, Rep. Prog. Phys. **56**, 1496 (1993).

V. E. Kravtsov and I. V. Lerner, Phys. Rev. Lett. **74**, 2563 (1995).

V. E. Kravtsov, I. V. Lerner, B. L. Altshuler, and A. G. Aronov, Phys. Rev. Lett. **72**, 888 (1994).

V. E. Kravtsov and A. D. Mirlin, Pis'ma Zh. Eksp. Teor. Fiz. **60**, 645 (1994).

P. A. Lee and T. V. Ramakrishnan, Rev. Mod. Phys. **57**, 287 (1985).

M. L. Mehta, *Random Matrices* (Academic Press, Boston, 1991).

G. Montambaux, Phys. Lett. A **233**, 430 (1998).

M. Schreiber and M. Ottomeier, J. Phys.: Condens. Matter **4**, 1959 (1992).

L. Schweitzer and I. Kh. Zharekeshev, J. Phys.: Condens. Matter **7**, L377 (1995).

L. Schweitzer and I. Kh. Zharekeshev, J. Phys.: Condens. Matter **9**, L441 (1997).

B. I. Shklovskii, B. Shapiro, B. R. Sears, P. Lambrianides, and H. B. Shore, Phys. Rev. B **47**, 11487 (1993).

K. Slevin and T. Ohtsuki, Phys. Rev. Lett. **78**, 4085 (1997).

D. J. Thouless, Phys. Rep. **13 C**, 93 (1974).

A. Weymer and M. Janssen, Ann. Physik (Leipzig), **7**, 159 (1998).

E. P. Wigner, Proc. Cambridge Philos. Soc. **47**, 790 (1951).

I. Kh. Zharekeshev, M. Batsch, and B. Kramer, Europhys. Lett. **34**, 587 (1996).

I. Kh. Zharekeshev and B. Kramer, Phys. Rev. B **51**, 17239 (1995).

I. Kh. Zharekeshev and B. Kramer, Jpn. J. Appl. Phys. **34**, 4361 (1995).

I. Kh. Zharekeshev and B. Kramer, Phys. Rev. Lett. **79**, 717 (1997).

I. Kh. Zharekeshev and B. Kramer, Ann. Phys. (Leipzig) **7**, 442 (1998).

I. Kh. Zharekeshev and B. Kramer, Computer Phys. Commun. **121/122**, 502 (1999).

Multiple Light Scattering in Nematic Liquid Crystals

D.S. Wiersma, A. Muzzi, M. Colocci, and R. Righini

European Laboratory for Non-linear Spectroscopy, and Istituto Nazionale per la Fisica della Materia, Largo E. Fermi 2, 50125 Florence, Italy,
e-mail: wiersma@lens.unifi.it.

Abstract. We have studied the propagation of light waves in large liquid crystal samples in a monodomain nematic phase. The strong scattering combined with the partial ordering of the nematic phase leads to an anisotropic diffusion process. Or better, the light waves perform an anisotropic random walk in these systems. We have studied the anisotropy in the diffusion constant for light in nematics, by means of time-resolved transmission experiments.

1 Introduction

In recent years various interesting interference effects have been recognized in light which is multiply scattered from complex media such as colloidal suspensions, semiconductor powders, or even common white paint[1]. For instance, it was found that the interference between counterpropagating waves in disordered structures gives rise to enhanced backscattering. The phenomenon is known as coherent backscattering or weak localization[2]. Later, more interference effects were found like the spatial correlations in the intensity transmitted through random media[3]. Inspired by solid state physics, many parallels were found between the multiple scattering of electrons and multiple scattering of light waves, for instance the photonic Hall effect and optical magneto resistance[4], Anderson localization of electromagnetic waves[5], and universal conductance fluctuations[6]. Important applications of multiple light scattering include medical imaging[7] and diffusing-wave spectroscopy (DWS)[8].

2 Liquid Crystals

Liquid crystals in the nematic phase are strongly scattering materials that also give rise to coherent backscattering effects[9], but they differ fundamentally from common isotropic random media. The nematic phase of a liquid crystal is characterized by a global aligment of the molecules in a direction called the nematic director $\mathbf{n}(\mathbf{r})$, and an otherwise translational disorder. (See Fig. 1.)

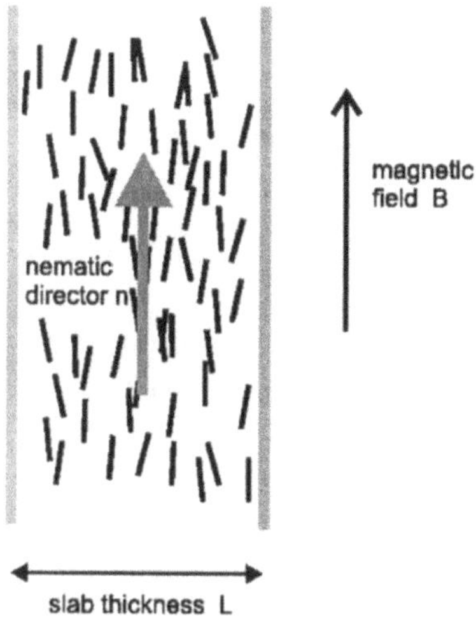

Fig. 1. Schematic representation of a liquid crystal sample in the nematic phase. The molecules have a global alignment in a direction called the nematic director, denoted by **n**. The sample depicted here has a slab geometry with the nematic director aligned along the slab, also referred to as planar alignment. A magnetic field is present along the slab to assure a homogenous monodomain phase. The nematic director is subject to local fluctuations which, for thick (several mm) slabs, give rise to multiple light scattering.

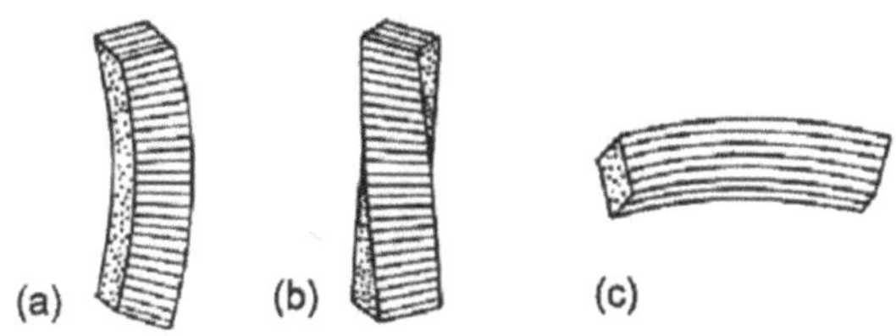

Fig. 2. The three modes of distortion of the nematic director, respectively splay (a), twist (b), and bend (c). The line pattern indicates the direction of the director.

The strong opacity of the nematic phase comes about from local fluctuations in the nematic director that elastically scatter light[10]. One can distiguish three modes of distortion of the nematic director called splay, twist, and bend distortions. (See Fig. 2.) These three distortion modes give rise to a complicated scattering function that depends on the direction of propagation of the incoming and outgoing wave with respect to the nematic director, and on the polarization of these waves. The scattering cross section for elastic scattering from the fluctuations of the nematic director is given by[11, 10]:

$$\sigma = V(\frac{\Delta\epsilon k^2}{4\pi})^2 \sum_{\alpha=x,y} \frac{k_b T(i_\alpha f_z + i_z f_\alpha)^2}{K_\alpha q_\perp^2 + K_3 q_{//}^2 + K_1 \xi^{-2}}. \tag{1}$$

Here $\mathbf{q} = \mathbf{k_f} - \mathbf{k_i}$ is the scattering wavevector with $\mathbf{k_i}$ the initial and $\mathbf{k_f}$ the final wavevector of the scattered light wave, K_1, K_2 and K_3 are the Franck elastic constants for the splay, twist and bend deformations respectively, $\mathbf{i}$ and $\mathbf{f}$ are unit vectors representing the initial and final polarization direction, V is the scattering volume, k_b the Boltzman constant, T is the temperature, and $\Delta\epsilon = \epsilon_{//} - \epsilon_\perp$, with $\epsilon_{//}$ and $\epsilon_\perp$ the dielectric constants for an electromagnetic field parallel respectively perpendicular to the nematic director. The projections $q_\perp$ and $q_{//}$ are the components of $\mathbf{q}$ respectively perpendicular and parallel to the nematic director. The length ξ is called the magnetic coherence length and expresses the distance over which the fluctutions are correlated. The magnetic coherence length is given by:

$$\xi = \sqrt{K_1/\chi_a B^2}, \tag{2}$$

where χ_a is the diamagnetic susceptibility and B the strength of an external magnetic field. This external magnetic field is used in practise to obtain a monodomain nematic phase. In the experiments described in this paper, rather large (several mm thickness) samples are used which without some external force would be very difficult to obtain in a monodomain phase.

From Eq. 1, we see that the scattered intensity is zero for conserving polarization channels, which means that the scattered light will be of opposite polarization compared to the incoming light. This interesting property of light scattering from nematics is expected to be averaged out however in the limit of diffusive transport.

Furthermore we see from Eq. 1 that the cross section diverges for $\mathbf{q} = 0$ at zero magnetic field. For isotropic systems, the scattering mean free path ℓ_{sc} is usualy defined as the average distance between two scattering events, and is inversely proportional to the scattering cross section. In a nematic, the scattering mean free path ℓ_s will depend on the direction of propagation and will have a component parallel and perpendicular to $\mathbf{n}$ denoted by $\ell_{s\,//}$ and $\ell_{s\,\perp}$ respectively. However, due to the above described divergence fo the scattering cross section, the scattering mean free path looses its meaning at zero magnetic field.

A more useful length is the transport mean free path ℓ which in isotropic systems is defined as the average distance over which the propagating light wave looses its memory of the original propagation direction. In isotropic systems where the single scattering function is strongly peaked in a certain direction (like in the case of a random collection of Mie scatterers), the transport mean free path can differ appreciably from the scattering mean free path. Only in the case of a completely isotropic scattering function (like in the Rayleigh limit), the transport mean free path is equal to the scattering mean free path. In many cases, the propagation of light in random systems can be mapped on a random walker with average step length given by the transport mean free path. For multiple light scattering in nematics, the transport mean free path is particularly usefull as it does not diverge at zero magnetic field[12, 13].

Due to the anisotropy of the scattering cross section, also the transport mean free path will depend on the direction in which the random walker is moving. This anisotropy is a manifestation of the partial order of the nematic phase. A light wave propagating through a large and monodomain nematic phase will perform an anisotropic random walk, of which both average step length (the transport mean free path ℓ) and velocity of propagation will be anisotropic. The transport mean free path ℓ will have the values $\ell_\perp$ and $\ell_\|$, respectively perpendicular and parallel to the nematic director. Likewise the transport velocity v will have the perpendicular and parallel components $v_\perp$ and $v_\|$. Anisotropic multiple light scattering was just recently observed by Kao el al. in an elegant experimental study on multiple scattering of light in nematic liquid crystals[14, 15]. Quite some inspiring theoretical work on light transport in these systems has been done as well[12, 13, 14, 15, 16, 17, 18, 19, 20, 21].

3 Time-resolved experiments

The observation of anisotropic light diffusion by Kao *et al.* was achieved in a static experiment. The difference between static and dynamic (time-resolved) experiments on multiple light scattering is fundamental as different properties of the system are probed in the two cases. In a static experiment one principally measures the average step length of the random walk that the light waves perform. In a time-resolved experiment one can measure the time evolution of the diffusion process as described by the diffusion constant D. The transport mean free path can be related to the diffusion constant via a velocity, called the transport velocity. This transport velocity for light waves in disordered systems behaves in a complicated but interesting way, and only recently was well understood[22]. In liquid crystals the difference between time-resolved and static experiments is even more important, as, apart from the diffusion constant and the mean free path, also the transport velocity is anisotropic. In this paper we describe the, to our knowledge, first time-resolved experiments on anisotropic multiple scattering of light, that allowed us to observe the anisotropy in the diffusion constant for light waves.

For isotropic media one can make the diffusion approximation, which means that one describes the transport of the energy density of the light by a common diffusion equation with diffusion coefficient D. The diffusion constant can be related to the transport mean free path and transport velocity via: $D = 1/3v\ell$. The only parameter which discribes the dynamics of the diffusion process is the diffusion constant, in contrast to the random walker where we have two relevant parameters: the mean free path and the transport velocity. In a diffusion picture, the relevance of the mean free path is only apparent in the boundary conditions via z_0 (see section 4). In the anisotropic case the simplification from random walker to diffusion process is even more drastic as both the mean free path and the transport velocity have different values for different directions of propagation with respect to the nematic director, which will depend on polarization as well.

For an anisotropic medium the diffusion equation reads:

$$\frac{\partial W(\mathbf{r},t)}{\partial t} = \nabla \cdot \mathbf{D} \nabla W(\mathbf{r},t) + S(\mathbf{r},t), \tag{3}$$

with $W(\mathbf{r},t)$ the energy density and $S(\mathbf{r},t)$ a source function. Choosing the nematic director along one of the coordinate axis, the distinct elements of the diffusion tensor will be $D_\perp$ and $D_\parallel$, which now can be related to the transport mean free path and transport velocity as:

$$D_\parallel = \frac{1}{3} v_\parallel \ell_\parallel, \qquad D_\perp = \frac{1}{3} v_\perp \ell_\perp. \tag{4}$$

We have measured the diffusion constant by recording the time-evolution of the diffuse transmission of short laser pulses through a sample with a slab geometry[23]. The slab was oriented in the x-y plane and the laser pulse incident along z. The laser beam was narrow (1mm) and the transmitted diffuse light through the slab was recorded around $x = y = 0$. The incident laser beam could be translated in x and y. We followed two strategies to observe the anisotropy. For zero translation of the incident beam, the time-evolution of the transmitted light only depends on the zz component of D. One can therefore measure $D_\perp$ and $D_\parallel$ by orienting the nematic director either parallel (planar alignment) or perpendicular (homeotropic alignment) to the plane of the slab. If the incident beam is translated over Δx or Δy, the transmitted intensity will also depend (weakly) on the xx or yy components of D, which allows to measure $D_\perp$ and $D_\parallel$ in principle in one measurement.

The time evolution of the diffusely scattered light in transmission was recorded with an optical gating technique as often used in time-resolved fluorescence spectroscopy. (See Fig. 3.) The probe pulse is an ultrashort (<1ps) light pulse at 405 nm that is incident from the front interface of a sample with slab geometry. A small part of the transmitted diffuse light is mixed in a non-linear crystal (BBO) to generate the sumfrequency of the probe light with a gate pulse at a different wavelength (810 nm). The sumfrequency light (at 270 nm) is separated from the probe and gate light and detected with a photo multiplier tube (PMT). Due to the limited acceptance angle of the phase matching process in the BBO and the diafragm ('D' in Fig. 3), only the small fraction of diffuse light which exits the sample around the direction of the slab normal and at position $x = y = 0$ is monitored.

The sample was contained in a glass cell of 25 mm diameter and thicknesses of 6.3 mm and 7.9 mm. The glass cell was temperature controlled with

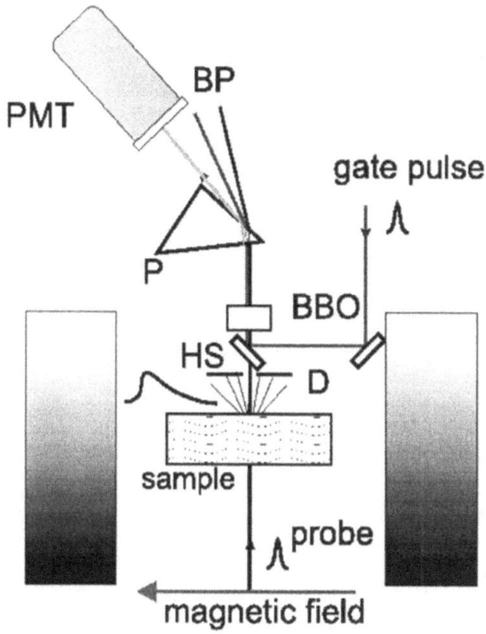

Fig. 3. Principle of the setup used to perform time-resolved measurements. A short probe pulse is incident from below. Part of the diffuse transmission is selected by a diafragm (diameter 1 mm), and mixed with a short gate pulse in a non-linear crystal (BBO), to generate a sum frequency signal that is then monitored by a photo multiplier tube (PMT). By changing the time delay between probe and gate, the temporal profile of the diffuse transmission is recorded. Time resolution of the setup $\approx$ 1 ps. HS: harmonic seperator, BBO: non-linear optical crystal (Beta-Barium Borate), D: diafragm, P: prism, BP: 270nm band pass filter, PMT: photo-multiplier tube.

an accuracy of better than 1 K, and placed in a magnetic field of 0.5 T, generated by an electromagnet with 10 cm diameter poleshoes to assure homogeneity of the field. We used the liquid crystal *p*-penthyl-*p*'-cyanobiphenyl (5CB), which is nematic at room temperature and which has its nematic-isotropic phase transition at 308 K. For every experiment, the sample was heated to 318 K and let cool down overnight in the magnetic field to obtain monodomain samples. We found that for achieving a homogeneous homeotropic alignment cooling slowly was important while the monodomain planar aligment could be achieved also within tens of minutes.

The results of the time resolved measurements for planar and homeotropic alignment of the director, keeping the incoming beam fixed at $x = y = 0$, are shown in Fig. 4. The scattering is stronger for the planar alignment (upper dataset) than for the homeotropic alignment (lower dataset). Therefore,

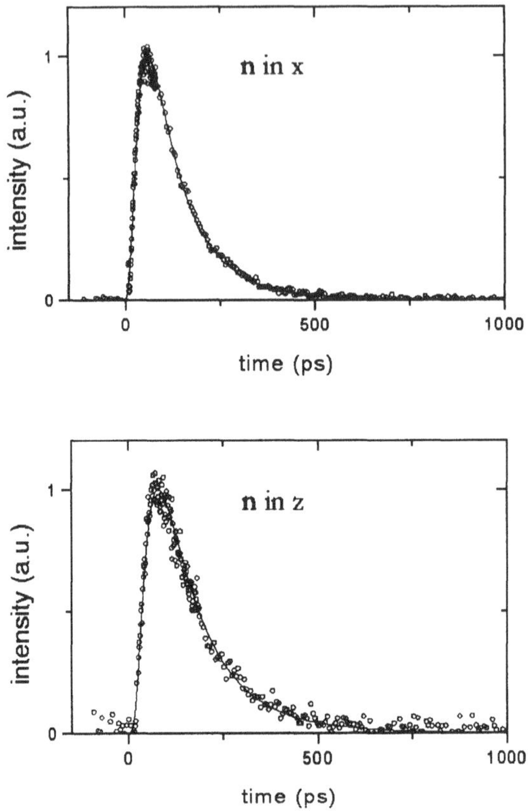

Fig. 4. Time evolution of the transmission of a short probe pulse through liquid crystal 5CB in the nematic phase, in a planar alignment (upper) and a homeotropic alignment (lower). Incident polarization vertical, observed polarization horizontal. (T = 300 K, B = 0.5 T.) The background signal (few percent of maximum) was detected at negative delay time (before presence of probe) and substracted. The solid lines are the theoretical curves as calculated from Eq. (6), taking into account internal reflection via Eq. (5) (refractive index contrast 1.15, $R = 0.211$[24]). From the planar alignment we find: $D_{zz} = D_{\perp} = 3.62 \cdot 10^4$ m^2/s and from the homeotropic alignment: $D_{zz} = D_{\parallel} = 4.56 \cdot 10^4$ m^2/s, which leads to an anisotropy in the diffusion constant of: $D_{\parallel}/D_{\perp} = 1.26$.

in order to have about the same optical thickness in both cases, we chose the physical thickness of the planar aligned sample L = 6.3 mm and of the homeotropic aligned sample L = 7.9 mm. We have performed all experiments with both vertical and horizontal incoming polarization and found no polarization dependence in the diffuse transmission.

4 Interpretation

If we assume that the diffusion approximation is valid for the dynamic behaviour of light transport in nematics, we can use the solution of Eq. (3) to describe our time-resolved transmission data. The solid line in Fig. 4, is the theoretical curve as obtained by solving Eq. (3), using the boundary conditions of a slab geometry: $W(\mathbf{r},t) = 0$ at $z = -z_0$ and $z = L + z_0$, with L the physical thickness of the slab, and using the source function $S(\mathbf{r},t) = \delta(x)\delta(y)\delta(z-\ell_z)\delta(t)$. The distance z_0 is called the extrapolation length and depends on the refractive index mismatch between sample and surrounding medium. For isotropic media z_0 is known[24], and is of the order of the transport mean free path. We conjecture that we can use the same expression for z_0 as given in Ref. [24], substituting ℓ by ℓ_z, the value of ℓ in the direction perpendicular to the slab. This yields:

$$z_0 = 2/3\,\ell_z\,(1+R)/(1-R), \tag{5}$$

with R the average reflectivity at the sample interface and ℓ_z the transport mean free path in z. We assume that the incident pulse is fully scattered at a depth ℓ_z and, for symmetry reasons, that the last scattering event takes place at $z = L - \ell_z$. The time evolution of the transmitted intensity is given by Fick's law ($I_{\mathrm{tr}} = -D_{zz}\nabla W(\mathbf{r},t)\,|_{z=L-\ell_z}$) and reads after solving Eq. (3):

$$I_{\mathrm{tr}} = \frac{I_0 \exp\left(-\Delta x^2/4D_{xx}t\right)\exp\left(-\Delta y^2/4D_{yy}t\right)}{\pi^{3/2}(4t)^{5/2}\sqrt{D_{xx}D_{yy}D_{zz}}} \times$$

$$\sum_{n=-\infty}^{+\infty} A\exp\left(-A^2/4D_{zz}t\right) - B\exp\left(-B^2/4D_{zz}t\right), \tag{6}$$

with $A = (1-2n)(L+2z_0) - 2(z_0+\ell_z)$ and $B = (2n+1)(L+2z_0)$, and where Δx and Δy denote the shift of the incoming beam in x and y respectively. Note that in the limit of long t, the transmitted intensity falls off as an exponential with time constant $\tau = (L+2z_0)^2/D_{zz}\pi^2$.

From Fig. 4 we see that there is excellent agreement between data and theory. From a fit of Eq. 6 to the data we find for the planar alignment: $D_{zz} = D_\perp = 3.62\,(\pm 0.15)\cdot 10^4\ \mathrm{m}^2/\mathrm{s}$, and for the homeotropic alignment: $D_{zz} = D_\| = 4.56\,(\pm 0.18)\cdot 10^4\ \mathrm{m}^2/\mathrm{s}$. The anisotropy in the diffusion constant is therefore: $D_\|/D_\perp = 1.26 \pm 0.07$. These results are in agreement with the theoretical predictions as e.g. in Ref. [13]. The value that we find in our time-resolved experiments differs as expected from the anisotropy as found by Kao et al. in static experiments[14].

To check the consistency of our results, we have performed time resolved transmission measurements with a translated input beam ($x, y \neq 0$), in which case the transmitted intensity depends on both D_{zz} and either D_{xx} or D_{yy}. By choosing a suitable orientation of the director with respect to the direction of the translation, one, in principle, could measure this way both $D_\perp$ and

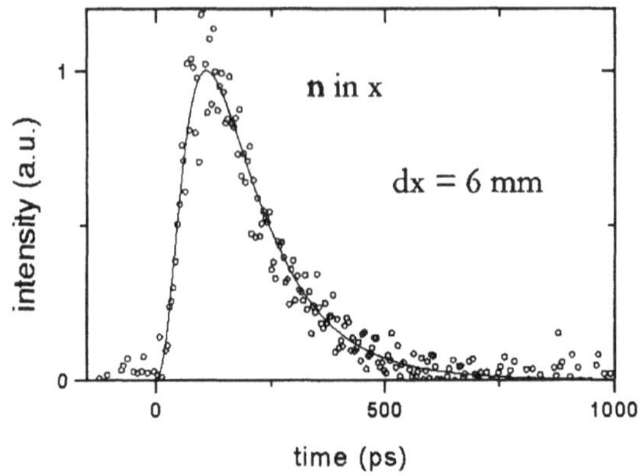

Fig. 5. Time evolution of the transmission of a short probe pulse through nematic 5CB in a planar alignment (with **n** in x); input beam translated 6 mm in x. ($T = 300$ K, $B = 0.5$ T, $L = 6.3$ mm.) Translating the input beam allows to observe contributions to the signal from both D_{zz} and D_{xx}. Solid line: theoretical curve from Eq. (6). The best fit to the data yields: $D_{zz} = D_{\|} = 4.40\,(\pm 0.50) \cdot 10^4$ m^2/s, which is consistent with the measurements presented in Fig. 2.

$D_{\|}$ at the same time. The dependence on D_{xx} and D_{yy} is weak however. Both D_{xx} and D_{yy} will influency mainly the rising slope of the signal, while the decay at long times is almost completely determined by D_{zz}. In Fig. 5 we have plotted the results for the director oriented along x and the input beam translated to $x = 6$ mm. The solid line is again the theoretical curve from Eq. (6). To check the consistency of our data we used the previously determined value of $D_{\perp} = 3.62 \cdot 10^4$ m^2/s. In that case we find for the parallel component of **D**: $D_{\|} = 4.40\,(\pm 0.50) \cdot 10^4$ m^2/s, which is consistent with the data measured in the homeotropic geometry.

5 Conclusions

We have measured the anisotropy in the diffusion constant for light transport in large monodomain nematic liquid crystals. The anisotropy was observed in direct time-resolved transmission experiments and it was found to be $D_{\|}/D_{\perp} = 1.26 \pm 0.07$. The temporal behaviour of the transmission data is well described by diffusion theory. It would be interesting however, to find deviations from diffusion theory in future studies especially in the comparison with static data because an anisotropic random walker with an anisotropic step length and anisotropic velocity is a more complete description of light propagation in nematics than an anisotropic diffusion approach with just an anisotropic diffusion constant.

It is a pleasure to thank Bart van Tiggelen, Holgar Stark, and Arjun Yodh for many stimulating discussions and Cecilia Gambi and Donatella Senatra for use of the electromagnet. This work was financially supported by the E.C. under contract number ERBFMGECT950017. DSW was supported by Marie Curie research grant no. ERBFMBICT972107.

References

1. See for instance: P. Sheng, *Introduction to Wave Scattering, Localization, and Mesoscopic Phenomena* (Academic Press, San Diego, 1995); POAN research group, *New Aspects of Electromagnetic and Acoustic Wave Diffusion* (Springer, Berlin, 1998).
2. Y. Kuga and A. Ishimaru, J. Opt. Soc. Am. A **8**, 831 (1984); M.P. van Albada and A. Lagendijk, Phys. Rev. Lett. **55**, 2692 (1985); P.E. Wolf and G. Maret, Phys. Rev. Lett. **55**, 2696 (1985).
3. N. Garcia and A.Z. Genack, Phys. Rev. Lett. **63**, 1678 (1989); M.P. van Albada, J.F. de Boer, and A. Lagendijk, Phys. Rev. Lett. **64**, 2787 (1990).
4. G.L.J.A. Rikken and B.A. van Tiggelen, Nature **381**, 54 (1996); A. Sparenberg, G.L.J.A. Rikken, and B.A. van Tiggelen, Phys. Rev. Lett. **79**, 757 (1997).
5. R. Dalichaouch, J.P. Armstrong, S. Schultz, P.M. Platzman, and S.L. McCall, Nature **354**, 53 (1991); A.Z. Genack and N. Garcia, Phys. Rev. Lett. **66**, 2064 (1991); D.S. Wiersma, P. Bartolini, A. Lagendijk, and R. Righini, Nature **390**, 671 (1997).
6. F. Scheffold and G. Maret, Phys. Rev. Lett. **81**, 5800 (1998).
7. A. Yodh and B. Chance, Physics Today **48-3**, 34 (1995).
8. G. Maret and P.E. Wolf, Z. Phys. B **65**, 409 (1987); D.J. Pine, D.A. Weitz, P.M. Chaikin, and E. Herbolzheimer, Phys. Rev. Lett. **60**, 1134 (1988).
9. D.V. Vlasov, L.A. Zubkov, N.V. Orekhova, and V.P. Romanov, Pis'ma Zh. Eksp. Teor. Fiz. **48**, 86 (1988) [JETP Lett. **48**, 91 (1988)]; H.K.M. Vithana, L. Asfaw, and D.L. Johnson, Phys. Rev. Lett. **70**, 3561 (1993).
10. P.G. de Gennes and J. Prost, *The Physics of Liquid Crystals* 2nd edition (Oxford, New York, 1993); S. Chandrasekhar, *Liquid Crystals* (Cambridge Univ. Press, Cambridge, 1977); A.Y. Val'kov and V.P. Romanov, Zh. Eksp. Teor. Viz. **82**, 1777 (1982) [Sov. Phys. JETP **56**, 1028 (1983)].
11. D. Langevin, Solid State Comm. **14**, 435 (1974); D. Langevin and M.-A. Bouchiat, J. Phys. (Paris) C **1**, 197 (1975).
12. H. Stark and T.C. Lubensky, Phys. Rev. E **55**, 514 (1997).
13. B.A. van Tiggelen, A. Heiderich, and R. maynard, Mol. Cryst. Liq. Cryst. **293**, 205 (1997).
14. M.H. Kao, K.A. Jester, A. Yodh, and P.J. Collins, Phys. Rev. Lett. **77**, 2233 (1996).
15. H. Stark, M.H. Kao, K.A. Jester, T.C. Lubensky, A.G. Yodh, and P.J. Collins, J. Opt. Soc. Am. A **14**, 156 (1997).
16. A.Y. Val'kov, L.A. Zubkov, A.P. Kovshik, and V.P. Romanov, Pis'ma Zh. Eksp. Teor. Viz. **40**, 281 (1984) [Sov. Phys. JETP lett. **40**, 1065 (1983)].
17. V.P. Romanov and A.N. Shalaginov, Opt. Spectrosc. (USSR) **64**, 774 (1988).

18. B.A. van Tiggelen, R. Maynard, and A. Heiderich, Phys. Rev. Lett. **77**, 639 (1996).
19. H. Stark and T.C. Lubensky, Phys. Rev. Lett. **77**, 2229 (1996).
20. B.A. van Tiggelen, Mol. Cryst. Liq. Cryst. bf 321, 197 (1998).
21. H. Stark, Mol. Cryst. Liq. Cryst. (1998).
22. M.P. van Albada, B.A. van Tiggelen, A. Lagendijk, and A. Tip, Phys. Rev. Lett. **66**, 3132 (1991).
23. D.S. Wiersma, A. Muzzi, M. Colocci, and R. Righini, Phys. Rev. Lett. **83**, issue 21 (1999).
24. A. Lagendijk, R. Vreeker, and P. de Vries, Phys. Lett. A **136**, 81 (1989); J.X. Zhu, D.J. Pine, and D.A. Weitz, Phys. Rev. A **44**, 3948 (1991).

Two Interacting Particles in a Two-Dimensional Random Potential

M. Ortuño and E. Cuevas

Departamento de Física, Universidad de Murcia, E-30071 Murcia, Spain

Abstract. The localization properties of two interacting particles in a two dimensional random potential are calculated by three different numerical procedures. These procedures focus on the nearest level spacing distribution, the inverse participation ratio of the states, and the two-particle decay length in long bars, respectively, and incorporate scaling techniques for the extrapolation of the data. Both short-range and long-range Coulomb interactions are considered. We always find a localized to extended states transition independently of the particle statistics and of the type of interaction. For low disorders, the interaction strongly mixes most unperturbed states, even those corresponding to electrons far apart.

1 Introduction

The combined effects of disorder and interactions in electronic systems have attracted a lot of attention during the last two decades (Lee and Ramakrishnan (1985)). Recent experiments in two-dimensional (2D) systems show clear indications that strong electron-electron interactions partially suppress the quantum interference effects responsible for localization (Kravchenko et al. (1996), Popović et al. (1997)). At the same time, the scaling theory of localization including the effects of interactions predicts that a 2D system may remain metallic even in the limit of zero temperature (Finkelstein (1983), Castellani et al. (1998)). The solution of the full problem is extremely difficult and so we focus on the simplest related problem, that of just two interacting particles in a random potential.

The statistical properties of the energy spectra of disordered systems are closely related to their localization properties (Altshuler and Shklovskii (1986), Shklovskii et al. (1993)). In the localized regime, the normalized inter-level spacings s are distributed according to the Poisson law $P_{\mathrm{P}}(s) = \exp(-s)$, while in the metallic regime the spacings follow Wigner-Dyson statistics

$$P_{\mathrm{W}}(s) = \frac{\pi}{2} s \exp\left(-\frac{\pi}{4} s^2\right) . \tag{1}$$

The study of the energy spectrum has already provided important results about the metal-insulator transition in different systems (Shklovskii et al. (1993), Schweitzer and Zharekeshev (1995)).

In this paper we use three different approaches to study the localization properties of two interacting particles in a random environment: i) the study of the spectral statistics, ii) the analysis of the inverse participation ratio and iii) the calculation of the two-particle decay length in long bars. The second section introduces the model used and the following three sections are dedicated to the three different numerical procedures employed. Finally, in section 6 we draw some conclusions.

2 Model

We consider both bosons and spinless fermions on a sample of size $L \times M$ described by the standard Hamiltonian

$$H = t\sum_{i,j}(a^{\dagger}_{i,j+1}a_{i,j} + a^{\dagger}_{i+1,j}a_{i,j} + \text{h.c.}) + \sum_{i,j}\epsilon_{i,j}a^{\dagger}_{i,j}a_{i,j} + H_1 \equiv H_0 + H_1 \ , \quad (2)$$

where the operator $a^{\dagger}_{i,j}$ ($a_{i,j}$) creates (destroys) an electron at site (i,j) of a square lattice and $\epsilon_{i,j}$ is the energy of this site chosen randomly between $(-W/2, W/2)$ with uniform probability. The hopping matrix element t is taken equal to -1 and the lattice constant equal to 1. We consider a long-range (LR) Coulomb interaction and a short-range (SR) interaction. For the LR case, the interaction Hamiltonian H_1 is given by

$$H_1 = U \sum_{i,j>k,l} \frac{a^{\dagger}_{i,j}a_{i,j}a^{\dagger}_{k,l}a_{k,l}}{|\boldsymbol{r}_{i,j} - \boldsymbol{r}_{k,l}|} \ , \quad (3)$$

while for the SR case we choose a nearest-neighbor (on-site) interaction Hamiltonian for spinless fermions (bosons).

3 Level statistics

In the first two procedures, we use square samples of size $L \times L$ and periodic boundary conditions. We restrict our investigation to the two-electron Hilbert subspace spanned by the basis of $N = L^2(L^2-1)/2$ antisymmetric products of one-electron states $|\psi_{i,j;k,l}\rangle = \frac{1}{\sqrt{2}}(a^{\dagger}_{i,j}a^{\dagger}_{k,l} - a^{\dagger}_{k,l}a^{\dagger}_{i,j})|0\rangle$, where $|0\rangle$ is the vacuum state. The Hamiltonian matrix is numerically diagonalized using a Lanczos tridiagonalization method (Cullum and Willoughby (1985)). The strength U of the interactions varies between $0-10$ and the disorder strength ranges the interval $(0.2, 16)$. The system sizes considered go from $L = 6$ to 20, and the number of random realizations is such that for a given triad of $\{U, L, W\}$ the number of studied eigenvalues in the range $(-1, 1)$ was kept around 2.5×10^4.

We use the following scaling variable (Cuevas (1999))

$$\eta(L, W) = \frac{\text{var}(s) - 0.273}{1 - 0.273} \ , \quad (4)$$

where $\text{var}(s) = \langle s^2 \rangle - \langle s \rangle^2$, and 0.273 and 1 are the variances of Wigner-Dyson and Poisson distributions, respectively. In this way, η ranges from 0 for Wigner-Dyson to 1 for Poisson distributions.

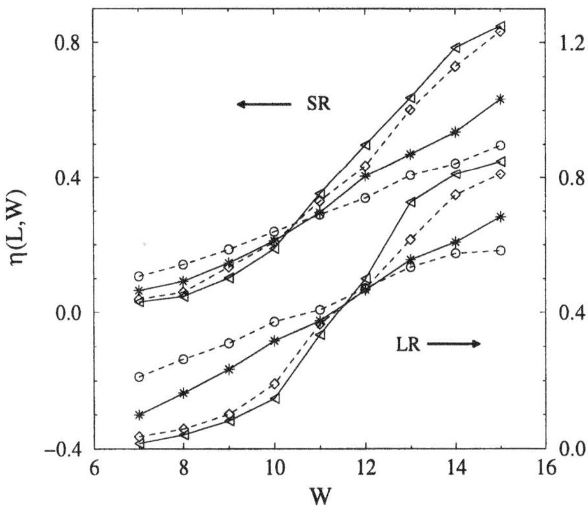

Fig. 1. Scaling variable η as a function of W for a LR and a SR interaction with $U = 1$. System sizes: $L = 6$ (○), 9 (⋆), 16 (◇) and 20 (▽).

Figure 1 shows the disorder dependence of η for both a LR (lower curves, right axis) and a SR interaction (upper curves, left axis) with $U = 1$ for different system sizes: $L = 6$ (○), 9 (⋆), 16 (◇) and 20 (▽). The curves for different sizes cross at a common point, which corresponds to the interaction-driven delocalization transition. We determine the critical disorder W_c and the critical exponent ν by minimizing the χ^2 statistic of the fitting function

$$\eta(L, W) = \eta_c + \sum_n A_n (W - W_c)^n L^{n/\nu} , \tag{5}$$

where we have truncated the series at $n = 4$ and have assumed that the correlation length near W_c diverges as $\xi(W) = \xi_0 |W - W_c|^{-\nu}$, where ξ_0 is a constant. The critical disorder found with this procedure for spinless fermions is $W_c = 11.8 \pm 0.2$ for a LR interaction and $W_c = 10.4 \pm 0.2$ for a SR interaction, and the corresponding critical exponent are $\nu = 1.2 \pm 0.2$ and $\nu = 1.3 \pm 0.2$. Note the strong similarity in the behavior of LR and SR systems.

We found that W_c increases monotonously with U starting from very small values for both LR and SR interactions. In the non-interacting limit we did not find any critical behavior and all states are localized, as expected.

4 Inverse participation ratio

The results in the previous section indicate that for small disorders a large fraction of the two-particle state energies must be perturbed by the interaction. In principle, this contradicts the standard believe that in the thermodynamic limit only a small fraction of the states can be appreciably affected by the interaction, specially in the SR case in which we focus on this section. In order to clarify this question we study the inverse participation ratio for the SR interaction.

We have obtained the eigenvectors ϕ around the center of the band of the Hamiltonian (2) and have calculated their inverse participation ratio in the non-interacting basis, defined as

$$R = \sum_{\alpha} \left|\langle \phi | \phi_{\alpha}^{0} \rangle\right|^{4} , \tag{6}$$

where ϕ_{α}^{0} denotes a two-electron state in this basis. For an infinite system $R = 0$ in the extended regime and increases with disorder up to 1 in the localized regime. Delocalization in real space is related to the inverse participation ratio with respect to the site basis, instead of the non-interacting basis. We have first checked the existence of interaction induced delocalization through the inverse participation ratio in the site basis. However, for quantitative purposes, it is better to use the non-interacting basis, as in Eq. (6), to study the transition since one separates in this way the enhancement in the wavefucntions due to the interaction from this due to the decrease of the disorder (Cuevas and Ortuño (1999)).

In Figure 2 we show the average value of R as a function of size and disorder for a SR interaction with $U = 1$ in both 1D (solid symbols) and 2D (empty symbols) systems. For a given disorder strength and length, averaging is done over 10 disorder realizations in a small energy window containing around 10 eigenvectors. For 1D we see a systematic increase of $\langle R \rangle$ with system size for all values of the disorder, as one expects in the absence of critical behavior. On the contrary, in the 2D case the curves for different sizes cross, indicating the presence of a transition. There is a systematic shift of the apparent critical disorder to lower values as the sizes increase. This is a typical feature of finite size effects, which in our case is relatively pronounced due to the finite slope of the asymptotic behavior near the crossing point.

5 Two-particle decay length

In this section we calculate the decay length of two particles moving together along the longitudinal direction of long bars of size $L \times M$ (Ortuño and Cuevas (1999)). In this case we use lateral periodic boundary conditions. We focus on the two-particle GF of the Hamiltonian (2)

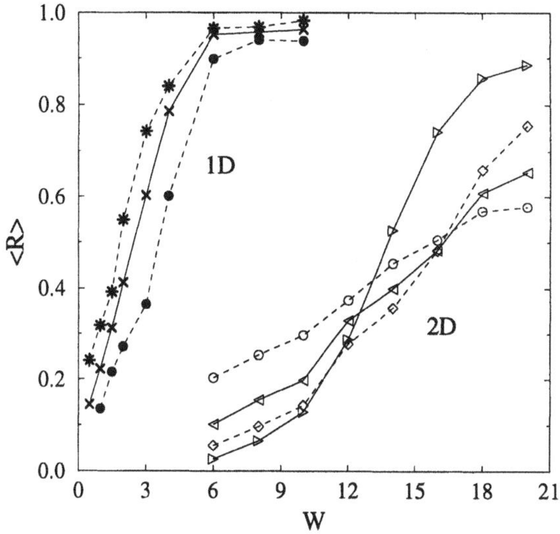

Fig. 2. Average inverse participation ratio $\langle R \rangle$ as a function of W for a SR interaction with $U = 1$ for 1D: $L = 50$ (•), 100 (×), 150 (∗) and 2D: $L = 6$ (○), 8 (◁), 10 (◇) and 16 (▷).

$$G = (E - H_0 - H_1)^{-1} , \tag{7}$$

which satisfies Dyson's equation $G = G_0 + G_0 H_1 G$, where G_0 is the two-particle GF in the absence of interactions. The eigenvectors and eigenvalues of the one-particle problem are enough to construct G_0. von Oppen et al. (1996) noted that for a local interaction we can obtain G very efficiently by projecting onto the subspace of doubly occupied sites. We will refer with a tilde to the matrices restricted to this subspace. Solving Dyson equation for $\widetilde{G}$, and taking into account that $\widetilde{H_1} = U\widetilde{I}$, we obtain

$$\widetilde{G} = (\widetilde{I} - U\widetilde{G}_0)^{-1}\widetilde{G}_0 . \tag{8}$$

This expression can be evaluated exactly by inverting matrices of range equal to the system size, $L \times M$.

Let us call $\widetilde{G}(m_1, n_1; m_2, n_2)$ to the matrix element of the GF between an initial (doubly occupied) site of coordinates (m_1, n_1), and a final (doubly occupied) site of coordinates (m_2, n_2). For a given strip of size $L \times M$ we calculate the following trace

$$\ln \mathrm{Tr}\, |\widetilde{G}(l)|^2 \equiv \langle \ln \sum_{i,j} |\widetilde{G}(1, i; l, j)|^2 \rangle , \tag{9}$$

with $l \leq L$, and where $\langle\,\rangle$ denotes an average over the disorder realizations. We ensure that L is large enough to get a linear exponential decay of the trace as a function of l, for any disorder W and width M considered. Once we reach the exponential regime, we fit the data in this regime to a straight line, whose slope α is related to the two-particle decay length ξ_M through $\xi_M = -2/\alpha$.

Finite-size scaling analysis (MacKinnon and Kramer (1983)) states that the renormalized decay length ξ_M/M is a function of a single parameter ξ/M,

$$\xi_M/M = f(M/\xi) . \tag{10}$$

The scaling parameter ξ is the two-particle localization length in the localized regime, and the two-particle correlation length in the extended regime.

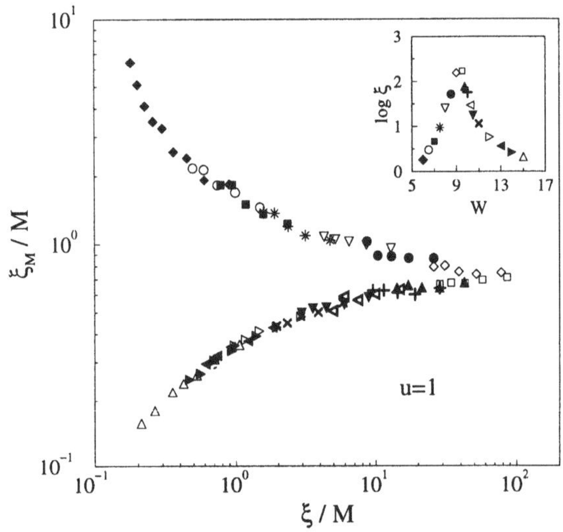

Fig. 3. Log-log plot of ξ_M/M as a function of ξ/M. Inset: disorder dependence of the scaling parameter ξ.

In Figure 3 we show the renormalized decay length ξ_M/M as a function of ξ/M on a double logarithmic scale. We have overlapped all points on one curve within the accuracy of the data, by shifting the raw data horizontally by a disorder dependent amount, which is determined by a least-square fit procedure (MacKinnon and Kramer (1983)). The on-site energy is $U = 1$ and the disorder energies range between $W = 6$ and 15. All data were obtained by averaging over a number of disorder realizations ranging between 300, for the largest M, and 1000, for $M = 2$. We consider the center of the band ($E = 0$), and a length $L = 62$. Fitting the data set for $W = 15$ and M between 4 and 10 to the form $\xi_M = \xi + A/M$ we obtain the localization length for this disorder $\xi(15) = 2.1 \pm 0.1$, which enables us to establish the absolute scale of

$\xi(W)$. The existence of two branches is a clear indication of a transition. In the inset of Figure 3 we plot the disorder dependence of $\log \xi$ for $U = 1$ at the center of the band ($E = 0$).

To obtain the critical disorder W_c and the critical exponent ν, we performed an statistical analysis for ξ_M/M similar to the one used in section 3 for η. The critical disorder found for bosons with this method is $W_c = 9.3 \pm 0.2$, and the corresponding critical exponent is equal to $\nu = 2.4 \pm 0.5$.

6 Conclusions

Since we first saw a transition for two-particle states in two-dimensional disordered systems, many authors have confirmed our results by different methods (Römer et al. (1999), Shepelyansky (1999)). As most unperturbed states correspond to electrons far apart, which should experience a very small interaction, one could expect negligible changes in the overall properties of these systems in the thermodynamic limit. To avoid the paradox between this and the numerical results, some authors have claimed that the behavior for LR and SR interactions are qualitatively different. However, our results clearly indicate very similar features for both type of interactions, which rules out any explanation of the transition based on the long-range character of the interaction.

In fact, Song and von Oppen (1999) already noticed that a SR interaction in 1D appreciably affects the two-particle states when the particle distance amply exceeds the one-particle localization length. In two-dimensions this effect is far more dramatic. For the largest systems that we can handle, we have checked that in situations where many unperturbed states do not appreciably feel the interaction, we still find practically all states delocalized over the whole sample. For low disorders, the eigenstates of the Hamiltonian are a superposition of many non-interacting states, and so there is no mapping between them. It does not make sense to say that states with electrons far apart will be slightly affected by the interaction. We believe that orthogonality between states is an important delocalizing mechanism. Orthogonality to states with electrons close together, which are likely to become delocalized, creates an effective potential acting on the other states.

We would like to stress that our results are not directly applicable to real interacting systems with a finite density of electrons. We only consider highly excited states in the center of band, and cannot construct in a self-consistent way some kind of Fermi sea out of the type of states considered.

Acknowledgements

We would like to thank the Spanish DGES, project number PB96-1118, and Fundación Séneca for financial support.

References

Altshuler B.L. and Shklovskii B.I. (1986): Zh. Eksp. Teor. Fiz. **91**, 220 [Sov. Phys. JETP **64**, 127 (1986)]

Castellani C., Di Castro C. and Lee P.A. (1998): Phys. Rev. B **57**, R9381

Cuevas E. (1999): Phys. Rev. Lett. **83**, 140

Cuevas E. and Ortuño M. (1999): Ann. Phys. (Leipzig), in press

Cullum J.K. and Willoughby R.A. (1985): *Lanczos Algorithms for Large Symmetric Eigenvalue Computations* (Birkhauser, Basel)

Finkelstein A.M. (1983): Zh. Eksp. Teor. Fiz. **84**, 168 [Sov. Phys. JETP **57**, 97 (1983)]

Kravchenko S.V., Simonian D., Sarachik M.P., Mason W. and Furneaux J.E. (1996): Phys. Rev. Lett. **77**, 4938

Lee P.A. and Ramakrishnan T.V. (1985): Rev. Mod. Phys. **57**, 287

MacKinnon A. and Kramer B. (1983): Z. Phys. B **53**, 1

Ortuño M. and Cuevas E. (1999): Europhys. Lett. **46**, 224

Popović D., Fowler A.B. and Washburn S. (1997): Phys. Rev. Lett. **79**, 1543

Römer R.A., Leadbeater M. and Schreiber M. (1999): Ann. Phys. (Leipzig), in press

Schweitzer L. and Zharekeshev Kh. (1995): J. Phys.: Con. Mat. **7**, L377

Shepelyansky D.L. (1999): cond-mat/9902246

Shklovskii B.I., Shapiro B., Sears B.R., Lambrianides P. and Shore H.B. (1993): Phys. Rev. B **47**, 11487

Song P.H. and von Oppen F. (1999): Phys. Rev. B **59**, 46

von Oppen F., Wettig T., and Müller J. (1996): Phys. Rev. Lett. **76**, 491

Part VIII

Mesoscopic Superconductors, Nanotubes and Atomic Chains

Paramagnetic Meissner Effect in Mesoscopic Superconductors

J.J. Palacios

Dept. de Física Teórica de la Materia Condensada, Universidad Autónoma de Madrid, Cantoblanco, Madrid 28049, Spain

Abstract. We calculate the saddle points or energy barriers that prevent the vortex escape and entrance in superconducting mesoscopic disks. As a result of the metastability of the vorticity, the system can exhibit paramagnetic response. From the height of the barriers we conclude that the transition points between vortex configurations with different vorticity observed in the experiments cannot be explained either on the basis of energetic instabilities or thermal activation.

1 Introduction

The interest in understanding the creation and annihilation mechanisms and, in general, the stability of vortices in superfluids has been recently boosted by a series of technological advances in both mesoscopic superconductors (Moshchalkov et al. 1995, Geim et al. 1997, Geim et al. 1998, Bolle et al. 1999) and atomic condensates (Butts et al. 1999). Most of the proposals for the creation of vortices in atomic condensates face, at the present time, severe technological difficulties. On the contrary, mesoscopic superconductors in magnetic fields are already proving to be an ideal scenario where the detection and even manipulation of vortices at the individual level is becoming more and more feasible (Moshchalkov et al. 1995, Geim et al. 1997, Geim et al. 1998, Bolle et al. 1999). A good example, although not the only one, of single-vortex sensitivity can be found in the Hall magnetometry measurements performed on mesoscopic Al disks by Geim et al. (1997,1998). Both field-cooled (FC) and constant temperature (CT) magnetization measurements provide evidence of the quantization of the vorticity of the order parameter. When the system is kept out of equilibrium, it can show paramagnetic response both in the CT and FC cases, whereas, as expected, equilibrium measurements *always* exhibit diamagnetism.

The magnetic response of a mesoscopic type-II superconducting disk is determined, to a large extent, by vortex escape and entrance barriers. Whether the system exhibits a conventional diamagnetic response or a not-so-conventional paramagnetic response depends on the history of the measurement process and relies on the existence of these barriers. These barriers are, however, poorly understood. The only models known to make quantitative predictions are based on the London theory (Bean et al. 1964) which neglects the core of the vortex and only considers the magnetic interaction and the interaction due to the currents with the surface. This model is valid close to the lower critical field H_{c1}, but, in general, the core of the vortex plays a dominant role, particularly close H_{c2} where surface superconductivity, which pushes away normal cores, develops.

In this work we present the basics of a method to calculate vortex escape and entrance barriers close to H_{c2} that fully takes into account vortex cores. These barriers correspond to saddle points of the Ginzburg-Landau energy functional that separate local minima with different vorticity or topological charge Q. We calculate the magnetization associated with the local minima and, comparing with the magnetization measurements in Geim's experiment, we conclude that the energetic instabilities do not suffice to explain the observed transition points between states with different Q. Thermal activation and, most likely, quantum tunneling mechanisms determine the transition points.

2 The projected Ginzburg-Landau functional

We start from the Ginzburg-Landau functional for the Gibbs free energy difference between the normal and superconducting states in an external magnetic field H:

$$G = \int d\mathbf{r} \left[\alpha |\Psi(\mathbf{r})|^2 + \frac{\beta}{2} |\Psi(\mathbf{r})|^4 + \frac{1}{2m^*} \left| \left(-i\hbar\boldsymbol{\nabla} - \frac{e^*}{c}\mathbf{A}(\mathbf{r}) \right) \Psi(\mathbf{r}) \right|^2 + \frac{[h(\mathbf{r}) - H]^2}{8\pi} \right], \tag{1}$$

where $\Psi(\mathbf{r})$ is the order parameter or Cooper pair wave function, $h(\mathbf{r}) = \boldsymbol{\nabla} \times \mathbf{A}(\mathbf{r})$, and α and β are the condensation and interaction energy parameters, respectively. Numerical minimization procedures have been used in the past (Schweigert et al. 1998) to find global and even local minima of the Ginzburg-Landau functional applied to mesoscopic superconducting disks. However, saddle points or energy barriers, which are essential for the analysis of the stability of the local minima, cannot be obtained from these methods. Before going into the details of how to overcome this problem, a few comments are in order. Thin Al disks behave like Type-II superconductors in a perpendicular magnetic field and, for sufficiently high intensities, one can consider a uniform magnetic induction B. Moreover, the order parameter is expected to lie almost entirely in the "lowest Landau level" subspace (Palacios 1998):

$$\Psi(\mathbf{r}) = \sum_{L=0}^{\infty} C_L \Psi_L(\mathbf{r}). \tag{2}$$

This subspace is spanned by normalized eigenfunctions of the linearized differential Ginzburg-Landau equations that are characterized by an angular momentum L, are nodeless in the radial direction, and are subject to the boundary condition of zero current through the surface. $C_L \equiv |C_L|e^{i\phi_L}$ are complex coefficients and we are considering the thickness of the disk to be smaller than the coherence length so that the system becomes effectively two-dimensional. The projected Ginzburg-Landau functional for the Gibbs free energy difference between the normal and superconducting states is

$$G = \sum_{L=0}^{\infty}(\alpha + \epsilon_L)|C_L|^2 + \frac{\beta}{2} \sum_{L_1,L_2,L_3,L_4=0}^{\infty} C^*_{L_1}C^*_{L_2}C_{L_3}C_{L_4} \int d\mathbf{r}\, \Psi^*_{L_1}\Psi^*_{L_2}\Psi_{L_3}\Psi_{L_4} + (B-H)^2/8\pi \tag{3}$$

where ϵ_L is the kinetic energy of the L component, and the second term represents the "interaction" between Cooper pairs. It is convenient for our purposes to express the projected functional in the following way:

$$\begin{aligned} G = {} & (B-H)^2 + \sum_{i=1}^{N} \alpha[1 - B\epsilon_{L_i}(B)]|C_{L_i}|^2 \\ & + \quad \pi\alpha^2\kappa^2 BR^2 \times \Bigg[\sum_{i=1}^{N} I_{L_i}(B)|C_{L_i}|^4 + \sum_{j>i=1}^{N} 4I_{L_iL_j}(B)|C_{L_i}|^2|C_{L_j}|^2 + \\ & \sum_{k>j>i=1}^{N} 4\delta_{L_i+L_k,2L_j}\cos(\phi_{L_i} + \phi_{L_k} - 2\phi_{L_j}) \\ & I_{L_iL_jL_k}(B)|C_{L_i}||C_{L_j}|^2|C_{L_k}| + \\ & \sum_{l>k>j>i=1}^{N} 8\delta_{L_i+L_l,L_j+L_k}\cos(\phi_{L_i} + \phi_{L_l} - \phi_{L_j} - \phi_{L_k}) \\ & I_{L_iL_jL_kL_l}(B)|C_{L_i}||C_{L_j}||C_{L_k}||C_{L_l}|\Bigg], \end{aligned} \tag{4}$$

where the energy is expressed in units of $H_{c2}^2V/8\pi$ (V being the volume of the disk), $\epsilon_L(B)$ is now the energy of the quantum state L expressed in units of $\hbar\omega_c/2$ ($\omega_c = e^*B/m^*c$), R is the radius of the disk expressed in units of the coherence length $\xi(T)$ and B and H are given in units of $H_{c2}(T)$. The interactions appear in $I_L(B) \equiv \int d\mathbf{r}\, \Psi_L^4$, which can be interpreted as the interaction between Cooper pairs occupying the same quantum state L, in $I_{L_iL_j}(B) \equiv \int d\mathbf{r} \Psi^2_{L_i}\Psi^2_{L_j}$, accounting for the interaction between Cooper pairs occupying different quantum states and in $I_{L_iL_jL_k}(B) \equiv \int d\mathbf{r} \Psi_{L_i}\Psi^2_{L_j}\Psi_{L_k}$ and $I_{L_iL_jL_kL_l}(B) \equiv \int d\mathbf{r}\, \Psi_{L_i}\Psi_{L_j}\Psi_{L_k}\Psi_{L_l}$, accounting for correlations. This form exhibits in a clear way all the terms that are relevant when a finite number N of components are considered.

3 Results and discussion

We restrict our discussion to a disk of radius $R = 5\xi(0)$ and an effective $\kappa = 1$. These parameters approximately correspond to those of the largest disk in Geim's experiment (Geim et al. 1998). Stationary solutions, either local minima or saddle points, of the projected Ginzburg-Landau functional are characterized by a given set of N components $\{L_1, L_2, \ldots, L_N\}$ (Palacios 1998). For the disk considered here, $N \leq 3$ for all the stationary solutions. Figure 1 shows the magnetization associated with the local minima. These local minima are characterized by a vorticity or topological charge Q which coincides with L_N. Different curves correspond to different values of Q. Along these curves the topological charge distributes itself in a variety of ways. For large Q ($Q \geq 12$), the local minimum is always a giant vortex with $L = Q$ whereas, for smaller Q, a multiple-vortex structure (Palacios 1998, Schweigert et al. 1998) appears at low enough fields.

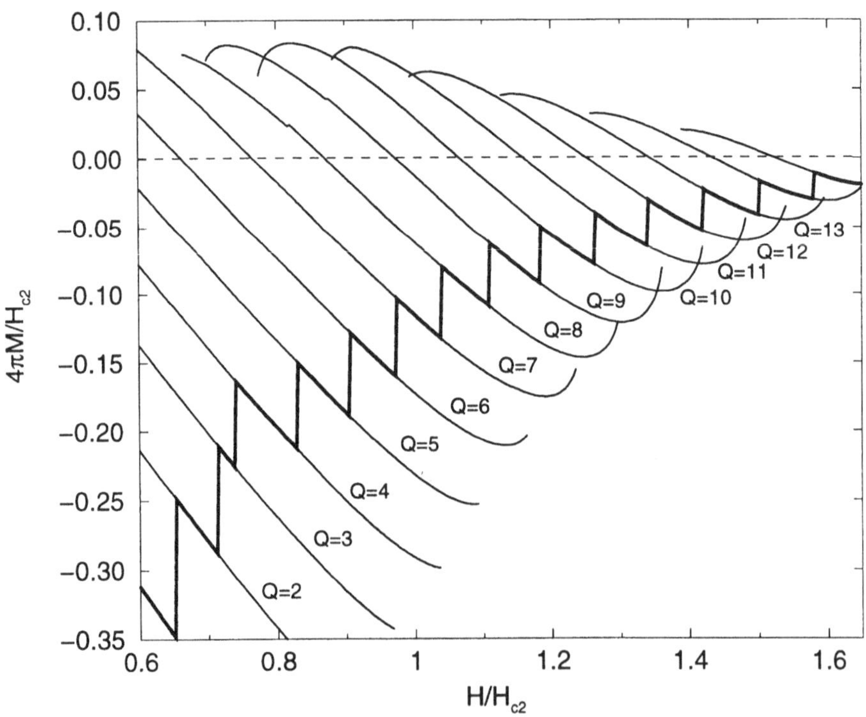

Fig. 1. Field dependence of the magnetization associated with different local minima (characterized by the topological charge Q) for a disk of radius $R = 5\xi$ and $\kappa = 1$. The equilibrium magnetization, i.e., the magnetization of the global minimum is represented by the thick solid line.

For illustration purposes we discuss $Q = 9$ in what follows. The energy of local minima (solid lines) and that of vortex entrance and exit barriers (dashed lines) are shown in Fig. 2. The local minimum on the high-field side corresponds to a giant vortex with $L = 9$. This is separated from neighboring local minima $L = 10$ and $L = 8$ by saddle points which are the energetically unstable stationary solutions $\{9, 10\}$ and $\{8, 9\}$, respectively. At the high-field end of the curve the energy of the local minimum merges with the energy of the saddle point $\{9, 10\}$. In the presence of dissipation this will drive the system toward the neighboring local minimum $L = 10$. Towards lower values of H, we cross the critical field where the multiple vortex solution in the form of a ring, $\{0, 9\}$, becomes energetically favorable (Palacios 1998, Schweigert et al. 1998). There is no barrier separating both solutions and a weak second order phase transition takes place. There are saddle point solutions, $\{0, 8, 9\}$ and $\{0, 9, 10\}$, separating the local minimum $\{0, 9\}$ from the local minima $\{0, 8\}$ and $\{0, 10\}$ on neighboring curves in Fig. 1, respectively. There is also a barrier, $\{0, 1, 9\}$, separating the local minimum $\{0, 9\}$ from the local minimum $\{1, 9\}$ which becomes energetically favorable as we move towards lower fields (see Fig. 2). These structural barriers are typically one or two orders of magnitude smaller than those separating states with different Q and the $\{0, 9\}$ solution could jump to the solution $\{1, 9\}$ in the presence of fluctuations (we have chosen this possibility for Fig. 1). At the low-field end of the curve, the saddle point separating the local minimum $\{1, 9\}$ from the local minimum $\{1, 8\}$ (i.e., the $\{1, 8, 9\}$ stationary solution) merges with the local minimum $\{1, 9\}$ and this becomes energetically unstable in favor of $\{1, 8\}$ (the same applies to the solution $\{0, 9\}$).

The height of the barriers, either for vortex entrance (high-field side) or vortex escape (low-field side), is several orders of magnitude the experimental temperature, becoming comparable only at both extremes of each curve for any Q. The experimental magnetization curves (Geim et al. 1998) are similar to the theoretical ones, but there are important discrepancies. The derivative of the magnetization with respect to the field changes sign close to the low- and high-field ends of the curves. Furthermore, neighboring curves even get to cross each other (see Fig. 1). This is never seen in the experiment (Geim et al. 1998) which seems to indicate that a vortex can escape or enter the disk before the barrier disappears and even before the thermal activation is effective. The macroscopic quantum tunneling of a single vortex could explain this discrepancy (Blatter et al. 1991).

Whatever mechanism may be responsible for the vorticity change at the low-field end of the curves, it does not seem to preempt the appearance of the paramagnetic response in the CT measurements. The origin of this paramagnetic response lies simply on the fact that the total supercurrent always ends up reversing the direction in which it flows on decreasing the field before the change in the topological charge takes place. This sign change in the response occurs approximately when the dominant eigenfunction in the

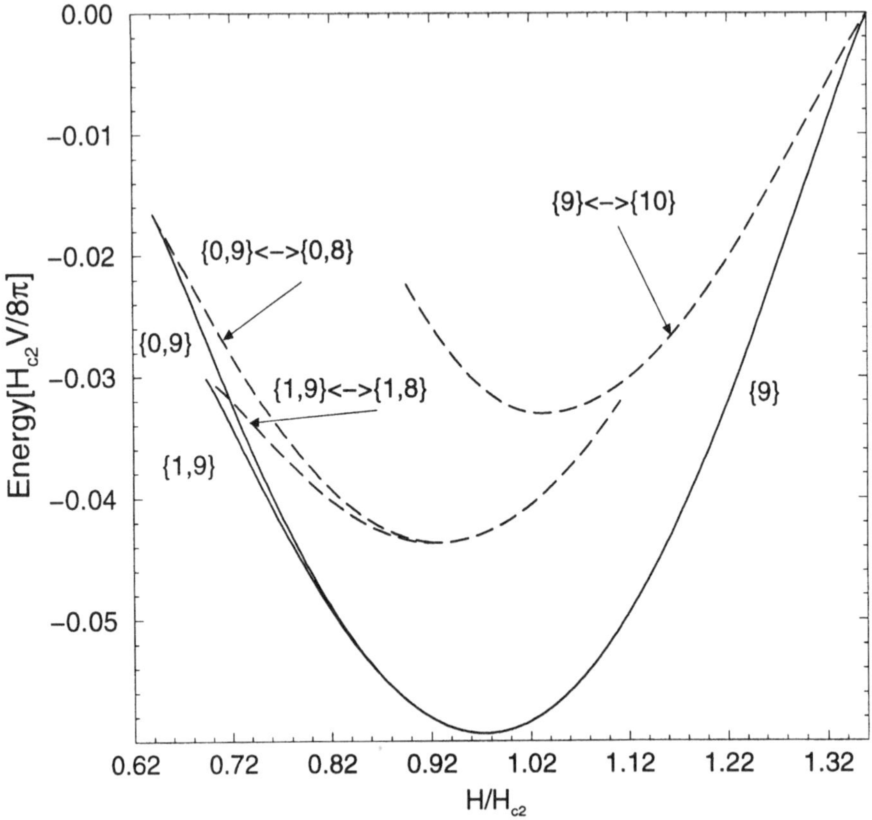

Fig. 2. Energy of the local minimum $Q = 9$ (solid line) compared to the energy of the saddle point (dashed line) that separates it from the closest neighboring local minimum $Q = 10$ (high field side) and $Q = 8$ (low-field side). Notice the two possible local minima at low fields described in the text and their corresponding escape barriers.

expansion of the order parameter, $L_N = Q$, crosses the minimum of the band structure $\epsilon_L(B)$ and reverses its group velocity (Palacios 1998). Notice that detector effects need not be invoked (Schweigert et al. 1998) to account for this paramagnetic response.

Even if the energetic instability on the diamagnetic side is preempted by some relaxation process as discussed above, the magnetization associated with the global minimum (thick line in Fig. 1) is not likely to be observed for increasing field without intentional relaxation. It has been suggested in the literature that surface roughness is responsible for the destruction of the saddle points associated to the surface barrier. The saddle points preventing the escape or entrance of vortices have the same origin and surface roughness should affect them similarly. Thus, in our view, there are no reasons for the system to follow the ground state and it is expected to continue along the metastability curve until the vortex entrance rate increases to typical measurement times.

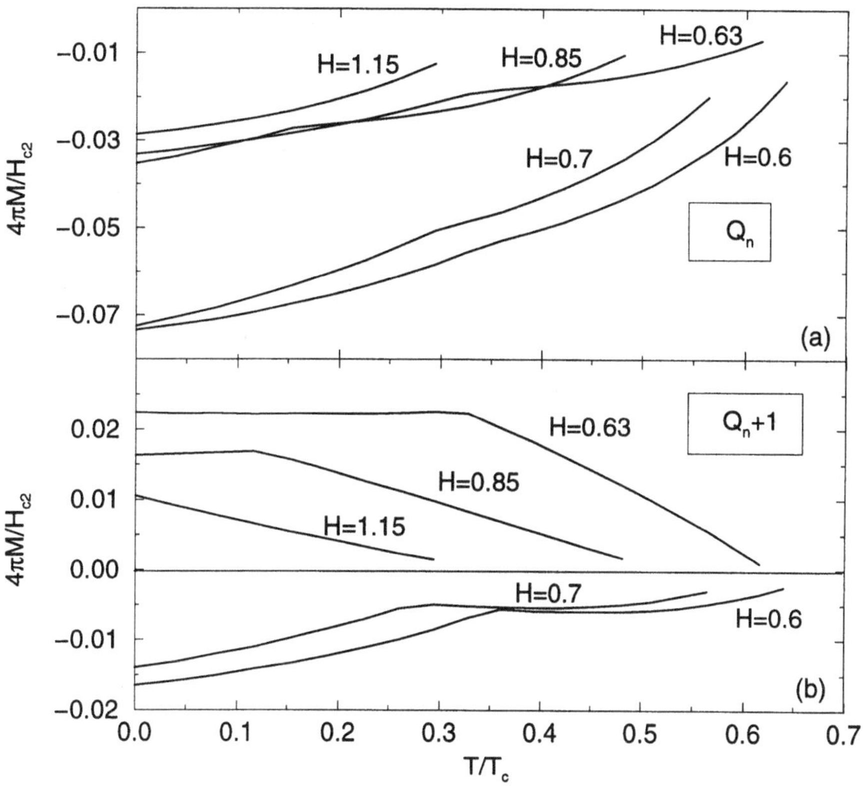

Fig. 3. (a) FC magnetization for a disk of radius $R = 5\xi(0)$ and $\kappa = 1$ at different values of the external magnetic field. Each curve corresponds to the topological charge Q_n of the giant vortex that nucleates at the highest critical temperature. (b) The same as in (a), but for a topological charge $Q_n + 1$.

Finally, we would like to comment on the FC measurements. The current understanding of the FC results is summarized in a work by Moshchalkov et al. (1997) which attributes the FC paramagnetic response to a flux-compression phenomenon. Figure 1 shows the magnetization as a function of temperature for different values of the magnetic field. The usual phenomenological temperature scaling of the parameters in the Ginzburg-Landau functional (1) has been considered (Moshchalkov et al. 1997). Each curve in Fig. 1(a) corresponds to the topological charge Q_n of the giant vortex that nucleates at the highest possible critical temperature for each chosen field. Q_n is maintained along the different curves down to $T = 0$ due to presence of the energy barriers discussed above which prevent the change of vorticity. The response is always diamagnetic in clear contrast with the FC data. Figure 3(b) shows magnetization curves for a topological charge $Q_n + 1$. Alternating paramagnetic and diamagnetic behaviors are obtained as a function of H and a low-temperature saturation of the paramagnetic response occurs due to explosion of the giant vortex as suggested by Moshchalkov et al. 1997). This behavior is in remarkable agreement with the FC data (Geim et al. 1998) which seems to suggest that either thermal fluctuations close to the critical temperature or surface roughness favor the nucleation of giant vortices with a higher topological charge than that expected from plain Ginzburg-Landau theory. Notice that, in our approach, the magnetic field is uniform in space which suggests that flux compression (Moshchalkov et al. 1997, Geim et al. 1998) is not essential as far as the existence of paramagnetism is concerned. Still, the origin of the paramagnetic response in the FC measurements remains an open issue.

This work has been funded by MEC of Spain under contract No. PB96-0085.

References

V. V. Moshchalkov *et al.*, Nature **373**, 319 (1995).
A. K. Geim et al. , Nature **390**, 259 (1997).
A. K. Geim et al. , Nature **396**, 144 (1998).
See for instance C. A. Bolle *et al.*, Nature **399**, 43 (1999) and references therein.
D. A. Butts and D. S. Rokshar, Nature **397**, 327 (1999).
C. P. Bean and J. B. Livingston, Phys. Rev. Lett. **12**, 14 (1964); A. L. Fetter, Phys. Rev. B **22**, 1200 (1980).
V. A. Schweigert, F. M. Peeters, and P. S. Deo, Phys. Rev. Lett. **81**, 2783 (1998).
J. J. Palacios, Phys. Rev. B **57**, 10873 (1998); *ibid.*, Phys. Rev. B **58**, 5948 (1998); *ibid.*, Physica B **256-258**, 610 (1998).
See for instance G. Blatter, V. B. Geshkenbein, and V. M. Vinokur, Phys. Rev. Lett. **66**, 3297 (1991) and references therein.
V. V. Moshchalkov, X. G. Qiu, and V. Bruyndoncx, Phys. Rev. B **55**, 11 793 (1997).

Novel 0D Devices: Carbon-Nanotube Quantum Dots

L. Chico[1,2], M.P. López Sancho[2], and M.C. Muñoz[2]

[1] Departamento de Física Aplicada, Facultad de Ciencias del Medio Ambiente, Universidad de Castilla-La Mancha, Avda. Carlos III s/n, 45071 Toledo, Spain
[2] Instituto de Ciencia de Materiales de Madrid, Consejo Superior de Investigaciones Científicas, Cantoblanco, 28049 Madrid, Spain

Abstract. We have designed a carbon-based quantum dot and a quantum barrier by introducing pentagon-heptagon defects in a carbon nanotube, in a way that is equivalent to combine two carbon-nanotube metal-semiconductor junctions. The quantum dot system has completely confined electronic states, so it behaves as an ideal 0-dimensional (0D) device, presenting well-separated discrete levels. By varying the distance between pentagon-heptagon pairs, the number and characteristics of the discrete levels can be modified. The inclusion of topological defects in a nanotube can also produce a quantum barrier. The study of the properties of such structure throws light onto the complex nature of topological barriers.

1 Introduction

Carbon nanotubes or, more precisely, tubular fullerenes, were imagined by Saito et al. (1992) as structures derived from spherical fullerenes C_{60} and C_{70}, by equatorially adding rings of carbon atoms (for a review, see Ebbesen, 1996). At the end of 1991 a theoretical work by Mintmire, Dunlap, & White (1992) predicted that rolling up a graphene sheet in a cilyndrical shape of nanometric diameter could yield for certain geometries a quasi-one-dimensional system with a carrier density similar to that of metals, and much higher than that of graphite. At the same time, Iijima (1991) found tubular fullerenes analyzing by electron microscopy the deposit on the graphite electrodes employed in the synthesis of C_{60}. These tubes consisted of multiple shells, where carbon atoms were arranged with respect to the tube axis with various degrees of helicity within the same nanotube. Band structure calculations by Hamada, Sawada, & Oshiyama (1992) predicted that an ideal single-wall nanotube (SWNT) would exhibit variations in electronic transport from metallic to semiconducting, depending on the diameter and chirality of the nanotube. Such theoretical predictions could not be verified until 1996, when Thess et al. (1996) achieved the production of SWNTs at high yield and structural uniformity. Recently, experimental measurements by Wildöer et al. (1998) and Odom et al. (1998) with scanning tunnelling microscopy and spectroscopy on individual SWNTs related directly electronic structure with differences in chirality, confirming the earlier theoretical predictions.

Since a variation of the chirality of the nanotubes may change their electronic character from metallic to semiconducting, the next theoretical step was to consider changes in the helicity within a single tube, thus making a heterostructure on a single molecule. Such a change in chirality can be achieved by the introduction of topological defects as pentagon-heptagon pairs, allowing the formation of a heterojunction between two semiconductors, two metals or between a semiconductor and a metallic tube (Chico et al. 1996; Saito, Dresselhaus, & Dresselhaus 1996; Lambin et al. 1995). A year after these theoretical proposals, Collins et al. (1997) experimentally found a carbon nanotube which presented a near-perfect rectifying behavior consistent with the existence of localized on-tube devices, as theoretically predicted. Furthermore, Tans et al. (1998) reported a field-effect transistor with only one SWNT as the active element, demonstrating that devices based on single molecules are possible. These kind of devices based on single nanotubes could be the first step towards the molecular electronics based on carbon.

We propose carbon-nanotube-based 0D devices achieved by introducing pentagon-heptagon (5-7) defects on the hexagonal bond network of an otherwise perfect nanotube, in a fashion that can be viewed as the combination of two carbon-nanotube metal-semiconductor junctions. One of the possibilities would be to sandwich a slice of a metallic tube between two semiinfinite semiconducting nanotubes. Such a system behaves as a quantum dot (Chico, López Sancho, & Muñoz 1998). The other case would be to join a semiconducting nanotube slab to a pair of metallic tubes, building a quantum barrier. Both systems have 0-dimensional features, and their properties are intimately related to the intringuing characteristics of the topological barriers.

In fact, transport measurements performed by Bockrath et al. (1997) in SWNT ropes show evidence of resonant tunnelling through quantized levels, and quantum-dot (QD) behavior has also been observed in SWNTs (Tans et al. 1997). In these experiments energy quantization is due to the presence of the metallic contacts needed to perform the measurements; so the origin of such quantized levels is different to that of the quantized states induced by the presence of topological defects, which are proposed here.

2 Model and method

As mentioned above, the 0-dimensional structures are constructed by combining two carbon-nanotube metal-semiconductor junctions. As a model system, we have chosen for the building blocks of the 0D-structures a (6,4) and a (5,5) tubes. The (6,4) tube is a semiconductor with a 1.2 eV gap, whereas the (5,5) tube is metallic. They have large enough radii so that curvature-induced hybridization effects are not important. A (6,4) tube can be joined to a (5,5) tube forming a pentagon-heptagon pair defect at the interface between them. So joining a slab of a (5,5) metallic tube to two semiconducting (6,4) tubes at its ends amounts to place two antiparallel pentagon-heptagon defects at the ends of the (5,5) metallic slab, as shown in Fig. 1. This system behaves as a quantum dot, as we will show later. We denote the QD structures constructed in this way as $(6,4)/m(5,5)/(6,4)$, where m denotes the number of (5,5) unit cells placed between the semiinfinite (6,4) nanotubes.

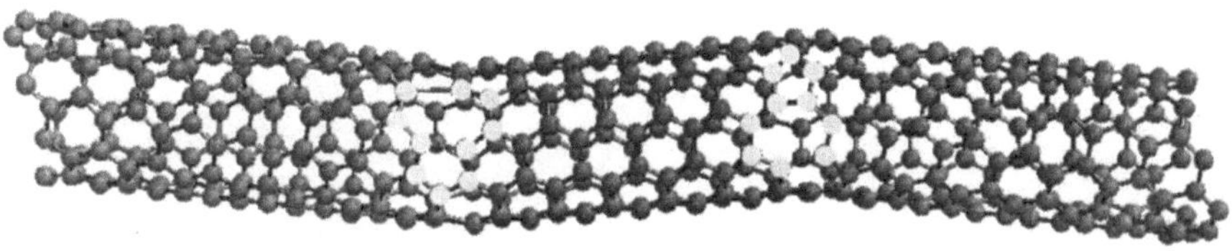

Fig. 1. Atomic structure of a (6,4)/6(5,5)/(6,4) carbon-nanotube quantum dot. Light-grey atoms form part of the pentagon-heptagon defects.

There are 20 C atoms in the unit cell of the (5,5) metallic tubule, while the number increases to 152 for the (6,4) semiconductor tube. All C atoms contained in the structures have 3-fold coordination, but the lattice connectivity changes at the (5,5)/(6,4) interfaces. The choice of the constituent tubes were made in order to prevent the occurrence of interface-induced states in the semiconductor gap. Interface states, originated by topological defects, have been predicted in various nanotube junctions (Chico et al. 1996).

The $(6,4)$ and $(5,5)$ tubules are described by a π-band TB approximation. Only nearest-neighbor interactions are considered, taking the hopping parameter $V_{pp\pi} = -2.66$ eV (Blase et al. 1994). For the energy range of interest, around the $(6,4)$ semiconductor gap, this model describes adequately the electronic structure of the nanotubes. The QD electronic properties are determined using a Green function (GF) matching approach (García-Moliner et al. 1992) along with a transfer matrix technique (López Sancho et al. 1984). These methods allow us to calculate the exact GF of any complex finite or infinite system formed by joining different media, avoiding size or supercell effects which appear in slab calculations. The GF of the complete structure is obtained from the bulk GFs of the constituent tubes and the interface hamiltonians. Localized and extended states are directly obtained from the GF.

3 Confined QD electronic states

Fig. 2 shows the dispersion relations of the $(5,5)$ and $(6,4)$ infinite nanotubes around the Fermi level, E_F, which is at 0 eV. The energy bands display the electron-hole symmetry characteristic of perfect nanotubes within the π-band approximation. In the $(5,5)$ tube the two bands crossing at E_F are responsible of its metallic character, while in the (6,4) tubule a gap of 1.2 eV develops at this energy range.

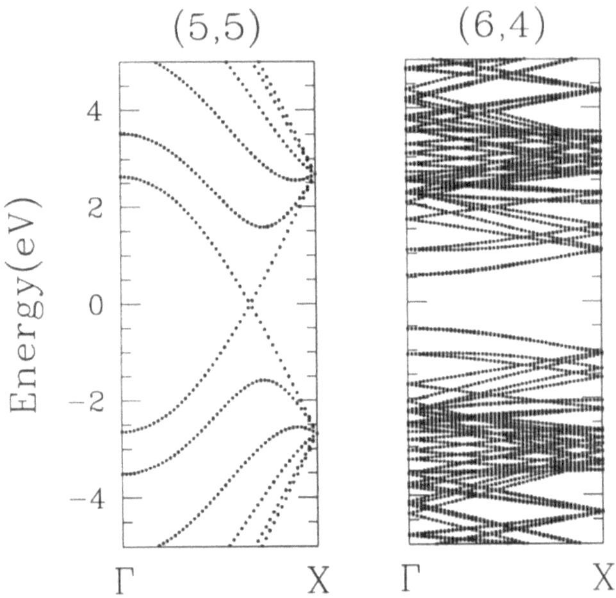

Fig. 2. Band structures of the (5,5) and (6,4) carbon nanotubes around the Fermi energy.

First, we will analize the $(6,4)/m(5,5)/(6,4)$ system. In Figs. 3 and 4 we represent the averaged local density of states (LDOS) at the unit cells forming the $(6,4)/(5,5)$ left interface for the $m = 6$ and $m = 7$ quantum dots respectively. Because of the arrangement of the 5-7 pair defects, the averaged LDOSs are symmetric with respect to the center of the m(5,5) slab, so it suffices to show the LDOS at one interface.

Both $m = 6$ and $m = 7$ QD structures present four sharp peaks in the energy range of the (6,4) semiconductor gap. Due to the fact that topological defects break the electron-hole symmetry, their energy positions are not symmetric with respect to the Fermi level ($E_F = 0$ eV) (Tamura et al. 1994). All the discrete states found in both structures show larger averaged LDOS at the interface unit cell corresponding to the (5,5) nanotube. Since the LDOS represents the squared amplitude of the wavefunction, Figs. 3 and 4 also illustrate the spatial localization of the discrete levels, which are mostly confined within the (5,5) tube, although their wavefunction amplitude spreads out to the (6,4) cells forming the junctions, as shown in the figures. The discrete nature and spatial localization of the (5,5) tube-derived states un-

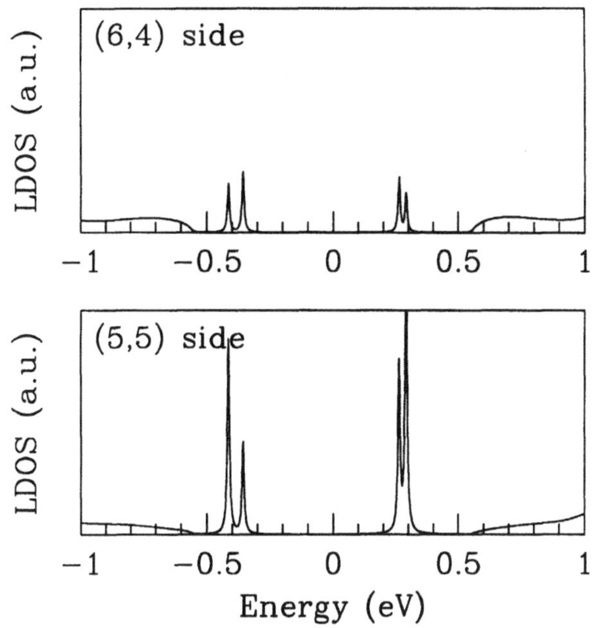

Fig. 3. LDOS at an interface of the (6,4)/6(5,5)/(6,4) QD averaged to the respective unit cells.

ambigously demonstrates that they are produced by quantum confinement. Therefore, both the $(6,4)/6(5,5)/(6,4)$ and $(6,4)/7(5,5)/(6,4)$ structures behave as quantum dots. But, at variance with ordinary QDs made of compositionally different materials, in a carbon-based QD, quantum confinement is due to changes in the network topology.

As shown in Figs. 3 and 4, the number of discrete states appearing in the $(6,4)$ semiconductor gap is the same for both $m = 6$ and $m = 7$ structures. However, the energy levels and the energy spacing between levels depend on m. Due to the complex nature of the confining barriers, it is essential to take into account the complete structure to find out the energy distribution of discrete states, given that the type and arrangement of the defects causing the barriers may influence the relative energy of the discrete QD levels. In our approach, we calculate the GF of the infinite system, which allows us to consider not only the energy dependence on m, but also the precise nature of the confining barriers. Nevertheless, in order to understand the origin of the QD states, one could in principle model the quantum dot as an m(5,5) slab surrounded by infinite barriers. In such a case, the allowed states would

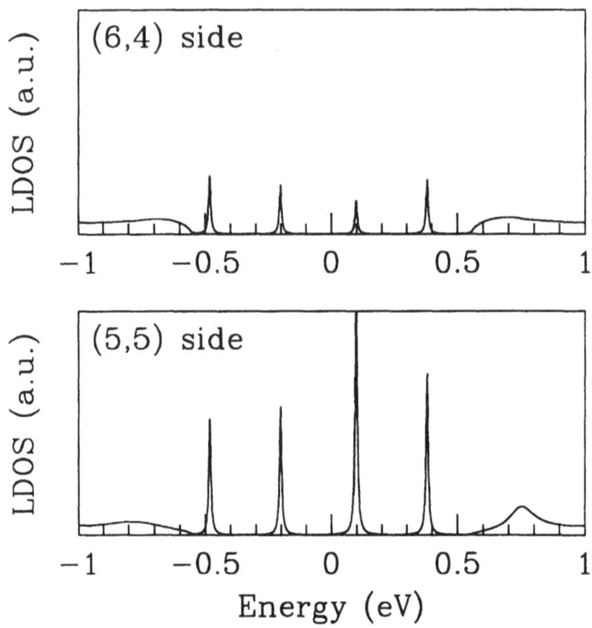

Fig. 4. LDOS at an interface of the (6,4)/7(5,5)/(6,4) QD averaged to the respective unit cells.

correspond to those bulk levels of the perfect (5,5) tube which result from sampling the bands at m evenly-spaced k-points. For a single band this size-quantization condition gives rise to two series of quantum well-like discrete states consecutively ordered and starting each series from a band edge. In the energy range of interest, i.e. around the Fermi energy, the perfect (5,5) nanotube presents two π-bands which cross at $k_F = \frac{2}{3}\frac{\pi}{a}$. Therefore, the QD discrete levels shown in Figs. 3 and 4 arise from quantization of these two π-bands.

We have investigated the size dependence of the QD energy spectrum, up to $m = 25$. As can be expected, when m increases, the number of discrete levels increases. However, there is not a simple relationship between the number of QD states appearing in the $(6,4)$ semiconductor energy gap and the value of m, since the series of quantum states stemming from two π-bands are intermingled. Electrons and holes have an analogous size dependence, although their confinement is different. In fact, the energies of electron-derived states are closer to the Fermi level -see Fig. 3- than those of the corresponding hole states. This indicates that the energy barriers caused by the 5-7 pair defects are higher for electrons. Furthermore, due to the singular electronic properties of carbon nanotubes, the QD energy spectra of all the structures investigated belong to one of two different categories, either the QD levels appear in doublets with almost degenerate energies, or the energy spacing between discrete levels is almost constant. The first type occurs for m multiple of 3 and is due to the fact that the QD states arise from k-points almost equidistant from $k_F = \frac{2}{3}\frac{\pi}{a}$ where the two π-bands cross in the perfect (5,5) nanotube. The second case has its origin in the linear dispersion relation of the two π-bands, which results in almost constant energy differences between contiguous states. In the range of m values investigated, the energy spacing between discrete levels is $\approx$ 150 meV, much larger than the thermal broadening at room temperature. Moreover, the wavefunction of discrete states remains localized in the (5,5) slab for large values of m. Thus, in the QD structures proposed in this work, electrons are 3-dimensionally confined in strongly localized discrete bound states, which open enormous possibilities for the technological applications of carbon-nanotube-based quantum dots.

4 Carbon-nanotube barriers

Now we consider a carbon-nanotube structure in which a slice of a semiconducting tube is placed between two metallic semiinfinite tubes. Choosing the same building blocks employed for the QD studied above, we denote such a structure as (5,5)/m(6,4)/(5,5), where m indicates the number of unit cells of the (6,4) tube forming the barrier. This can be achieved in a similar way to that employed for the QD system, i.e., by joining the unit cells of the two kinds of tubes, so two antiparallel 5-7 pair defects are formed at the interfaces between the semiconducting slice and the metallic tubes.

The (6,4) tube unit cell length along the tube axis is 18.6 Å, so a barrier composed of only one (6,4) unit cell has a considerable size, given that the gap of this tube is 1.2 eV. We have calculated the conductance of a (5,5)/1(6,4)/(5,5) system using the Landauer approach. Within this scheme, the conductance of a multichannel system is given by

$$\Gamma = \frac{2e^2}{h}\mathrm{Tr}(t^\dagger t)$$

where t is the transmission matrix from either the left or the right, as defined by Fisher and Lee (1981). For a perfect (5,5) metallic nanotube, the conductance given by this expression is $2\frac{2e^2}{h}$, since this tube has two channels at the Fermi energy. In Fig. 5 we show the conductance of the (5,5)/1(6,4)/(5,5) structure as a function of the energy.

It is remarkable that the presence of a barrier of almost 20 Å length and 0.5 eV height for electrons at the Fermi level does not suppress the conductance completely. This is due to the complex nature of the topological barriers formed by the 5-7 defects. In a normal heterostructure formed by

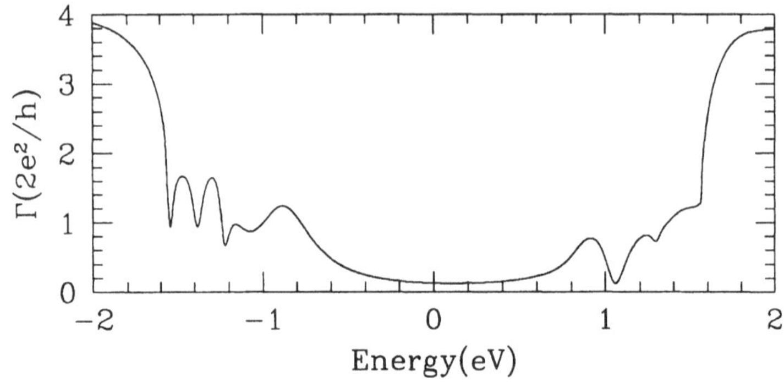

Fig. 5. Conductance of a (5,5)/1(6,4)/(5,5)) barrier.

the combination of compositionally different materials, the change from one medium to other is local, whereas in these carbon-nanotube structures, made with only one element, the change from one type of nanotube to other is noticed when comparing the position of second-nearest neighbors, given that the three-fold coordination of the first nearest neighbors is preserved in the 5-7 defects. In this sense, the topological barriers have a complex, non-local nature, which manifests in the transport properties of these systems.

5 Outlook

The singular electronic properties of carbon-based 0D systems may be important in future nanoelectronics: a carbon nanotube QD with metallic contacts can behave as a one-electron transistor, where Coulomb blockade effects due to occupation of these strongly localized discrete levels are expected. Furthermore, the atomic-like character of their DOS should result in a great enhancement of the energy selectivity and resonance effects in optical transitions, while the almost constant separation of energy levels can lead to nonlinear optical properties. The properties of the topological barriers have to be explored; their non-local nature might be of use for building coupled quantum dots based on carbon nanotubes, which could be of interest not only for optical applications, but for the design of quantum computers.

References

Blase X., Benedict L.X., Shirley E.L. & Louie S.G. M.L. (1994): Phys. Rev. Lett. 72, 1878

Bockrath M., Cobden D.H., McEuen P., Chopra N.G., Zettl A., Thess A., & Smalley R.E. (1997): Science 275, 100
Chico L., Crespi V., Benedict L.X., Louie S.G., & Cohen M.L. (1996): Phys. Rev. Lett. 76, 971
Chico L., López Sancho, M.P,& Muñoz M.C. (1998): Phys. Rev. Lett. 81, 1278
Collins P.G., Zettl A., Bando H., Thess A., & Smalley R.E. (1997): Nature 386, 474
Ebbesen, T.W. (1996) Physics Today, 26
Fisher, D. S. and Lee, P.A. (1981): Phys. Rev. B 23, 6851
García-Moliner F. and Velasco V.R. (1992): *Theory of Single and Multiple Interfaces*, World Scientific, Singapore
Hamada N. ,Sawada S., & Oshiyama A. (1992): Phys. Rev. Lett. 68, 1579
Iijima, S. (1991): Nature 354, 56
Lambin P. et al. (1995): Chem. Phys. Lett. 245, 85
López Sancho M.P., López Sancho J.M., and Rubio J. (1984): J. Phys. F: Metal Phys. 14, 1205
Mintmire, J.W., Dunlap, B.I., & White, C.T. (1992): Phys. Rev. Lett. 68, 631
Saito, R, Dresselhaus, G.D., & Dresselhaus M.S. (1996): Phys. Rev. B 53, 2044
Tamura R., & Tsukada M. (1994): Phys. Rev. B 49, 7697
Tans, S.J. et al. (1997): Nature 386, 474
Thess, A. et al. (1996): Science 273, 583
Wildöer, J.W.G. et al. (1998): Nature 391, 59
Odom, T.W. et al. (1998): Nature 391, 62

Atomic-Size Conductors

Nicolás Agraït

Laboratorio de Bajas Temperaturas
Departamento de Física de la Materia Condensada C-III
Universidad Autónoma de Madrid, E-28049 Madrid, Spain
Instituto Universitario de Ciencia de Materiales "Nicolás Cabrera"

Abstract. The electronic transport properties of a metallic conductor of atomic dimensions differ from those of other mesoscopic systems: the wave functions of the electrons not only maintain phase coherence over the whole system but are also very sensitive to the local atomic arrangement. As a result electronic and mechanical properties are strongly coupled in these systems. The chemical nature of the atoms of the system is also of fundamental importance in determining the properties of the system.

1 Introduction

A metallic system of atomic dimensions differ from other mesoscopic conductors because not only the dimensions of the contact L is smaller than the elastic and inelastic mean free paths of the electrons, l_e and l_i, but also L is of the order of electron wavelength, which in metals is typically of the order of an atomic diameter. At this scale a full quantum description of transport becomes necessary.

Interest in these systems has been triggered by the experimental posibilities opened by scanning tunneling microscopy (STM) and related techniques like mechanically controlled break-junctions (MCBJ)(for recent review see van Ruitenbeek 1997). It has been shown that atomic contacts, or more precisely, atomic constrictions, and even atomic chains of up to seven atoms in length can be formed (Yanson et al 1998, Onishi et al 1998). For these small systems electronic transport properties cannot be considered independently of mechanical properties since these play an essential role in the formation of the contacts.

We will review some of the fundamental theoretical and experimental aspects of metallic constrictions.

2 The conductance of a constriction

Depending on the size of the constriction L with respect to the electronic mean free path ℓ, three different regimes are possible.

If the constriction is much larger than ℓ, the electrons move through the constriction diffusively and the conductance depends on the resistivity of the material ρ. This is Maxwell's regime. In the limiting case of a contact the conductance is $G_M = 2a/\rho$, where a is the contact radius.

If the constriction is smaller than ℓ, the electrons will traverse it ballistically and the conductance, which can be computed semiclassically, is independent of ℓ and ρ. This is Sharvin's regime. For a contact, $G_S = 3\pi a^2/4\rho\ell = (2e^2/h)(\pi a/\lambda_F)^2$, where λ_F is the Fermi wavelength (Sharvin 1965).

If the constrictions is of atomic dimensions, in the case of a metal $\lambda_F \sim a$, we are in the quantum regime and the conductance must be computed using Schrödinger equation. The simplest way is to use a free-electron model. In this model the geometry of the constriction is modelled as a hard-wall potential and the conductance is computed in terms the scattering transmission of each eigenmode of the constriction. The conductance is computed through Landauer formula (Landauer 1970) in terms of the transmission of the quantum channels of the contact

$$G = \frac{2e^2}{h} \sum_n T_n, \tag{1}$$

where T_n is the transmission of the n channel.

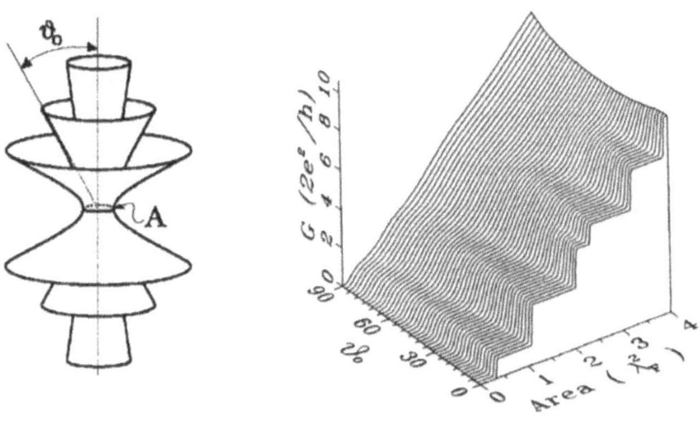

Fig. 1. The conductance of a 3d axisymmetric hyperbolic constriction for different aperture angles.

The results for 3d constrictions (Torres et al 1994) are shown in Fig. 1. For relatively long constrictions the conductance varies in a stepwise manner as the contact size is varied continuously. The plateaus are at integer numbers of the quantum of conductance $G_0 = 2e^2/h$. For shorter contacts this structure is almost entirely washed out: the conductance presents very small oscillations rather than well defined steps. This situation is similar to that of 2DEG (Tekman and Ciraci 1991). For larger contacts the semiclassical Sharvin formula is recovered.

The free electron model has some limitations that become clear for contacts of atomic dimensions. In these cases the constriction is rough in the scale of the wavelength of the electrons and its dimensions cannot be varied continuously. A tight-binding method, which makes possible to define atomic structures and can be combined with non-equilibrium Green's functions methods to calculate transport properties, is more adequate. The electronic structure (that is the band structure or chemistry) of the different materials can also be taken into account easily (Cuevas et al 1998).

3 Experimental technique

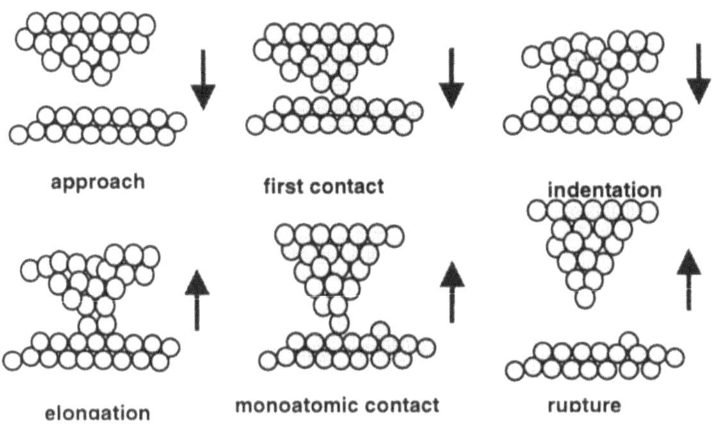

Fig. 2. Formation of an atomic-size contact.

Metallic constrictions of atomic size can be fabricated using scanning tunneling microscopy (STM), or mechanically controlled break junctions (MCBJ) (for a recent review see van Ruitenbeek 1997). In the second method a thin wire with a notch is stretched in a controlled manner (picometer precision). The wire deforms plastically at its weaker spot (the notch), and eventually breaks. If the pulling is performed in a clean enviroment (UHV) the junction will stay clean and the contact could be reconstituted by reapproaching the freshly exposed surfaces. The STM experiment is similar, but since at the start the tip and sample are separeted, the process starts by first making a large contact. This contact is then pulled as in the MCBJ case (see Fig.2). The main advantages of STM with respect to MCBJ are its versatility, the posibility of changing the position of the contact to a different spot on the sample, and the possibility of having heterogeneous contacts by making tip and sample of different materials. The weakest points of STM are the possible contamination of the contacting surfaces before contact, and the mechanical stability of the junction. For certain materials (Au, Al, Pb) making and breaking repeatedly the contact solves this problem: the freshly exposed surfaces are clean, as in the MCBJ. The problem of mechanical stability can be overcome with special design for the STM.

Most of the experiments in atomic constrictions are performed at low temperatures (liquid helium). At these temperatures contacts can be very stable since there is no thermal drift, and electronic properties are sharply defined.

Since the constrictions are formed by plastic deformation, the mechanical properties of the constriction cannot be completely separated from the transport properties. Besides the constriction can be in a state of high stress and this affects transport (see below). Mechanical properties, like the force to form a contact or its elastic constant can be measured by mounting either the tip or the sample on some kind of force sensor, whose compliance must be much smaller than that of the constriction in order not to affect the measurements (Agraït et al 1995, Rubio et al 1996). A rigid cantilever (far more rigid than those use in conventional AFM) is necessary (Fig. 3).

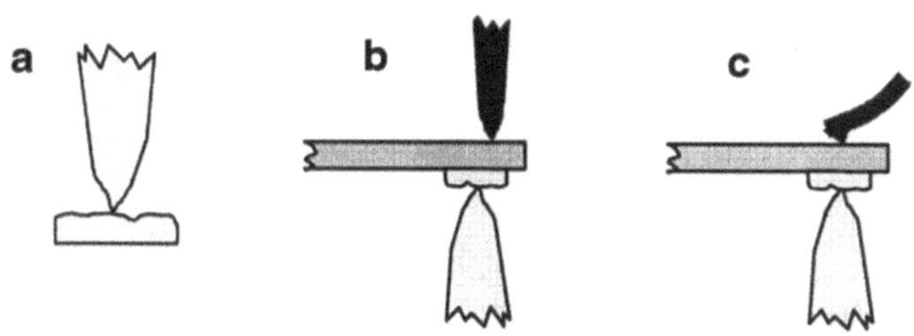

Fig. 3. Different STM setups: (a) Conventional STM; (b) STM with STM as a force sensor; (c) STM with AFM as a force sensor.

4 Large constrictions

Understanding the experimental results for relatively large constrictions (of the order of 100 quantum units or about a hundred atoms in cross-section) is essential for understanding the experimental results on smaller contacts. At this scale the mechanical properties show some differences with macroscopic properties while the electronic properties can be described semiclassically.

In Fig. 4 we can see the conductance and force vs displacement for a constriction of relatively large size. As the constriction is pulled the conductance decreases stepwise as the cross-section becomes narrower (Agraït et al 1995). The cause of these steps is readily seen in the force vs displacement curves: the plateaus in conductance correspond to linear elastic changes in force, while the jumps in conductance are related to irreversible mechanical relaxations, as demonstrated in molecular dynamics simulations (Landman et al 1990). At any given instant in the evolution of the constriction, the conductance gives the minimal cross-section and the slope of the elastic stages gives the elastic constant of the constriction. For any given constriction the elastic constant of the constriction depend on the elastic properties of the material (Young's modulus) and the geometry of the constriction (see Fig. 5). The local crystalline orientation is also important since the elastic properties are quite anisotropic.

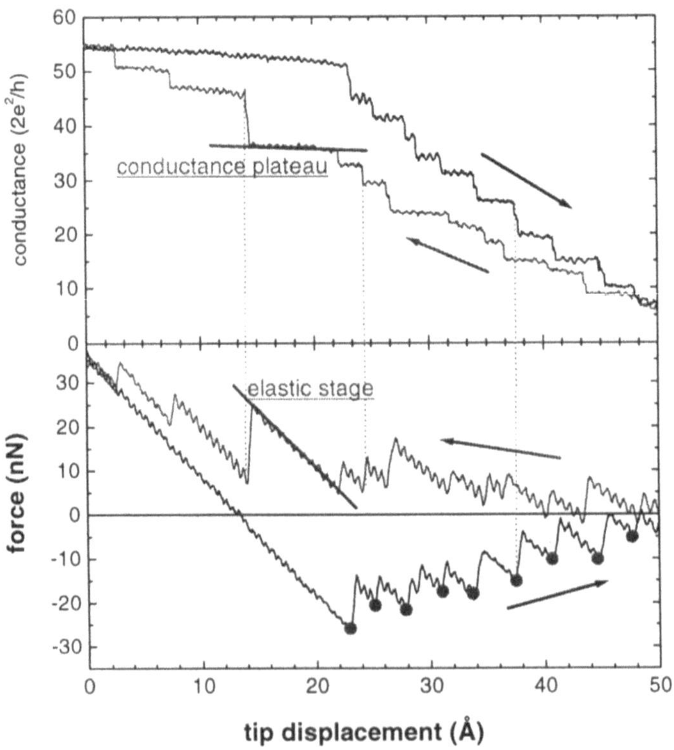

Fig. 4. Force and conductance as the constriction is stretched, for a relatively large Au constrictions.

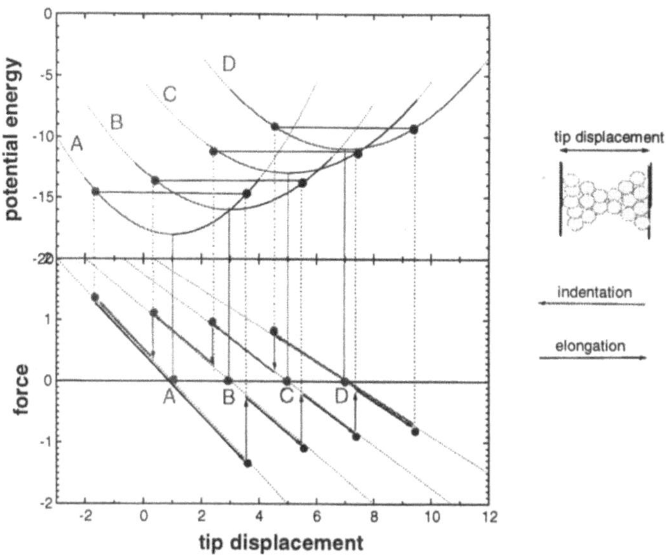

Fig. 5. Simple model for contact deformation: deformation about the equilibrium configurations, A, B, C, and D, is elastic, with a different elastic constant for each configuration.

A given configuration of the constriction deforms elastically up to a certain maximal force. Beyond this force, which can be related to a yielding strength, the constriction changes to a new configuration, that is, it deforms plastically. The yielding strength of a microscopic constriction is found to be many times larger (about 20 times for Au and Pb) than that of a macroscopic constriction. These values are of the order of magnitude expected for the ideal strength of a solid in the absence of dislocations (Agraït et al 1994, Agraït et al 1996). This mechanical properties of contacts and metallic structures of very small dimensions are of great interest, both applied and fundamental, since many important phenomena like adhesion, friction, involve macroscopic bodies in contact which involve regions of microscopic dimensions.

From the electronic point of view, even relatively large contacts cannot be completely understood in terms of the free-electron model: one would expect the slope of a conductance plateau to be zero (simplest model) or rather slightly negative (downwards as constriction is stretched) since materials typically contract as they are stretched (this is given by Poisson modulus). However in Fig. 6, the slope is clearly material dependent (upwards for Al, downwards for Au). This is due to the different electronic nature of the materials Sánchez-Portal et al 1997. Note that this deviation from the free-electron model is important even simple metals like Au and Al.

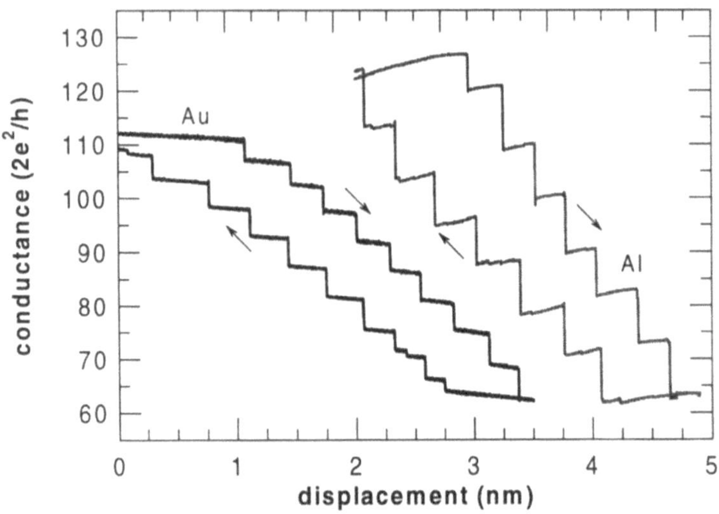

Fig. 6. The conductance vs displacement curves are characteristic of each material. In Al the slope of the plateaus is positive, while in Au it is negative.

5 Shape of the constriction

The shape of the constrictions formed experimentally can be deduced from the conductance measurements, using the fact that the conductance is related to the narrowest cross section of the constriction and that this magnitude is measured as a function of the relative tip-sample displacement. A simple model that assumes that only this narrowest part, which is the weakest, will deform plastically and that volume is conserved (see Fig. 7), gives a reliable shape for the constriction (Untiedt et al 1997).

6 Atomic-size contacts

Fig. 8 shows the conductance and force vs displacement for a contact of atomic size. The contact is pulled up to the rupture point, which is signaled by a sudden decrease of both conductance and force. The smallest constriction before rupture consists, typically, of a single atom. After breaking the junction is in the tunneling regime.

The correlation between plateaus and elastic deformation, and sudden conductance changes and force relaxations is very clear, as in the case of larger contacts. The steps observed in the current should not be confused with steps due to conductance quantization, as argued by some authors (for references, see van Ruitenbeek 1997). They have a mechanical origin: they result from plastic deformation of the constriction.

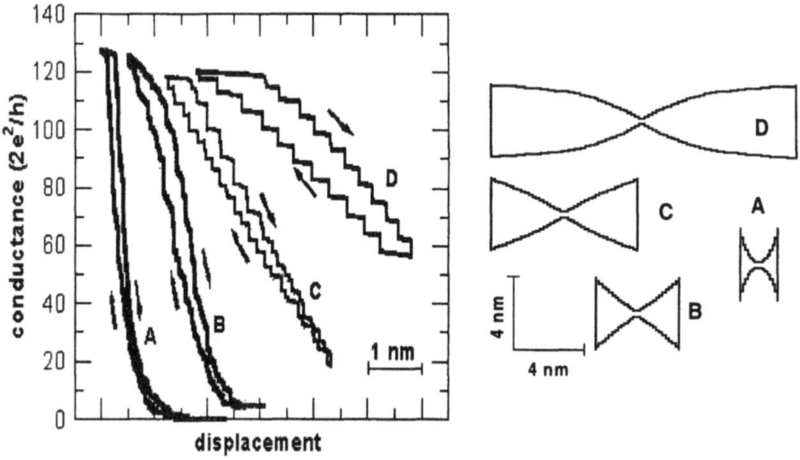

Fig. 7. Evolution of the conductance of constrictions of Au of different shapes.

A situation of special interest is the one-atom contact. In this case, the atomic structure of the constriction is specially simple and amenable to detailed calculation. In Fig. 9 the conductance of several one-atom contacts of Au, Al, and Pb is shown. For Au the conductance is $2e^2/h$ and does not change with strain. However for Al and Pb the conductance changes elastically with strain. This is similar to the case of larger contacts discussed above.

In order to understand this behavior, it is necessary to go beyond the free-electron model. A tight-binding model in which it is possible to take into account the difference in electronic structure of the different materials is more adequate. As shown by Cuevas et al 1998, these variations in the conductance reflect the existence of more than one channnel in the constriction, and the variation of their transmission with strain. The number of quantum channels in a one-atom constriction depends on the chemical nature of the atom: for Au there is only one channel; for Al there are 3; and for Pb there are 3 or 4. This number of possible channels is related to the chemical valence (Scheer et al 1998).

7 Atomic chains

In experiments on gold occasionally during the contact breaking process the atoms in the contact form a stable chain of single atoms, up to 7 atoms long, freely suspended between two electrodes (Yanson et al 1998). The maximum value of the conductance of this chain is $2e^2/h$ showing that it has one single quantum mode (see Fig. 10). Fluctuations to lower values correspond to a reduced transmission probability for this mode as a result of backscattering.

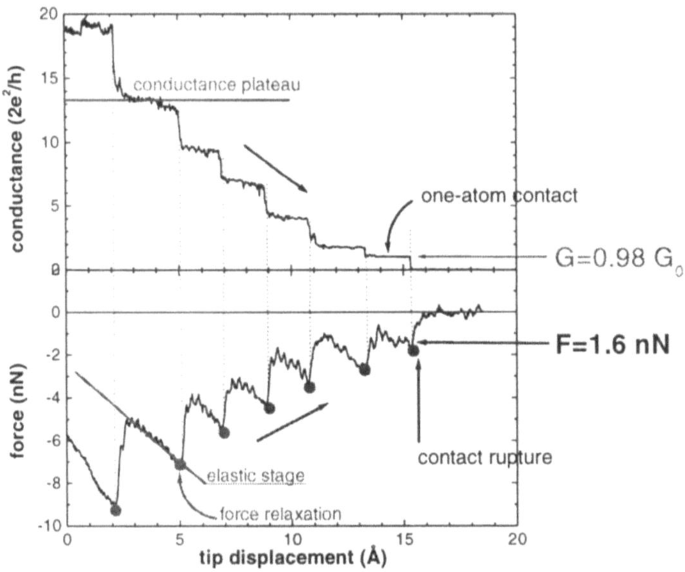

Fig. 8. Force and conductance for a small constriction.

The probability of formation of such chains is low, but once formed they have been held stable for longer than one hour, and can sustain enormous current densities (up to 8×10^{14} A/m^2) since most of the power is dissipated in the electrodes far away from the contact. This makes them suitable candidates for future applications in atomic electronic circuits. Mechanical manipulations of such chains is also possible and may bring new insight into the rapidly developing field of nano-tribology.

Such chains, which constitute the ultimate metallic wires, are true one-dimensional conductors for which theory predicts many unusual properties. The electronic properties are expected to be those of a perfect one-dimensional conductor, where the interaction with the lattice may lead to a Peierls transition, and due to the interaction between electrons the Fermi liquid description should break down. Experiments in course of realization, aim at studying one-dimensional excitations in the chain, such as phonons.

8 Point-contact spectroscopy in atomic-size contacts

At low temperatures the conductance of atomic-size contacts is voltage-dependent due to inelastic backscattering of electrons with phonons. Since at these temperatures there are almost no phonons present, the main process is backscattering by phonon emmission. This is a local probe for phonons,

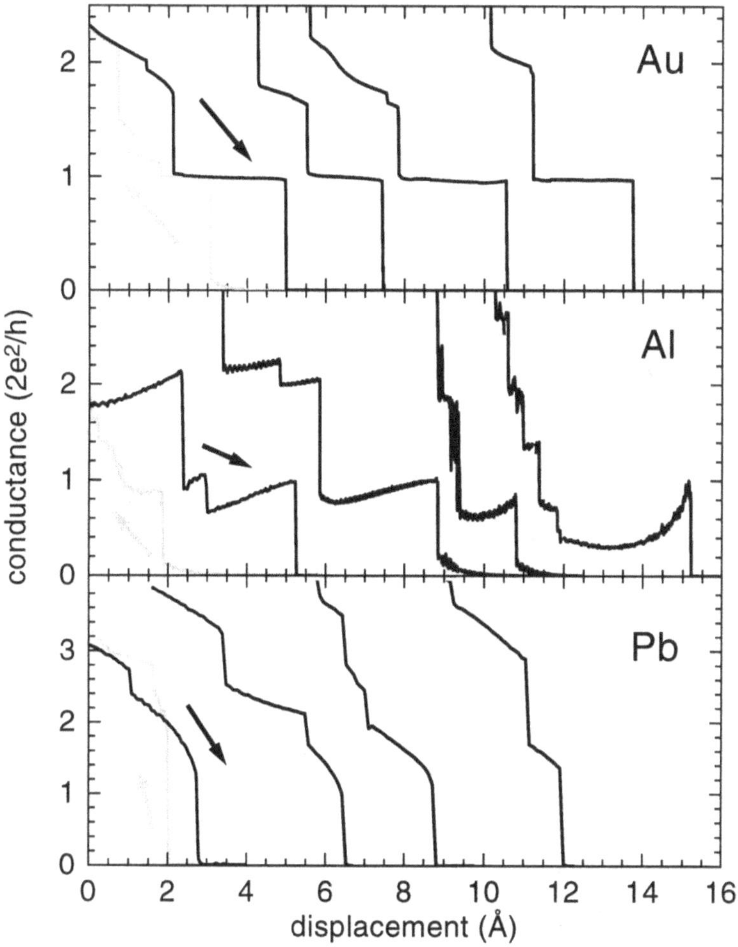

Fig. 9. One-atom contacts of Au, Al, and Pb: the conductance changes in a characteristic way for each material.

because only scattering events taking place in the immediate vicinity of the contact have an effect in the conductance. The inset of Fig. 11 shows the derivative of the conductance which is related to the phonon density of states Jansen et al 1980. The volume probed by the electrons is of atomic dimensions.

Atomic-size contacts typically show also oscillations in conductance Ludolph et al 1999, as those observed in Fig. 11, due to quantum interference with impurities even in samples with low impurity concentrations. Transmission through these small contacts depends not only on the local atomic

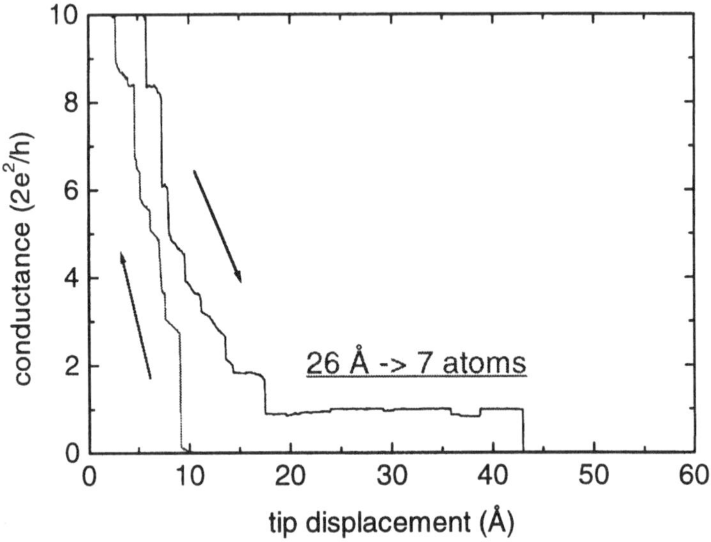

Fig. 10. atomic chain.

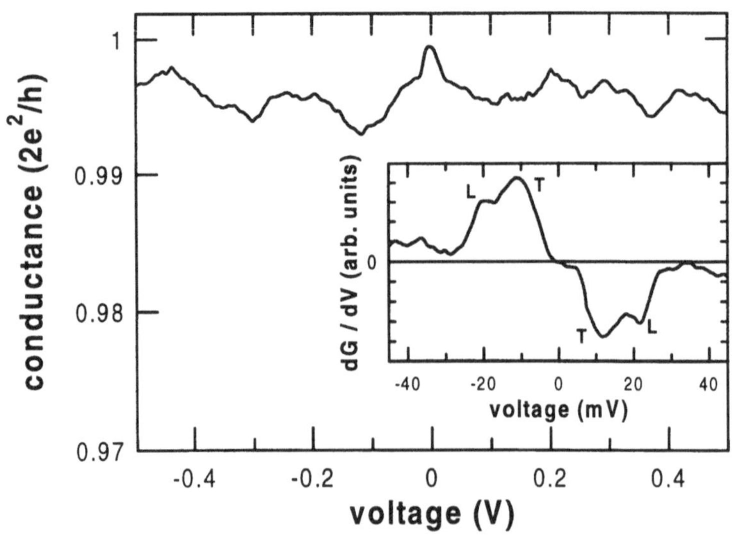

Fig. 11. Conductance and its derivative for a one-atom contact.

structure at the contact but also on the distribution of impurities or defects within a coherence length of the contact.

References

J.M. van Ruitenbeek (1997): in Mesoscopic Electron Transport, NATO-ASI Series E: Appl. Sci., Vol. 345 (eds Sohn, L.L., Kouwenhoven, L.P. & Schön, G.) 549-579 (Kluwer Academic Publishers).
A.I. Yanson, G. Rubio Bollinger, H.E. van den Brom, N. Agraït and J.M. van Ruitenbeek (1998): Nature **395**, 783.
H. Onishi, Y. Kondo, and K. Takayanagi (1998): Nature **395**, 780.
R. Landauer (1970): Phil. Mag. **21**, 863.
J.A. Torres, J.I. Pascual, J.J. Sáenz (1994): Phys. Rev. B **49**, 16581.
E. Tekman and S. Ciraci (1991): Phys. Rev. B **43**, 7145.
Yu. V. Sharvin (1965): Zh. Eksp. Teor. Fiz. **48**, 984 [Sov. Phys. JEPT **21**, 655 (1965)].
J.C. Cuevas, A. Levy Yeyati and A. Martín-Rodero (1998): Phys. Rev. Lett. **80**, 1066.
N. Agraït, G. Rubio and S. Vieira (1995): Phys. Rev. Lett. **74**, 3995.
G. Rubio, N. Agraït, and S. Vieira (1996): Phys. Rev. Lett. **76**, 2302.
N. Agraït, J.G. Rodrigo, G. Rubio, C. Sirvent and S. Vieira (1994): Thin Solid Films **253**, 199.
N. Agraït, G. Rubio and S. Vieira (1996): Langmuir **12**, 4505.
U. Landman, W.D. Luedtke, N.A. Burnham, R.J. Colton (1990): Science **248**, 454.
D. Sánchez-Portal, C. Untiedt, J.J. Sáenz, J.M. Soler, and N. Agraït (1997): Phys. Rev. Lett. **79**, 4198.
C. Untiedt, G. Rubio, S. Vieira and N. Agraït (1997): Phys. Rev. B **56**, 2154.
T.N. Todorov, G.A.D. Briggs adn A.P. Sutton (1993): J. Phys.: Condens. Matter **5**, 2389.
J.C. Cuevas, A. Levy Yeyati, A. Martín-Rodero, G. Rubio Bollinger, C. Untiedt and N. Agraït (1998): Phys. Rev. Lett. **81**, 2990.
E. Scheer, N. Agraït, J.C. Cuevas, A. Levy Yeyati, B. Ludolph, A. Martín-Rodero, G. Rubio Bollinger, J.M. van Ruitenbeek and C. Urbina (1998): Nature **394**, 154.
A.G.M. Jansen, A.P. van Gelder and P. Wyder (1980): J. Phys. C: Solid State Phys. **13**, 6073.
B. Ludolph, M.H. Devoret, D. Esteve, C. Urbina and J.M. van Ruitenbeek (1999): Phys. Rev. Lett. **82**, 1530

Appendix I

Contributions Presented as Posters

Observation of Shell Structure in Sodium Nanowires

A.I. Yanson[1], I.K. Yanson[1,2], and J.M. van Ruitenbeek

[1] Kamerlingh Onnes Laboratorium, Leiden University, PO Box 9504, NL-2300 RA Leiden, The Netherlands

[2] B. Verkin Institute for Low Temperature Physics and Engineering, National Academy of Sciences, 310164, Kharkiv, Ukraine

Abstract. The fact that the rare gas atoms at the end of each row in the periodic table of the elements are exceptionally stable is explained by the closed-shell configuration of their electronic structure. In general, the quantum states of a system of particles in a finite spacial domain form a set of discrete energy eigenvalues, which are usually grouped into bunches of degenerate or close-lying levels, called shells [1]. For fermions this gives rise to a local minimum in the total energy of the system when all states of a given shell are occupied. Shell effects have previously been observed for protons and neutrons in nuclei and for clusters of metal atoms [2, 3, 4]. Here, we report the first observation of a shell effect in an open system, a metal nanowire. When recording the statistical distribution of conductance values observed while pulling off the contact between two bulk sodium metal electrodes, the histogram shows oscillations up to contacts larger than 100 atoms in cross section. The period follows the law expected for the electronic shell-closing effect similar to that determining the "magic numbers" in metal clusters [3, 4].

References

1. Balian, R. & Bloch, C. Distribution of eigenfrequencies for the wave equation in a finite domain: III. Eigenfrequency density oscillations. *Ann. Phys.(N.Y.)* **69**, 76-160 (1972).
2. Bohr, Å. & Mottelson, B.R. *Nuclear Structure*, Vol.II (Benjamin, Reading, MA, 1975).
3. de Heer, W.A. The physics of simple metal cluster: experimental aspects and simple models. *Rev. Mod. Phys.* **65**, 611-676 (1993).
4. Brack, M. The physics of simple metal clusters: self-consistent jellium model and semiclassical approaches. *Rev. Mod. Phys.* **65**, 677-732 (1993).

Strong Charge Fluctuations in the Single-Electron Box: A Quantum Monte Carlo Analysis

C.P. Herrero[1,2], G. Schön[1], and A.D. Zaikin[1,3]

[1] Institut für Theoretische Festkörperphysik, Universität Karlsruhe, D-76128 Karlsruhe, Germany
[2] Instituto de Ciencia de Materiales, C.S.I.C., Cantoblanco, 28049 Madrid, Spain
[3] Lebedev Physics Institute, Leninski pr. 53, 117924 Moscow, Russia

Abstract. Path-integral Monte Carlo simulations have been employed to study strong electron tunneling in the single-electron box (a small metallic island coupled to an electrode by a tunnel junction). Results will be presented for the free energy of this system, as well as for the average charge on the island, as a function of the tunneling strength, the temperature, and an external bias voltage. In much of the parameter range an extrapolation to the ground state ($T = 0$) is possible. Our results for the effective charging energy for strong tunneling will be compared with earlier theoretical predictions and Monte Carlo simulations.

Double Quantum Dots as Detectors of High-Frequency Quantum Noise in Mesoscopic Conductors

R. Aguado[1,2] and L.P. Kouwenhoven[1]

[1] Department of Applied Physics and DIMES, Delft University of Technology, P.O.Box 5046, 2600 GA, Delft, The Netherlands.

[2] Department of Physics and Astronomy, Rutgers University, Piscataway, NJ 08854-8019, USA.

Abstract. In this work we propose a measurement set-up for detecting quantum noise over a wide frequency range using inelastic transitions in a tunable two-level system as a detector. Recently, two device structures were realized that can be used as tunable two-level systems. In a superconducting single-electron transistor a Cooper-pair [1] and in a double quantum dot an electron [2] can make inelastic transitions between two discrete energy states. The transition rate for levels separated by an energy ϵ, is a measure of the spectral density of the fluctuations in the enviroment at a frequency $f = \epsilon/h$. In our set-up the frequency-resolving detector consists of a double quantum dot which is capacitively coupled to the leads of a nearby mesoscopic conductor. The inelastic current through the double quantum dot is calculated in response to equilibrium and non-equilibrium current fluctuations in the nearby conductor, including vacuum fluctuations at very low temperatures. As a specific example, the fluctuations across a quantum point contact are discussed.

References

1. Y. Nakamura, C. D Chen and J. S. Tsai, Phys. Rev. Lett. **79**, 2328(1997).
2. T. Fujisawa *et al*, Science **282**, 932 (1998).

Large Wigner Molecules and Quantum Dots

C.E. Creffield[1], W. Häusler[2], J.H. Jefferson[3], and S. Sarkar[1]

[1] Dept. of Physics, King's College London, Strand, London, WC2R 2LS

[2] Universität Freiburg, Fakultät für Physik, Hermann-Herder-Str. 3, D-79104 Freiburg, Germany

[3] DERA, Electronics Sector, St. Andrews Road, Malvern, Worcs. WR14 3PS

Abstract. The low-lying eigenstates of a system of two electrons confined within a two-dimensional quantum dot with a hard polygonal boundary are obtained by means of exact diagonalisation. The transition from a weakly correlated charge distribution for small dots to a strongly correlated 'Wigner molecule' for large dots is studied, and the behaviour at the crossover is determined. The quasi-crystalline structure found in large dots suggests that the low energy states of the system may be mapped to an effective charge-spin lattice model, as was recently proposed in Ref.1, and the effectiveness of this procedure is investigated by comparison with the results from exact diagonalisation. It is found that the effective model predicts the correct ordering of energy levels, and gives a reasonable first approximation to the size of the energy spacings. The model can be further refined to account for the detailed behaviour of the low energy levels by including spin-flip processes previously neglected in the derivation of the effective model. We conclude that this approach is a useful method to obtain the low energy spectrum of few-electron quantum dots.

References

1. John H. Jefferson and Wolfgang Haeusler, Phys. Rev. B 54, 4936 (1996)

Fundamental Problems for Universal Quantum Computers

T.D. Kieu and M. Danos

CSIRO MST, Private Bag 33, Clayton South MDC, Victoria 3169, AUSTRALIA

Abstract. Several fundamental problems and restrictions for universal quantum computers are pointed out. Firstly, it is shown that the halting of universal quantum computers is incompatible with the constraint of unitarity of the quantum dynamics. Secondly, the consequences of the second law of thermodynamics and the existence of time constants associated with the generation of initial states of quantum systems are shown to impose severe limits on the capabilities of quantum computers.

Kondo Photo-Assisted Transport in Quantum Dots

R. López[1], G. Platero[1], R. Aguado,[1] and C. Tejedor[2]

[1] Instituto de Ciencia de Materiales (CSIC), 28049 Madrid, Spain

[2] Universidad Autonoma de Madrid. Cantoblanco, Spain

Abstract. The competition between electron-electron interaction and quantum mechanical hybridization between the delocalized electrons in a non-magnetic metal and the unpaired electrons of a magnetic impurity leads to the Kondo effect [1]. It has been predicted [2, 3] and recently observed experimentally [4, 5] that the transport at low temperatures through a quantum dot (QD) coupled by tunneling barriers to two leads containing Fermi liquids is governed by a Kondo-like singularity existing in the quasiparticle density of states (DOS) at the Fermi level ϵ_F of the leads. The physics behind such a behavior is well described by the low energy excitations of the Anderson Hamiltonian. When one electron at ϵ_F, becomes scattered by the dot its wave function suffers a phase shift which is proportional to the exact QD displaced charge (DC). As a consequence of this Friedel-Langreth (FL) theorem [7] the conductance takes the value $\mathcal{G} = 2e^2/h\, sin^2(\pi\langle\delta(n_\sigma)\rangle)$ where $\langle\delta(n_\sigma)\rangle$ is the DC [2, 3].
In this work we have generalized the Friedel-Langreth sum rule [8] to the case with AC potentials at finite temperature. We study the photoassisted transport through QD taking especial care in the fulfillment of this generalized rule. New features are found in the transport properties due to the AC potentials. We study the behaviour of the density of states in presence of AC potentials at temperature below and above of the Kondo temperature. The effect of the AC potential on both the quantum dot density of states and the linear conductance shows the importance of using a theory which describes intradot finite interaction and nonperturbative effects at finite temperature. In conclusion, our model considers, for the first time, finite U as well as charge conservation through the extension of the FL sum rule at finite temperature.

References

1. A. C. Hewson: *The Kondo problem to Heavy Fermions* (Cambridge University Press, Cambridge, 1993).
2. T. K. Ng and P. A. Lee, Phys. Rev. Lett., **61**, 1768 (1988).
3. L. I. Glazman and M. E. Raikh, JETP Lett., **47**, 452 (1988).
4. D. Goldhaber-Gordon, *et al*, Nature, **391**, 156 (1998).
5. Sara M. Cronenwett, *et al*, Science, in press.
6. H. Kajueter and G. Kotliar, Phys. Rev. Lett., **77**, 131 (1996).
7. D. C. Langreth, Phys. Rev. **150**, 516 (1966).
8. R. López, R. Aguado, G. Platero, C.Tejedor, Phys.Rev. Lett. **81**, (1998).

Shot Noise and Coherent Multiple Charge Transfer in Superconducting Quantum Point-Contacts

J.C. Cuevas, A. Martín-Rodero, and A.L. Yeyati

Departamento de Física Teórica de la Materia Condensada C-V,
Universidad Autónoma de Madrid, E-28049 Madrid, Spain

Abstract. We analyze the shot noise in a voltage biased superconducting quantum point-contact. Results are presented for the single channel case with arbitrary transmission. In the limit of very low transmission it is found that the effective charge, defined from the noise-current ratio, exhibits a step-like behavior as a function of voltage with well defined plateaus at integer values of the electronic charge. This multiple charge corresponds to the transmitted charge in a Multiple Andreev Reflection (MAR) process. This effect gradually disappears for increasing transmission due to interference between different MAR processes.

References

1. This work has been published in Phys. Rev. Lett. **82**, 4086 (1999).

Evidence for Ising Ferromagnetism and First-Order Phase Transitions in the Two-Dimensional Electron Gas

V. Piazza[1], V. Pellegrini[1], F. Beltram[1], W. Wegscheider[2], M. Bichler[2], T. Jungwirth[3,4], and A.H. MacDonald[3]

[1] Scuola Normale Superiore and Instituto Nazionale per la Fisica della Materia, Pisa, Italy
[2] Walter Schottky Institute, Munich, Germany
[3] Department of Physics, Indiana University, Bloomington, Indiana 47405
[4] Institute of Physics ASCR, Cukrovarnická 10, 162 00 Praha 6, Czech Republic

Abstract. The two-dimensional (2D) electron gas in the quantum Hall regime offers unique possibilities to study the impact of many-body correlations under well-controlled conditions. One of the fields to which quantum Hall effect studies are bringing new microscopic insight is ferromagnetism. At filling factor $\nu = 1$, for instance, the exchange interaction leads to an isotropic ferromagnetic ground state even in the limit of vanishing Zeeman energy[1]. Systems with two nearby 2D electron layers can develop at $\nu = 1$ ground states with spontaneous inter-layer coherence[2]. In these systems a pseudospin index can be associated to the layer and phase transitions between states with different pseudospin polarizations can be driven changing the level alignment of the two layers, e. g. by applying a bias potential or an in-plane magnetic field. This configuration led to the observation of soft collective excitations associated to a continuous (second order) quantum phase transition by resonant inelastic light scattering[3].

While continuous quantum phase transitions were recently observed and extensively studied[1, 4], till now no clear evidence of first-order transitions between Ising-ferromagnetic and paramagnetic ground states was reported. We shall present experimental evidence of such kind of phase transition in the two-dimensional electron gas formed in a wide GaAs/AlGaAs quantum well in the quantum-Hall regime at $\nu = 2$ and 4. Theoretical calculations based on local-density and Hartree-Fock approximations will also be presented providing quantitative agreement with experimental results.

Experimental evidence of a first-order phase transition was found in the longitudinal resistivity as a function of external magnetic and electric fields. Striking features associated to an anomalous evolution of integer quantum Hall minima developed at temperatures below 1 K when two Landau levels with opposite pseudospin (that originate from two different subbands) were brought close to resonance. Remarkably, at even values of n and low temperatures we observed a complete suppression of the quantum-Hall-state excitation gap correlating with the emergence of hysteretical behavior of the diagonal resistivity in up and down sweeps of the magnetic field. At these particular values of n, the two Landau levels with opposite pseudospin have also different spins. These observations demonstrate that crossing of these levels lead to the formation of easy-axis ferromagnetic states associated to a first-order phase transition. Detailed many-body calculations provide an unambiguous identification of the nature of the transition and allowed the determination of the exact properties of the electron ground states involved in the transition. We also investigated the temperature dependence of the observed hysteresis. These measurements displayed evidence of an additional finite-temperature transition at critical temperatures close to $T_C = 900$ mK. These additional results suggest that large domains of particular pseudospin orientation lead to the hysteretic behavior and indicate that the finite-temperature properties of the spin-polarized 2D electron gas are similar to those of a classical Ising ferromagnet.

References

1. S.M. Girvin and A.H. MacDonald in *Perspectives in Quantum Hall Effects* (Wiley, New York, 1996). pp. 161-224.
2. S.Q. Murphy, J.P. Eisenstein, G.S. Bobinger, L.N. Pfeiffer, and K.W. West, Phys. Rev. Lett. **72**, 728 (1994).
3. V. Pellegrini, A. Pinczuk, B. Dennis, A.S. Plaut, L.N. Pfeiffer, and K.W. West, Science **281**, 799 (1998).
4. S. Das Sarma, S. Sachdev, and L. Zheng, Phys. Rev. Lett. **79**, 917 (1997);

Mechanical Properties of Metallic One-Atom Quantum Point Contacts

G.R. Bollinger, N. Agraït, and S. Vieira

Laboratorio de Bajas Temperaturas Dpt. Fisica de la Materia Condensada,
C-III Universidad Autonoma de Madrid 28049-Madrid, Spain

Abstract. It is possible to fabricate nanoscopic metallic structures moving the atoms one by one with a scanning tunneling microscope. This technique can be combined with atomic force microscopy, allowing for a combined study of the electronic transport and the mechanical behaviour during the fabrication of these nanoscopic structures. Our experiments show that there is a close relationship between conductance and applied force during the formation of one-atom quantum point contacts. It is shown taht the mechanical process of stretching an atomic-sized metallic contact proceeds in a discrete sequence of structural transformations involving elastic and yielding stages.

We have also found that it is possible to obtain a chain of several single gold atoms from a one-atom contact of gold [2]. This new structure is a truly one-dimensional conductor and the conductance is independent of its length and has the same value as a single one-atom contact of gold [3]: one conductance quantum ($2e^2/h$).

References

1. G.Rubio, N.Agrait and S.Vieira, *Phys. Rev. Lett.* 76, 2302-5 (1996)
2. A.I. Yanson, G. Rubio Bollinger,H.E. van den Brom, N. Agraït, J.M. van Ruitenbeek, *Nature* 395, pp. 783-785, 1998.
3. E. Scheer, N. Agraït, J.C. Cuevas, A. Levy Yeyati, B. Ludoph, A. Martín-Rodero,
G. Rubio Bollinger, J.M. van Ruitenbeek,C. Urbina. *Nature* 395, pp. 154-157, 1998.

Nanosized Superconducting Constrictions in High Magnetic Fields

H. Suderow[1], E. Bascones[2], W. Belzig[3], S. Vieira[1], and F. Guinea[2]

[1] Laboratorio de Bajas Temperaturas, Universidad Autonoma de Madrid, Spain
[2] Instituto de Ciencia de Materiales de Madrid, CSIC, Spain
[3] Theoretical Physics. TU Delft, Netherlands

Abstract. Nanowires of lead between macroscopic electrodes can be produced experimentally by means of an STM. Magnetic fields higher than the critical field of the bulk may destroy the superconductivity in the electrodes, while the wire remains in the superconducting state[3]. The I-V characteristics can be described in terms of an effective gap and a pair breaking parameter, which depends on the field. The evolution of the order parameter, the density of states and the transport properties, as function of the applied field are experimentally and theoretically analyzed.

References

1. M. Poza, E. Bascones, J. G. Rodrigo, N. Agraït, S. Vieira and F. Guinea, Phys. Rev. B **58**, 11173 (1998).

Interaction-Induced Dephasing in Disordered Electron Systems

S. Sharov[1,2] and F. Sols[1]

[1] Departamento de Física Teórica de la Materia Condensada, Universidad Autónoma de Madrid, 28049 Madrid, Spain
[2] I.E.Tamm Department of Theoretical Physics, P.N.Lebedev Physics Institute, Leninskii pr. 53, 117924 Moscow, Russia

Abstract. The destructive effect of the electron-electron interaction on weak localization phenomena in disordered metals is studied. We associate this effect with a cutoff in the particle-particle diffusion propagator (Cooperon) in the presence of the interaction. This cutoff, also called dephasing rate, is evaluated diagrammatically in the lowest order of the perturbation theory in the screened Coulomb interaction and in the disorder strength. Unlike in previous studies, we explicitly take into account the interaction processes with energy exchange between the upper and the lower electron lines entering the Cooperon propagator and directly solve the integral equation obtained. Besides, special care is taken to treat the virtual processes with energy transfer larger than temperature (quantum fluctuations). We accurately keep all the contributions coming from them, not appealing to the known semi-classical procedure which reduces the effect of the interaction to that of a fluctuating external field. It is demonstrated that the quantum fluctuations only lead to a rescaling of the momentum and frequency dependences of the Cooper pole but do not affect its cutoff. Thus, we do not confirm the idea that "zero-point fluctuations" can dephase electrons. In contrast, the processes with energy transfer smaller than temperature do that. The temperature dependence of the dephasing rate due to such processes for quasi-1D and -2D conductors is analyzed.

Resonant Tunneling Through Three Quantum Dots with Interdot Repulsion

M.R. Wegewijs[1], Yu.V. Nazarov[1], and S.A. Gurvitz[2]

[1] Delft Institute of Microelectronics and Submicrontechnology (DIMES), Delft University of Technology, 2628 CJ Delft, The Netherlands
[2] Department of Particle Physics, Weizmann Institute of Science, Rehovot 76100

Abstract. Arrays of quantum dots have received an increasing amount of interest with the advance of fabrication technologies. At present mostly arrays of 2 coherently coupled dots (double dot) have been studied theoretically and experimentally. In this contribution we present new theoretical results on the resonant transport through a triple quantum dot connected to leads. We assume that the resonant states of each dot are ground states differing by the addition of an extra electron and that a large bias is applied to the leads. In such small structures Coulomb repulsion between electrons in different dots is important. Whereas in a double dot only one charging energy is of importance, in a triple dot we expect that the competition between nearest neighbor and next-nearest neighbor charging energies to affect transport through the structure. The addition energy spectrum of the three dots with interdot charging energies gives rise to many different regimes for resonant tunneling depending on the positioning of the chemical potentials in the leads. For the most interesting regimes we have calculated the stationary resonant current as a function of the tunnel rates to and from the leads and the parameters characterizing the coherent electronic state in the array. In the "free" electron regime interdot charging energies hardly affect transport properties (intradot charging is incorporated) and as many as 3 extra electrons can populate the array. In the Coulomb blockade regime all charging energies large enough to allow at most 1 extra electron in the array. In intermediate regimes a large difference in the finite interdot charging energies can suppress the current through many-electron states with 2 extra electrons by negatively affecting their coherence. This effect is not possible in a double dot.

We employ the density matrix approach[1],[2] to obtain analytical results in all parameters of our model. These include the coupling to the leads (which has a pronounced influence on the condition for a resonant peak and which is not present in a double dot), the interplay between this coupling and the interdot charging and a possible asymmetry of the array (which affects the coherent couplings and addition energies).

References

1. S.A. Gurvitz , Ya. S. Prager, Phys. Rev. B 53, 15932 (1996)
2. Yu. V. Nazarov, Physica B 189, 57 (1993)

Spin-Isospin Textures in Quantum Hall Bilayers at Filling Factor $\nu = 2$

B. Paredes[1], C. Tejedor[1], L. Brey[2] and L. Martín-Moreno[3]

[1] Departamento de Física Teórica de la Materia Condensada, Universidad Autónoma de Madrid, 28049 Madrid, Spain.
[2] Instituto de Ciencia de Materiales de Madrid (CSIC), Cantoblanco, 28049, Madrid, Spain.
[3] Departamento de Física de la Materia Condensada, ICMA(CSIC), Universidad de Zaragoza, Zaragoza 50015, Spain.

Abstract. We study the charged excitations of a double layer at filling factor 2 in the ferromagnetic regime. In a wide range of Zeeman and tunneling splittings we find that the low energy charged excitations are spin-isospin textures with the charge mostly located in one of the layers. As the tunneling between the layers increases, the *parent* spin texture in one layer increases and it induces, in the other layer, a *shadow* spin texture antiferromagnetically coupled to the *parent* texture. The increase of the quasiparticle spin can be observed experimentally.

Hall Resistance of a Two-Dimensional Electron Gas in the Presence of Magnetic Clusters with Large Perpendicular Magnetization

J. Reijniers,[1] A. Matulis,[2] and F.M. Peeters[1]

[1] University of Antwerp (UIA), Department of Physics, Belgium
[2] Semiconductor Physics Institute, Vilnius, Lithuania

Abstract. The Hall and magnetoresistance of a two-dimensional electron gas (2DEG) placed a small distance from a random distribution of identical perpendicular magnetized ferromagnetic clusters is studied. The magnetic clusters are modeled both by magnetic dipoles, and by thin magnetic disks. The electrons in the 2DEG are scattered by the magnetic field profiles as created by the magnetic clusters.

Although the average magnetic field is zero, we find a nonzero Hall resistance (R_{xy}), which increases with k_F, for small Fermi energies ($E_F = \hbar^2 k_F^2/2m$), but which tends to zero for higher energies. For magnetic disks we find resonances in both the Hall and the magnetoresistance (R_{xx}) as function of the Fermi wave vector.

The physical reason is that quasi bound electron states are formed in the non-homogeneous magnetic field profiles, and so electrons are trapped underneath the magnetic disk. Such resonances enhance R_{xx}, but reduce R_{xy}.

Superconductivity Under Magnetic Fields in Nanobridges of Lead

H. Suderow[1], A. Izquierdo[1], E. Bascones[2], F. Guinea[2], and S. Vieira[1]

[1] Laboratorio de Bajas Temperaturas, Departamento de Fisica de la Materia Condensada, Intituto de Ciencia de Materiales Nicolás Cabrera, Facultad de Ciencias, C-III, Universidad Autónoma de Madrid, 28049 Madrid-Spain
[2] Instituto de Ciencia de Materiales, Consejo Superior de Investigaciones Cientificas, Cantoblanco, 28049 Madrid-Spain

Abstract. We study the properties of superconducting bridges of Pb of nanoscopic dimensions fabricated by an STM under magnetic fields. The bridges behave as weak links of variable dimensions whose form and smallest contact radius can be varied in-situ. We find that these structures have upper critical fields several times the bulk critical field of Pb, forming at large fields an $N - s - junction - s - N$ system whose superconducting part s is of mesoscopic dimensions. We could find no evidence for the nucleation of vortices in these structures, in agreement with the estimation for the geometry which we can get from R-z curves and the model of Ref... Therefore, their phenomenology cannot be explained by the usual type II superconducting behavior. The physical properties of the system are then expected to be dominated by the pair breaking effect of the magnetic field. For instance, in the case of a bridge whose smallest contact radius is of the order of a single atom, we can follow the subharmonic gap structure as a function of the pair breaking parameter. By contrast, in bridges formed by two *long* cone like structures with a smallest contact of the order of 20Åor larger a new phenomenology appears. We observe two well defined conduction regimes as a function of the field and under strong current flow. In an intermediate field regime, strong peaks appear in the dV/dI curves which disappear at large fields, and at the highest fields, an new anomaly in dV/dI appears. This behavior can be understood in terms of nonequilibrium superconductivity by considering the magnetic field dependence of the quasiparticle charge imbalance length Λ_{Q*} through the pair breaking effect. At intermediate fields, our curves can be interpreted as an indication for the nucleation of phase slip centers within the bridge (of dimensions orders of magnitude smaller than usual phase slip wires), and at high magnetic fields the observed anomaly should be related to the contact resistance of the $N - s$ interface formed by the destruction of superconductivity in the bulk. The different magnetic field dependence of Λ_{Q*} and of the length of the superconducting part L_s makes the appearence of two regimes possible.

Effect of the Measurement on the Decay Rate of a Quantum System

B. Elattari[1,2] and S. Gurvitz[1]

[1] Weizmann Institute of Science, Department of Condensed Matter Physics 76100 Rehovot, Israel
[2] Université Chouaïb Doukkali, Fauclté des Sciences, El Jadida, Morocco

Abstract. We investigate how the measurement influences the decay of a quantum unstable system. The latter is represented by an electron tunneling from a quantum dot to a reservoir of empty states. The dot is coupled to a ballistic point-contact acting as a detector. The entire setup can be realized in actual mesoscopic experiments. Using a microscopic description of the whole system, we derived a new type of rate equations and calculated the energy distribution of the tunneling electron. We show that the continuous measurement of the unstable system does not influence its decay rate, while the energy distribution of the tunneling electron is strongly affected. This is in contradiction with rapidly repeated measurements, which are expected to slow down the decay rate (Quantum Zeno effect).

Statistics of Intensities in Surface Disordered Waveguides

A. García-Martín[1], J.J. Sáenz[1], and M. Nieto-Vesperinas[2]

[1] Departamento de Física de la Materia Condensada and Instituto de Ciencia de Materiales "Nicolás Cabrera",
Universidad Autónoma de Madrid, E-28049 Madrid, Spain.
[2] Instituto de Ciencia de Materiales, CSIC, Campus de Cantoblanco, 28049-Madrid, Spain.

Abstract. Numerical calculations of the statistical distributions of the transmited and reflected intensities from surface disordered waveguides are presented. An oscillating behavior of the enhanced backscattering and localization length as a function of the wavelength is predicted.

Although the transport is strongly non isotropic, the analysis of the probability distributions of the transmitted waves confirms in this configuration distributions predicted by Random Matrix Theory for volume disorder.

We use RMT to analytically deduce the probability densities of reflected waves in the localization regime. Numerical calculations of the coupling to backward modes are also put forward for comparison. Interestingly, the speckle distributions are found to be independent of the transport regime. The predicted probability densities reproduce accurately the numerical results.

References

1. A. García-Martín, J.A. Torres, J.J. Sáenz and M. Nieto-Vesperinas, Appl. Phys. Lett. **71** 1912 (1997).
2. A. García-Martín, J.A. Torres, J.J. Sáenz and M. Nieto-Vesperinas, Phys. Rev. Lett. **80**, 4165 (1998).
3. S.A. van Lagen, P.W. Brouwer and C.W.J. Beenakker, Phys. Rev. E **53**, R1344 (1996).
4. P. Pereyra and P.A. Mello, J. Phys. A **16** 237 (1983).
5. A. García-Martín, T. López-Ciudad, J.J. Sáenz and M.Nieto-Vesperinas, Phys. Rev. Lett. **81**, 329 (1998).

Optical Transmission Through Strong Scattering and Highly Polydisperse Media

J.G. Rivas[1], R. Sprik[1], C.M. Soukoulis[2], K. Busch[3], and A. Lagendijk[1]

[1] Van der Waals-Zeeman Instituut, Universiteit van Amsterdam, Valckenierstraat 65, 1018 XE Amsterdam, The Netherlands. rivas@wins.uva.nl

[2] Ames Laboratory and Department of Physics and Astronomy, Iowa State University, Ames, Iowa 50011.

[3] Institute for Theory of Condensed Matter, Department of Physics, University of Karlsruhe, P.O. Box 6980, 76128 Karlsruhe, Germany.

Abstract. Anderson localization refers to a break-down of the wave propagation in disordered scattering systems due to interference. As localization is essentially a wave phenomenon it should hold for all kind of waves *i.e.* electrons, electromagnetic and acoustic waves. For isotropic scatterers localization is established if $kl_s \leq 1$, where k is the wavevector in the medium and l_s is the scattering mean free path. To approach the localization transition, l_s can be reduced by using scatterers with a high refractive index, n, and a size such that the scattering cross section is maximal. An interesting class of materials are semiconductors that have very large refractive indexes and almost not absorption for wavelengths well below the energy of the band gap. Recently localization of light has been observed in GaAs powders [1], opening the possibility to new studies in this field. We investigate the infrared transmission through samples of randomly packed silicon powders. In the wavelength range $1.4\mu m$ to $2.5\mu m$ we analyze in detail the scattering properties and the effects of residual absorption. In this range we observe a nearly constant value of kl_s around 3.5. We attribute the non-variation of kl_s with the wavelength to the high polydispersity in the size of the Si particles. Due to the similar refractive indexes of GaAs and Si, it is surprising that we do not observe Anderson localization in the Si samples. An explanation could be a difference in the connectivity of the particles.

References

1. D.S. Wiersma, P. Bartolini, A. Lagendijk, and R. Righini, *Localization of light in a disordered medium*, Nature **390**, 671 (1997).

Interference in Random Lasers

G. van Soest, F.J. Poelwijk, R. Sprik, and A. Lagendijk

Van Der Waals-Zeeman Institute, University of Amsterdam, Valckenierstr. 65, 1018 XE Amsterdam NL

Abstract. A random laser is a disordered medium that both scatters and amplifies light. Important parameters in this system are the transport mean free path l, the size of the amplifying volume and the amplification length l_{amp}. Properties of light that is propagating through such a medium are modified by the presence of gain, giving rise to a range of surprising phenomena. One of these is the occurrence of a lasing threshold in the output power, pulse duration and spectral width of the emission [1]. These experiments have raised a large number of questions regarding the mechanism that gives rise to these phenomena. One of the most intriguing is how to connect the multiple scattering picture with the laser picture.

Recently, we have performed measurements and simulations investigating the dependence of the laser threshold on the size of the gain volume [3]. It was found that for a small amplifying volume the threshold is at a much higher pump intensity than for a large volume. This effect was reproduced very accurately in a simulation of diffusion in a two- component (amplifying/passive) medium.

Enhanced backscattering is affected by amplification in the multiple scattering medium [3]. As the amount of gain in the medium increases, the backscattercone gets narrower. This is due to the fact that longer light paths, the ones that make up the top of the cone, are amplified more strongly than shorter paths, contributing to the wings. In ref. 2, however, the laser threshold of the medium could not be reached. We present backscattering experiments in a material which allows us to cross the laser threshold. In this way we want to shed more light on how propagation in a multiply scattering medium is affected by gain.

References

1. See e.g. N. M. Lawandy et. al. Nature 368, 436-438 (1994)
2. G. van Soest, M. Tomita, and A. Lagendijk, Opt. Lett. 24, 306-308 (1999)
3. D. S. Wiersma, M. P. van Albada and A. Lagendijk, Phys. Rev. Lett. 75, 1739-1742 (1995).

Electron Patterns Under Bistable Electro-Optical Absorption in Quantum Well Structures

C.A. Velasco[1], L.L. Bonilla[1], V.A. Kochelap[2], and V.N. Sokolov[2]

[1] Universidad Carlos III de Madrid, Escuela Politécnica Superior, Avda Universidad 30, E-28911 Leganés, Spain
e-mail: cvelasco@math.uc3m.es, bonilla@ing.uc3m.es

[2] Institute of Semiconductor Physics, National Academy of Sciences, 252650 Kiev, Ukraine, e-mail: nika@div1.semicond.kiev.ua

Abstract. Heterostructures with electrically biased quantum wells show bistable absorption in both, hybrid electro–optical devices (SEED's) [1] and all-optical devices (wireless SEED's [2], and multiple quantum well structures [3,4]). We report a new phenomenon of formation of transversal electron patterns under bistable electro-optical absorption in wireless single and multiple quantum well heterostructures. The patterns consist of regions with different densities of two-dimensional photogenerated electron-hole plasma. This phenomenon results in patterning of optical absorption and light transmission.

We have formulated and analyzed a model addressing these self–sustained patterns which includes: self–consistent calculations of the wave functions and subband energies of photoexcited electrons and holes in a strongly biased quantum well, nonlinear interband light absorption, configuration of the electrostatic potential, its screening and the lateral motion of the bidimensional electron-hole plasma. The transversal characteristic length scale of the patterns is of the order of the ambipolar diffusion length of the two–dimensional plasma, $L_{\mathcal{D}}$. For large transversal dimensions of the QW layer most of the patterns consist of wide plateaus with high (low) absorption and plasma density and relatively narrow domains with low (high) absorption and plasma density. There is a strong coupling between the vertical and transversal degrees of freedom of photoexcited carriers due to the electrostatic interaction. Transversal redistributions of the plasma induce complex configurations (two-, or three–dimensional) of the electrostatic potential. For a finite transversal dimensions of the QW layer the patterns are strongly affected by the boundary conditions at the edges of the layer. Depending on the heterostructure design there are either electrically charged, or quasineutral patterns (for the latter case the electron concentration, n, is almost equal to the hole concentration, p). If the distance, d_c, between the QW and the electrodes (heavy doped regions of the structure to which the bias is applied) is considerably greater than $L_{\mathcal{D}}$, the patterns are quasineutral. In the opposite case, the patterns are charged. For all cases patterning of transmitted light is calculated.

References

1. D. A. B. Miller *et al.*, Appl. Phys. Lett. **45**, 13 (1984). A. L. Levin, IEEE J. of Quant. Electron. **29**, 655 (1993).
2. J. Couturier *et al.*, Appl. Phys. Lett. **64**(6), 742 (1994).
3. R. Merlin *et al.*, Surf. Science **41**, 9953 (1990).
4. V. A. Kochelap, L. L. Bonilla, V. N. Sokolov and C. A. Velasco, Phys. Stat. Solidi (b) **204**, 559 (1997).
   L. L. Bonilla, V. A. Kochelap and C. A. Velasco, J. Phys. C **31**, L539 (1998).
   C. A. Velasco, L. L. Bonilla, V. A. Kochelap and V. N. Sokolov, Microelectronic Eng. **43-44**, 153 (1998).
   V. A. Kochelap, L. L. Bonilla and C. A. Velasco, Semiconductor Physics, Quantum Electronics and Optoelectronics **1(1)**, 50 (1998).
   V. A. Kochelap, L. L. Bonilla and C. A. Velasco, J. Opt. B: Quantum Semiclass. Opt. **1**, 84 (1999).

Simulation of Mesoscopic Devices with Bohm Trajectories and Wavepackets

X. Oriols, J.J. Garcia, F. Martín, and J. Suñé

Departament d'Enginyeria Electrònica.
Universitat Autònoma de Barcelona
08193 Bellaterra SPAIN.
FAX:34 93 581 13 50.
Phone: 34 93 581 13 61.
E.mail: xoriols@cc.uab.es

Abstract. Following the idea that the great amount of information extracted from "realistic" Monte Carlo (MC) simulations helps to understand and improve the performance of eletron devices, we present a new proposal for the MC simulation of electronic transport in mesoscopic devices where quantum mechanical (QM) effects are important. In particular, we have developed a one-dimensional self-consistent quantum MC simulator to show the viability of our proposal by applying it to analyze static and dynamic properties of resonant tunneling diodes (RTD).

Our proposal can be explained (understood) from simple and intuitive physical ideas. We use Bohm trajectories to describe the quantum dynamics of electrons in the active region of the device. Among the various causal formulations of QM, the most widely known is the one due to Bohm [1,2] that assures that the measurable results of standard quantum mechanics are perfectly reproduced by averaging the Bohm trajectories with adequate relative weights. Our simulator [3] defines a QM window (QW) which includes the double-barrier of the RTD, and restricts the QM treatment (i.e. each electron associated to a Bohm trajectory) to this window. Outside the QW, where the potential changes smoothly in the scale of the de Broglie wavelength of the carriers, the classical MC technique is used to simulate the electron transport.

On the other hand, we will also explain that in spite of the simplicity (from a physical point of view) of our proposal, it can be demonstrated that our model provides a particular solution of the Liouville equation. In this regard, we will present a deconstruction of our proposal in terms of the density matrix [4]. So, we can conclude that our proposal is a simple and intuitive way for solving the Liouville equation, at the same level, as the classical MC method provides a simple and intuitive solution of the Boltzman equation. We will also show some results for a typical RTD. The obtained results qualitatively resemble those obtained with other approaches when no scattering is considered. In particular, within our proposal we can define a new phase-space distribution (positive defined everywhere) that is quite similar to the Wigner distribution function. In conclusion, we will present a RTD simulation based on causal trajectories and we will show several examples to discuss the profit of the information (static and dynamic) extracted from it and the viability of our proposal.

References

1. D.Bohm, Physical Review, 85 (2), 166 (1952)
2. X.Oriols, F.Martín and J.Suñé, Phys. Rev. A. 54(4) (1996)
3. X.Oriols, J.J.Garcia, F.Martín and J.Suñé, T.Gonzalez, J.Mateos and D.Pardo, Appl. Phys. Lett, 72, 806 (1998)
4. X.Oriols, J.J.Garcia, F.Martín and J.Suñé, T.Gonzalez, J.Mateos,D.Pardo.and O.Vanbesien, Semicond. Sci. and Tech., 14(6) (1999)

Chaotic Motion of Space Charge Monopole Waves in Semiconductors Under Time-Independent Voltage Bias

I.R. Cantalapiedra[1], M.J. Bergmann[2], S.W. Teitsworth[2], and L.L. Bonilla[3]

[1] Departamento de Física Aplicada, Universidad Politècnica de Catalunya, c/ Dr. Marañón, 44, 08028 Barcelona, Spain
[2] Department of Physics and Center for Nonlinear and Complex Systems, Box 90305, Duke University, Durham, NC 27708-0305
[3] Universidad Carlos III de Madrid, Escuela Politécnica Superior, 28911 Leganés, Spain

Abstract. A standard drift-diffusion model of space charge wave propagation in semiconductors has been studied numerically and analytically under dc voltage bias. For sufficiently long samples and appropriate voltage bias - such that the sample is biased in a regime of negative differential resistance - we find period doubling and chaos in the propagation of nonlinear fronts (charge monopoles of alternating sign) of electric field. The chaos is always low-dimensional, but it still has a complex spatial structure, lack of spatial coherence; this behavior can be interpreted using a finite dimensional asymptotic model (which is exactly derivable from the full model in the limit of infinitely long samples) in which the front (charge monopole) positions and the electrical current are the only dynamical variables.

References

1. I. R. Cantalapiedra, L. L. Bonilla, M. J. Bergmann, and S. W. Teitsworth, Phys. Rev. B **48**, 12278 (1993).
2. L. L. Bonilla, I. R. Cantalapiedra, M. J. Bergmann, and S. W. Teitsworth, Semicond. Sci. Technol. **9**, 599 (1994).
3. M. J. Bergmann, S. W. Teitsworth, L. L. Bonilla and I. R. Cantalapiedra, Phys. Rev. B **53**, 1327 (1996).
4. L. L. Bonilla and I. R. Cantalapiedra, Phys. Rev. E **56**, 3628 (1997).

Improving Electron Transport Simulation in Mesoscopic Systems by Coupling a Classical Monte Carlo Algorithm to a Wigner Function Solver

J. García-García, F. Martín, X. Oriols, and J. Suñé

Departament d'Enginyeria Electrònica. Universitat Autònoma de Barcelona 08193 BELLATERRA (Barcelona). Spain

Abstract. Because of its high switching speed, low power consumption and reduced complexity to implement a given function, resonant tunneling diodes (RTD's) have been recently recognized as excellent candidates for digital circuit applications [1]. Device modeling and simulation is thus important, not only to understand mesoscopic transport properties, but also to provide guidance in optimal device design and fabrication.

Several approaches have been used to this end. Among kinetic models, those based on the non-equilibrium Green function formalism [2] have gained increasing interest due to their ability to incorporate coherent and incoherent interactions in a unified formulation. The Wigner distribution function approach has been also extensively used to study quantum transport in RTD's [3-6]. The main limitations of this formulation are the semiclassical treatment of carrier-phonon interactions by means of the relaxation time approximation and the huge computational burden associated to the self-consistent solution of Liouville and Poisson equations. This has imposed severe limitations on spatial domains, these being too small to succeed in the development of reliable simulation tools.

Based on the Wigner function approach, we have developed a simulation tool that allows to extend the simulation domains up to hundreds of nanometers without a significant increase in computer time [7]. This tool is based on the coupling between the Wigner distribution function (quantum Liouville equation) and the Boltzmann transport equation. The former is applied to the active region of the device including the double barrier, where quantum effects are present (quantum window, QW). The latter is solved by means of a Monte Carlo algorithm and applied to the outer regions of the device, where quantum effects are not expected to occur.

Since the classical Monte Carlo algorithm is much less time consuming than the discretized version of the Wigner transport equation, we can considerably increase the simulation domains without paying a penalty in efficiency.

We have modeled this coupling by using the Monte Carlo distribution of carriers in k-space in the cells adjacent to the QW as boundary conditions for the step-by-step solution of the Liouville equation, while the Wigner distribution function at the edges of the QW dictates carrier injection to the classical regions.

By introducing in our tool a Poisson solver, necessary for self-consistency, we have simulated the I-V characteristic of RTD's with typical physical parameters. Realistic simulation boxes of 300nm have been considered. These are much higher than those of previous works, while the times required to achieve convergence are similar. The main qualitative features of actual devices are reproduced by means of our tool, i.e., oscillatory behavior and current plateau in the negative differential resistance region.

By analyzing the potential profile and electron density distribution at various applied bias it is seen that charge accumulation in the well is maximum at resonance and no spurious or discontinuities have been found at the boundaries of the QW, which reveals that the coupling model is well-behaved.

We are currently comparing the simulation results to experimental data provided by other authors. Our preliminary results seem to indicate that allowing physical parameters to slightly vary from nominal values (parameter dispersion is unavoidable at the length scales we are dealing with), a reasonable fit between simulated and experimental results is possible.

The main conclusion of the work is that within the framework of the Wigner distribution function, we have developed a tool that provides improvement over previous simulators since realistic device dimensions can be considered without efficiency degradation. This allows to obtain more accurate simulated results and make the tool a potential candidate to aid in RTD device design and fabrication.

This work has been supported by the Dirección General de Enseñanza Superior under contract PB97-0182.

References

1. P. Mazumder, S. Kulkarni, M. Bhattacharya, J.P. Sun and G.I. Haddad, Proc. IEEE, 86, 664 (1998)
2. R. Lake, G. Klimeck, R.C. Bowen and D. Jovanovich, J. Appl. Phys., 81, 7845 (1997).
3. W.R. Frensley, Rev. Mod. Phys., 62, 745 (1990).
4. K.L. Jensen and F.A. Buot, Phys. Rev. Lett., 66, 1078 (1991).
5. K.K. Gullapalli, D.R. Miller and D.P. Neikirk, Phys. Rev. B., 49, 2622 (1994).
6. B.A. Biegel and J.D. Plummer, Phys. Rev. B., 54, 8070 (1996).
7. J. García-García, F. Martín, X. Oriols and J. Suñé, Appl. Phys. Lett., 73, 3539 (1998).

Extended States in Correlated-Disorder GaAs/AlGaAs Superlattices

V. Bellani[1], E. Diez[2], R. Hey[3], G.B Parravicini[1], L. Tarricone[4], and F. Domínguez-Adame[5]

[1] INFM-Dipartimento di Fisica "A. Volta", Università di Pavia, I-27100 Pavia, Italy
[2] GISC-Departamento de Matemáticas, Universidad Carlos III, E-28911, Leganés, Madrid, Spain
[3] Paul Drude Institut für Festkörperelektronik, D-10117 Berlin, Germany
[4] INFM-Dipartimento di Fisica, Università di Parma, I-43100 Parma, Italy
[5] GISC-Departamento de Física de Materiales, Universidad Complutense, E-28040 Madrid, Spain

Abstract. We report [1] the first experimental evidence that spatial correlations inhibit localization of states in disordered low-dimensional systems, as previous theoretical calculations suggested [2,3] in contrast to the earlier belief that all eigenstates are localized. This has been done studying the dc vertical transport and photoluminescence (PL) in GaAs-AlGaAs superlattices (SL's) with intentional correlated disorder. The spectra are compared to those obtained in ordered and uncorrelated disordered superlattices. To verify this theoretical prevision we grew several n-i-n heterostructures (being i the undoped SL's) by molecular beam epitaxy. All SL's have 200 period and $Al_{0.3}Ga_{0.7}As$ barriers 3.2 nm thick. In the *Ordered-SL* all the 200 wells are identical with thickness 3.2 nm (hereafter referred to as A wells). In the *Random-SL*, 58 A wells are replaced by wells of thickness 2.6 nm (hereafter referred to as B wells) and this replacement is done randomly. The so-called *Random dimer-SL* is identical to the *Random-SL* with the additional constraint that the B wells appear only in pairs. X-ray diffraction experiments confirm that the dimer constraint intentionally introduced during sample growth is the only difference between the *Random* and *Random Dimer-SL*.

We measured [1] the vertical dc resistance of our sample at dark as a function of temperatures. The resistance of the *Random Dimer-SL* is very similar to the resistance of the *Ordered-SL* for temperatures below 50 K, and the small differences are due to the different miniband-width between the two. On the other hand, *Random-SL* shows a much higher resistance in this range of temperature. This is due to the presence of extended states in the *Random Dimer-SL* showing transport properties very similar to a *Ordered-SL*. According to theoretical studies [3], these extended states in *Random Dimer-SL's* are not Bloch-like, as occurs in *Ordered-SL's*. PL experiment confirm this interpretation. The PL peak of the *Ordered-SL* is at the lower energy among the three SL's. The PL peak of the *Random-SL* shifts towards higher energies compared with the other two samples. In this SL the intentional disorder introduced by the random distribution of thinner wells B (2.6 nm) localizes the electronic states [3]. The PL peak of the *Random Dimer-SL* is *red-shifted* with respect to the PL peak for the *Random-SL*. This red-shift of the PL peak is due to the formation of a miniband with tunnel process for carriers between the GaAs wells. The position of the electronic levels were calculated with the Kronig-Penney model and calculation show that the *Ordered-SL* and the *Random Dimer-SL* exhibit extended electronic states [3]. The experimental PL positions of the three SL's are in very good agreement with the calculated ones. This is completely consistent with the above interpretation of the transport experiments.

References

1. V. Bellani et al., Phys. Rev. Lett., **82**, 2159 (1999).
2. D. H. Dunlap et al., Phys. Rev. Lett. **65**, 88 (1990).
3. E. Diez et al., IEEE J.Quantum Electron. **31**, 1919 (1995).

Non-Linear Charge Dynamics in Semiconductor Superlattices

D. Sánchez[1], M. Moscoso[2], R. Aguado[1], G. Platero[1], and L.L. Bonilla[2]

[1] Instituto de Ciencia de Materiales (CSIC), 28049 Madrid, Spain
[2] Universidad Carlos III, 28911 Leganés, Spain

Abstract. In the last few years, vertical transport in n-doped weakly coupled semiconductor superlattices has been shown to exhibit non-linear phenomena such as domain formation, multistability, self-sustained current oscillations, and driven and undriven chaos [1]. The most succesful theoretical models combine discrete rate equations for the carrier density and the electric field in the different wells, and use reasonable boundary conditions which mimic the experimental setup [2]. Although these models explain rather well the observed phemomena, the boundary conditions turn out to have a strong effect on the dynamics of electric-field domains.

We study self-sustained current oscillations in weakly-doped superlattices by means of a self-consistent microscopic model of vertical sequential tunneling which includes boundary conditions in a natural way [3]. For highly doped injecting contacts, self-oscillations arise due to recycling and motion of domain walls which are charge monopoles. As the contact doping decreases, a new oscillatory mode due to recycling and motion of charge dipole waves appears. This mode has not been observed so far because the contact doping density is too high in the usual experimental setups. We predict that dipole-mediated oscillations dominate at low doping for which monopole-mediated oscillations disappear. There is an intermediate doping range where both oscillation modes coexist as stable solutions, and hysteresis between them is possible [4].

In addition, our model reproduce experimentally observed current spikes. They are due to the well-to-well hopping of domain walls and appear as a high-frequency oscillation superimposed to the natural current oscillation due to monopole dynamics. Our model makes a distinction between the average potential drops at barriers and wells which (together with the backward tunneling current present only at low electric fields) causes spikes. Thus spikes should also be present at high bias, as experimentally observed. Several preceding models needed to introduced disorder in the doping to obtain current spikes, in a less natural way than ours. [4].

References

1. H. T. Grahn *et al*, Phys. Rev. Lett. **67**, 1618 (1991);
   J. Kastrup *et al*, Phys. Rev. B **55**, 2476 (1997);
   Y. Zhang *et al*, Phys. Rev. Lett. **77**, 3001 (1996).
   K.J. Luo *et al*, Phys. Rev. Lett. **81**, 1290 (1998).

2. F. Prengel *et al*, Phys. Rev. B **50**, 1705 (1994);
L. L. Bonilla *et al*, Phys. Rev. B **50**, 8644 (1994);
O. M. Bulashenko and L. L. Bonilla, Phys. Rev. B **52**, 7849 (1995);
L. L. Bonilla *et al*, SIAM J. Appl. Math. **57**, 1588 (1997).
3. R. Aguado *et al*, Phys. Rev. B **55**, R16053 (1997).
4. D. Sánchez *et al*, (unpublised).

Time-Dependent Resonant Tunneling in the Presence of an Electromagnetic Field

P. Orellana[1] and F. Claro[2]

[1] *Departamento de Física,*
Universidad Católica del Norte ,
Angamos 0610, Casilla 1280, Antofagasta , Chile
[2] *Facultad de Física, Pontificia Universidad*
Católica de Chile,
Vicuña Mackenna 4860, Casilla 306,
Santiago 22, Chile

Abstract. We have analyzed the electronic current density and the charge stored in the well in a resonant tunneling double barrier heterostructure in the presence of an external electromagnetic field. We studied the processes of charging and discharging of the well in the boundaries of the bistability region, and we calculated the transient times as a function of the field parameters. We found that the width of the bistability region is reduced by the external field. Moreover, dynamical instabilities in the charge stored in the well may be triggered by the oscillating field.

The Interplay of Chaos and Dissipation in a Driven Double-Well Potential

S. Kohler[1](*), P. Hanggi[1], and T. Dittrich[2]

[1] Institut fur Physik, Universitat Augsburg
(*) present address:
Dpto. Fisica Teorica de la Materia Condensada,
Universidad Autonoma de Madrid
[2] Universidad de los Andes, Santa Fe de Bogota

Abstract. As a characteristic feature of the spectrum of a driven bistable potential, there exist crossings of chaotic quasienergy singlets with regular tunnel doublets. As the two partners of the tunnel doublet are have opposite symmetry, the crossing with one of them is avoided, the other is exact. Close to the avoided crossing, the tunnel splittings are drastically increased, resulting in a correspondingly higher tunnel rate (chaos-assisted tunneling) [1].

In the presence of dissipation, the tunneling becomes a transient phenomenon within the relaxation towards a quantum-mechanical attractor. The transient dynamics near a singlet-doublet crossing involves at least all three states in the crossing. Depending on temperature and parameters of the crossing, it can be qualitatively different from the familiar fading out of two-state tunneling [2].

As a model for dissipation, we couple the driven system to an ensemble of harmonic oscillators (Caldeira-Leggett model) and introduce a Markov approximation to the driven system, integrated within the Floquet formalism. This results in a master equation which describes the dissipative dynamics in the basis of the Floquet states of the conservative system.

We discuss both the coherent and the incoherent dynamics near singlet-doublet crossings in terms of a simple three-state model [1]. By comparison with exact numerical results, we identify its limitations. In particular, we investigate the duration of the coherent tunneling and the type of quantum-mechanical attractor, depending on temperatur and distance to the avoided crossing.

References

1. O. Bohigas, S. Tomsovic, D. Ullmo, Phys. Rep. 223, 44 (1993).
2. S. Kohler, R. Utermann, T. Dittrich, P. Hanggi, Phys. Rev. E 58, 7219 (1998).

Monte Carlo Simulation of Quantum Transport in Semiconductors Using Wigner Paths

A. Bertoni, J. García-García[2], P. Bordone, R. Brunetti, and C. Jacoboni

[1] Istituto Nazionale per la Fisica della Materia
Dipartimento di Fisica - Università di Modena,
Via Campi 213/A, I-41100 Modena, ITALY.
[2] Departament d'Enginyeria Electrònica
Universitat Autònoma de Barcelona,
08193 Bellaterra (Barcelona), SPAIN.

Abstract. Charge transport in mesoscopic semiconductor systems must be analyzed in terms of a quantum theory since nowadays typical dimensions of the physical structures are comparable with the electron coherence length. Theoretical approaches based on fully quantum mechanical grounds have been developed in the last decade with the purpose of analyzing the quantum electron-phonon interaction in electron transport. The Wigner function (WF) formalism is particularly suitable for the analysis of mesoscopic structures owing to its phase-space formulation that allows a natural treatment of space dependent problems with given boundary conditions. The Hamiltonian describing the system is [3]

$$\mathbf{H} = -\frac{\hbar^2}{2m}\nabla^2 + \sum_{\mathbf{q}} \mathbf{b}_{\mathbf{q}}^{\dagger}\mathbf{b}_{\mathbf{q}}\hbar\omega_{\mathbf{q}} + \sum_{\mathbf{q}} i\hbar F(\mathbf{q})\left(\mathbf{b}_{\mathbf{q}}e^{i\mathbf{qr}} - \mathbf{b}_{\mathbf{q}}^{\dagger}e^{-i\mathbf{qr}}\right) + V(\mathbf{r}) + e\mathbf{E}\cdot\mathbf{r}\,,$$

where the terms in the RHS describe, respectively: free electron evolution, free evolution of the phonon system, electron-phonon interaction, structure potential, constant uniform accelerating field. Due to the linearity of the Liouville equation, these various contributions can be independently developed in the equation of motion for the generalized WF including phonon variables [2], leading to an equation of the form:

$$f_W\left(\mathbf{r},\mathbf{p},\{n_{\mathbf{q}}\},\{n'_{\mathbf{q}}\},t\right) \;=\; \tilde{\mathcal{F}}(t_0)f_W + \int_{t_0}^{t} \mathrm{d}t' \left[\tilde{\mathcal{V}}(t') + \tilde{\mathcal{P}}(t')\right] f_W\,, \qquad (1)$$

where $\tilde{\mathcal{F}}$ is the operator describing the ballistic free evolution of the WF, while $\tilde{\mathcal{V}}(t')$ and $\tilde{\mathcal{P}}(t')$ are the operators accounting respectively for potential and phonon scattering at time t'. Eq. (1) is formally analogue to the Chambers transport equation for the Boltzmann distribution, so even the numerical solution technique and the physical interpretation may be quite similar to a semiclassical approach. This equation is in fact iteratively solved by Monte Carlo sampling, and the concept of Wigner paths is introduced [3, 3]; they are formed by ballistic flights during which the constant field $\mathbf{E}$ acts, interrupted by scattering events due to other fields or phonons, in analogy to the semiclassical case. The WF is supposed to be known inside the simulated device at a given initial time, and at device boundary at every time. The value of the WF at a given phase-space point at a time t can be constructed as the sum of the contributions due to a very large number of Wigner paths that terminate at that point and start at points (and times) in which the WF is known. These contributions are weighted by the proper quantum phase and by a factor due to the scattering mechanisms acting along each path.

A Monte Carlo algorithm has been developed for the case of homogeneous transport in bulk systems, and the effect of optical phonon scattering on macroscopic phenomenological parameters such as electron mobility is obtained. This can be done generating, during the simulation, only phonon scatterings, taking $\tilde{\mathcal{V}} = 0$ in eq. (1). In this way it is possible to obtain the WF (and consequently the current as a function of the applied field $\mathbf{E}$) at any desired time: this means to have the possibility to study non equilibrium conditions and transient phenomena.

A slightly different approach, based on the same formalism, allowed to obtain an I-V characteristic for a RTD device in presence of phonon scattering [3]. At present an algorithm including an arbitrary potential profile, that allows to simulate semiconductor heterostructures, is under study.

References

1. A. Bertoni, P. Bordone, R. Brunetti, and C. Jacoboni, *Wigner Function for Electron Transport in Mesoscopic Systems*, to appear in J. Phys.: Cond. Matt. (1999).
2. F. Rossi, C. Jacoboni, and M. Nedjalkov, Semicond. Sci. and Technol. **9** 934 (1994).
3. P. Bordone, M. Pascoli, R. Brunetti, A. Bertoni, and C. Jacoboni, Phys. Rev. B **59** 3060 (1999).

Transient Currents Through Quantum Dots

J.A. Vergés[1] and E. Louis[2]

[1] Instituto de Ciencia de Materiales de Madrid, Consejo Superior de Investigaciones Científicas, Cantoblanco, E-28049 Madrid, Spain.
[2] Departamento de Física Aplicada, Universidad de Alicante, Apartado 99, E-03080 Alicante, Spain.

Abstract. Kubo formula is used to get the time dependent current that flows through a quantum dot after switching on a small voltage bias. Technically, the calculation involves the evaluation of the linear response function for all frequencies and is, therefore, sensibly more expensive from the computational point of view than the evaluation of the d.c. conductance. Previous estimations of the transient current were done by Prigodin *et al.* in Phys. Rev. Lett. **72**, 546 (1994) for chaotic mesoscopic systems. Our numerical results are completely different from the purely inductive results given in the mentioned paper. Both the regular and the chaotic system show initially a linear increase of the conductance that grows well beyond its static value. Afterwards, it decreases in an oscillating fashion towards its stationary value. While oscillations quickly attenuate in the chaotic model, a power law decay is obtained for the ideal system. Apart for the rapid oscillations, the result can be modelled by a classic circuit having resistive, inductive and also capacitive elements. In principle, our result opens a straightforward experimental way allowing a clear distinction between chaotic and regular systems.

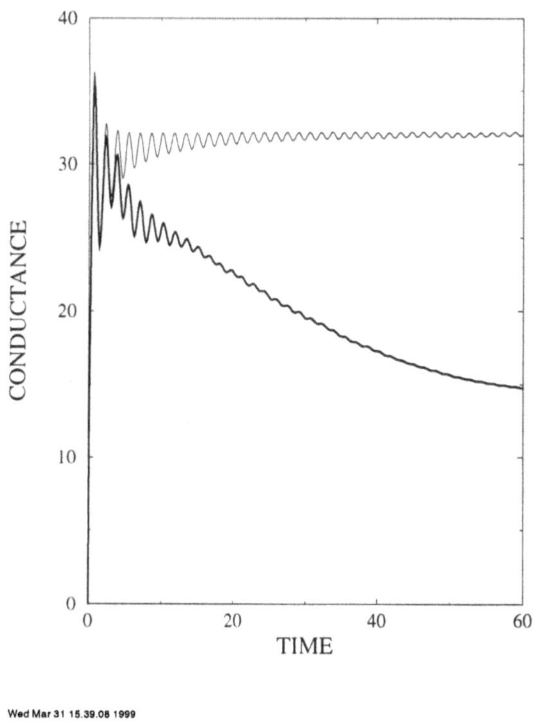

Fig. 1. Conductance (current over voltage) after switching on a small voltage bias calculated for a 32 × 32 dot with a small number of vacancies (billiard, thick line) compared with the result obtained for a regular system (thin line). Conductance is given in quantum conductance units and time is measured in $\hbar$ over hopping energy units.

Ultrafast Coherent Spectroscopy of the Fermi Edge Singularity

D. Porras, J. Fernández-Rossier, and C. Tejedor

Departamento de Física Teórica de la Materia Condensada, Universidad Autónoma de Madrid, Cantoblanco, 28049, Madrid, Spain.

Abstract. In this work we present a theoretical description of the transient response of the Fermi Edge Singularity (FES). We study the linear and the nonlinear response of an n-doped QW to laser pulses in the Coherent Control (CC) and Four Wave Mixing (FWM) Configurations. We calculate the FWM signal emitted by the sample when it is excited by pulses spectrally peaked around the FES by means of a bosonization formalism and show that the long time behavior of the nonlinear signal, for zero and nonzero temperature, is very similar to the linear case.

Self-Consistent Theory of Shot Noise Suppression in Ballistic Conductors

O.M. Bulashenko[1], J.M. Rubí[1], and V.A. Kochelap[2]

[1] Departament de Física Fonamental, Universitat de Barcelona
Av. Diagonal 647, 08208 Barcelona, Spain
[2] Department of Theoretical Physics, Institute of Semiconductor Physics,
Kiev 252028, Ukraine

Abstract. Shot-noise measurements become a fundamental tool to probe carrier interactions in mesoscopic systems [1]. A matter of particular interest is the significance of Coulomb interaction which may keep nearby electrons more regularly spaced rather than strictly at random and lead to the noise reduction. That effect occurs in different physical situations. Among them are charge-limited ballistic transport, resonant tunneling, single-electron tunneling, etc.

In this communication we address the problem of Coulomb correlations in *ballistic* conductors under the space-charge-limited transport conditions, and present for the first time a semiclassical self-consistent theory of shot noise in these conductors by solving analytically the kinetic equation coupled self-consistently with a Poisson equation. Basing upon this theory, exact results for current noise in a two-terminal ballistic conductor under the action of long-range Coulomb correlations has been derived. The noise reduction factor (in respect to the uncorrelated value) is obtained in a closed analytical form for a full range of biases ranging from thermal to shot-noise limits which describe perfectly the results of the Monte Carlo simulations for a nondegenerate electron gas [2]. The magnitude of the noise reduction exceeds 0.01, which is of interest from the point of view of possible applications.

Using these analytical results one may estimate a relative contribution to the noise from different groups of carriers (in energy space and/or real space) and to investigate in great detail the correlations between different groups of carriers. This leads us to suggest an electron energy spectroscopy experiment to probe the Coulomb correlations in ballistic conductors. Indeed, while the injected carriers are uncorrelated, those in the volume of the conductor are strongly correlated, as follows from the derived formulas for the fluctuation of the distribution function. Those correlations may be observed experimentally by making use of a combination of two already realized techniques: a hot-electron spectrometer [3,4] which allows one to analyze different energy groups of electrons collected at the contact and shot-noise measurements [5,6]. Such "shot noise reduction spectroscopy" allows one to measure the novel phenomena. In particular, we predict the (anti)correlation of the "tangent" electrons having the energy close to the potential barrier height, to all other electron energy groups collected at the receiving contact.

References

1. R. Landauer, Nature **392**, 658 (1998).

2. T. González *et al.*, Phys. Rev. B **56**, 6424 (1997).
3. J. R. Hayes, A. F. J. Levi, and W. Wiegmann, Phys. Rev. Lett. **54**, 1570 (1985).
4. M. Heiblum *et al.*, Phys. Rev. Lett. **55**, 2200 (1985).
5. M. Reznikov *et al.*, Phys. Rev. Lett. **75**, 3340 (1995).
6. A. Kumar *et al.*, Phys. Rev. Lett. **76**, 2778 (1996).

Transfer Matrix Formulation of Field-Assisted Tunneling

C. Pérez del Valle[1], S. Miret-Artés[1], R. Lefebvre[2,3], and O. Atabek[2]

[1] Instituto de Matemáticas y Física Fundamental, CSIC, Serrano 123, 28006 Madrid, Spain
[2] Laboratoire de Photophysique Moléculaire, Campus d'Orsay, 91405 Orsay, France
[3] Université Pierre et Marie Curie, 75231 Paris, France

Abstract. A transfer matrix technique is formulated to treat the scattering of an particle incident on a piecewise constant potential and interacting with an oscillatory field. An appropriate choice of parameters gives evidence of the inhibition of transmission at a resonance energy of the intermediate. The transmissivity as a function of a bias potential in a double-barrier structure is calculated.

References

1. Transfer matrix formulation of field-assisted tunneling" by C.Pérez del Valle, R.Lefebvre and O.Atabek, Phys.Rev.A59 3701 (1999).
2. Electron scattering assisted by an intense electromagnetic field : Exact solution of a simplified model " by R.A.Sacks and A.Szoke, Phys.Rev.A 40 5614 (1989).
3. Continuity Conditions for a time-dependent wavefunction" by R.Lefebvre, J.Mol.Structr. (THEOCHEM) (in press).

Two-Dimensional Gunn Effect

L.L. Bonilla[1], R. Escobedo[1], and F.J. Higuera[2]

[1] E.P.S. de la Universidad Carlos III de Madrid, 28911 Leganés, Spain
[2] E.T.S.I. Aeronáuticos, Pza. Cardenal Cisneros 3, 28040 Madrid, Spain

Abstract. Current self-oscillations in semiconductors with a region of negative differential resistivity in their current-field characteristic are known since J. B. Gunn's early experiments on n-GaAs samples in 1963. Most studied are Gunn self-oscillations in one-dimensional spatial configurations which appear when planar contacts are placed in bulk semiconductor samples: during each period of the current oscillation, a charge dipole wave is triggered at the injecting contact, moves and is annihilated at the receiving contact. Dynamics of planar dipole waves can be surprisingly rich for systems with one-dimensional geometry: besides periodic self-oscillations, under *dc* voltage bias there may appear period doubling, frequency blocking and intermittency routes to (low-dimensional) chaos.

We study by numerical and asymptotic methods the solutions of a widely used drift-diffusion model of the Gunn effect in a circular geometry (Corbino disks: a circular disk with one contact on its circumference and a point contact at its center). The result is that axisymmetric pulses of the electric field are periodically shed by an inner circular cathode for a dc voltage bias above a certain onset. These waves decay during their journey to the outer anode, which they may not reach. Meanwhile the current continuously increases and then abruptly decreases when a new wave is shed, in agreement with existing experimental results of Willing and Maan [1]. Depending on the bias, more complex patterns with multiple shedding of pulses at the cathode are possible [2].

References

1. B. Willing and J. C. Maan, Phys. Rev. B **49**, 13995 (1994).
   B. Willing, Ph.D. Thesis, University of Nijmegen (1994).
2. L. L. Bonilla, R. Escobedo, F. J. Higuera (preprint 1999).

An Explanation for Spikes in Current Oscillations of Doped Superlattices

A. Perales, M. Moscoso, and L.L. Bonilla

Escuela Politécnica Superior, Universidad Carlos III de Madrid,
Av. de la Universidad, 30. 28911 Leganés, Madrid, SPAIN

Abstract. The current oscillations in semiconductor superlattices under domain formation, which are due to a recycling of the domain boundary, contain additional spikes of much higher frequency than the fundamental oscillations. While the recycling of the domain boundary covers a larger number of periods of the superlattice, these spikes are due to a relocation of the domain boundary, which is a charge monopole, by one period. The number of spikes within one period of the fundamental oscillation can therefore serve as a measure for the number of periods, which are involved in the recycling motion. The theoretical model used in the simulations of the dynamics of the domain boundary has been outlined in Ref.[2]. In addition we have included a time delay in the drift term (proportional to the tunneling probability through a given barrier) so that the drift term is evaluated at a previous time, $(t - \tau_{tun})$ [1]. This accounts for the nonzero tunneling time τ_{tun} that is estimated as the scattering time due to the interaction between electrons and Coulombic impurities, interface defects and optical and acoustical phonons [3]. Delay effects occur when the average time that the monopole spends crossing a SL period (roughly, the oscillation period divided by number of wells) is comparable to the tunneling time. Our simulations show that the time delay results in high-frequency spikes similar to those experimentally observed [1]. In the following figure we show the time evolution of the current, (a), and the Electric-field profiles at the times depicted in the inset, (b).

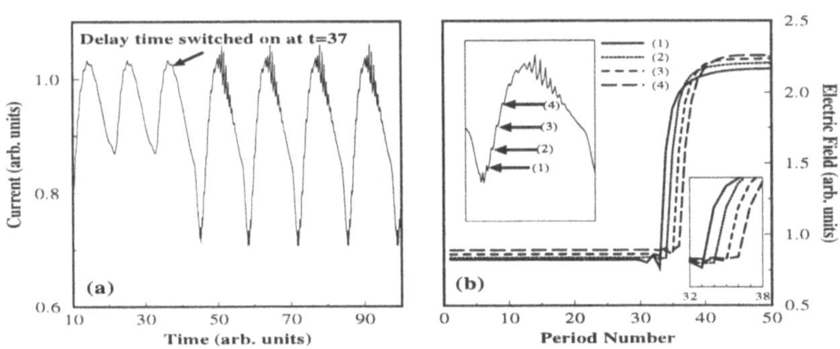

References

1. J.W Kantelhardt, H.T. Grahn, K.H. Ploog, M. Moscoso, A. Perales and L. L. Bonilla. Phys. Stat. Sol. (b) **204**, 500 (1997).
2. J. Kastrup R. Hey, K. H. Ploog, H. T. Grahn, L. L. Bonilla, M. Kindelan, M. Moscoso, A. Wacker, and J. Galán, Phys. Rev. B**55**, 2476 (1997).
3. R. Ferreira and G. Bastard. Phys. Rev. B**40**, 1074 (1989)

Beyond the Static Aproximation in a Mean Field Quantum Disordered System

F. González-Padilla[1] and F. Ritort[2]

[1] Universidad Carlos III de Madrid, 28911 Leganés, Spain
[2] Departament de Física Fonamental, Facultat de Física
Universitat de Barcelona, Diagonal 647, 08028 Barcelona, Spain

Abstract. We study the phase transition of a mean-field spin-glass model in a transverse field in which the interaction matrix is taken from a random orthogonal ensamble (ROM). This model behaves in a similar way to the Sherrinton-Kirkpatrick (SK) model in the classical regime: at low temperatures a first-order replica symmetry breaking phase transition appears induced by the collapse of the configurational entropy and the dominance of a very large number of metastable states which compensate the paramagnetic free energy (E. Marinari, G. Parisi and F. Ritort, J. Phys A **27**, 7647 1994) . The depletion of the transition in the presence of a transverse field has already been studied in presence of quantum fluctuations in the context of the static approximation. We go a little bit further and study these properties beyond the static approximation, using the method introduced by Miller and Huse to study the SK model (J. Miller and D. Huse, Phys. Rev. Lett. **70**, 3147, 1993). Our calculations suggest that the static approximation correctly predicts the order of the zero-temperature phase transition although fails in giving a precise estimate of the value of the critical field.

Quantum-Classical Crossover of the Escape Rate in a Spin System

X. Martínez-Hidalgo

Departament de Física Fonamental, Facultat de Física
Universitat de Barcelona, Diagonal 647, 08028 Barcelona, Spain

Abstract. Quite generally, a system in a metastable well is rendered unstable by thermal and quantum fluctuactions. At high temperature, the former dominate the decay rate, whereas the latter do so at low temperatures. There is a narrow region of temperature where the system shows a crossover between the thermal and the quantum regimes. This crossover resembles a phase transition. It can be first-order (with an abrupt change of regime) or second-order (when the escape rate changes smoothly). More complicated behaviour is possible as well.

Spin systems provide a highly nontrivial example of the general theory. For these systems, the type of crossover can be tuned by an external field. Our predictions can be tested experimentally in molecular nanomagnets, such as $Mn_{12}Ac$ and Fe_8, and in single domain particles.

Appendix II

List of Participants

List of Participants

N. Agraït — nicolas.agrait@uam.es
Universidad Autónoma de Madrid, Spain,

R. Aguado — raguado@physics.rutgers.edu
Rutgers University, USA,

T. Alarcón — tomas@ffn.ub.es
Universitat de Barcelona, Spain,

B. Altshuler — bla@research.nj.nec.com
Princeton University, USA,

S. Bailey — s.bailey@lancaster.ac.uk
Lancaster University, UK,

N. Barberan — nuri@ecm.ub.es
Universitat de Barcelona, Spain,

E. Bascones — lleni@quijote.icmn.csic.es
ICMM-CSIC, Spain,

V. Bellani — bellani@fisav.unipv.it
University of Pavia, Italy,

A. Bertoni — bertoni@teseo.unimo.it
Università di Modena - Instituto Nazionele di Fisica della Materia, Italy,

R.H. Blick — Robert.Blick@Physik.Uni-Muenchen.DE
Ludwig-Maximilians-University, Germany,

L.L Bonilla — bonilla@ing.uc3m.es
Universidad Carlos III de Madrid, Spain,

L. Brey — brey@naomi.icmm.csic.es
Instituto de Ciencia de Materiales CSIC, Spain,

O. Bulashenko — oleg@ffn.ub.es
Universitat de Barcelona, Spain,

M. Büttiker — buttiker@serifos.unige.ch
University of Geneva, Switzerland,

M.J. Calderón — maryjoe@eclipse.icmm.csic.es
Instituto de Ciencia de Materiales CSIC, Spain,

J. Calleja — calleja@uam.es
Universidad Autónoma de Madrid, Spain,

M. Casse — casse@polycnrs-gre.fr
CNRS Grenoble High Magnetic Field Laboratory, France,

L. Chico lchico@amb-to.uctm.es
Universidad de Castilla-La Mancha, Spain,
C.E. Creffield charles@feynman.ph.kcl.ac.uk
King's College London, UK,
T.D. Kieu kieu@mst.csiro.au
CSIRO, Australia,
S. De Franceschi silvano@qt.tn.tudelft.nl
Delft University of Technology, The Netherlands,
M. Devoret mdevoret@cea.fr
CEA-Saclay, France,
D.P. DiVincenzo divince@watson.ibm.com
IBM T. J. Watson Research Center, USA,
P. Dolby p.dolby@lancaster.ac.uk
Lancaster University, UK,
B. Elattari hbrahim@wicc.weizmann.ac.il
Weizmann Institute of Science, Israel,
A. Emperador agusti@ecm.ub.es
Universitat de Barcelona, Spain,
R. Escobedo escobedo@math.uc3m.es
Universidad Carlos III, Spain,
R. Folch roger@ecm.ub.es
Universitat de Barcelona, Spain,
J. García joan.g@cc.uab.es
Universitat Autònoma de Barcelona, Spain,
A. García-Martín a.garcia@uam.es
Universidad Autónoma de Madrid, Spain,
L. Garrido garrido@ffn.ub.es
Universitat de Barcelona, Spain,
T. Geisel geisel@chaos.gwdg.de
Max-Planck-Institut fur Stromungsforschung, Germany,
L. Glazman glazman@mnhepw.hep.umn.edu
University of Minnesota, USA,
J. Gòmez-Rivas rivas@wins.uva.nl
Universiteit van Amsterdam, The Netherlands,
G. Gomila gomila@axpmat.unile.it
Università di Lecce, Italy,
T. González tomasg@gugu.usal.es
Universidad de Salamanca, Spain,
M. González manelgp@ffn.ub.es
Universitat de Barcelona, Spain,
H.T. Grahn htg@pdi-berlin.de
Paul-Drude-Institut, Germany,
F. Guinea paco.guinea@uam.es
Instituto de Ciencia de Materiales CSIC, Spain,

P. Hänggi Hanggi@Physik.Uni-Augsburg.de
University of Augsburg, Germany,
M. Heiblum Heiblum@wis.weizmann.ac.il
Weizmann Institute of Science, Israel,
A. Hernández aurora@ecm.ub.es
Universitat de Barcelona, Spain,
C.P. Herrero ch@noe.icmm.csic.es
Instituto de Ciencia de Materiales CSIC, Spain,
A.P. Jauho antti@mic.dtu.dk
Technical University of Denmark, Denmark,
G. Kastrinakis kastr@phy.cam.ac.uk
Hitachi Cambridge Laboratory, UK,
V. Kochelap nika@div1.semicond.kiev.ua
Institut of Semiconductor Physics, Ukraine,
S. Kohler sigmund@nuria.fmc.uam.es
Universidad Autónoma de Madrid, Spain,
B. Kramer kramer@theorie1.physnet.uni-hamburg.de
Universtitaet Hamburg, Germany,
C. Lambert c.lambert@lancaster.ac.uk
Lancaster University, UK,
A.L. Yeyati aly@uamc11.fmc.uam.es
Universidad Autónoma de Madrid, Spain,
R. López rosalg@everest.icmm.csic.es
Instituto de Ciencia de Materiales CSIC, Spain,
E. Louis enrique@fisic1.ua.es
Universidad de Alicante, Spain,
A. Luzón luzon@montes.upm.es
Universidad Politécnica de Madrid, Spain,
A.H. MacDonald macd@gibbs.physics.indiana.edu
University of Bloomington, USA,
C. Marcus cmarcus@leland.stanford.edu
Stanford University, USA,
M.A. Martín-Delgado mardel@eucmax.sim.ucm.es
Universidad Complutense de Madrid, Spain,
J. Martínez-Hidalgo xavim@ffn.ub.es
Universitat de Barcelona,
L. Martín-Moreno lmm@posta.unizar.es
Universidad de Zaragoza, Spain,
K. Matveev matveev@phy.duke.edu
Duke University, USA,
P. Messina pm127@york.ac.uk
University of York, UK,
Y. Naveh ynaveh@ccmail.sunysb.edu
State University of New York, USA,

M. Ortuño moo@fcu.mu.es
Universidad de Murcia, Spain,
J.J. Palacios palacios@kim.fmc.uam.es
Universidad Autónoma de Madrid, Spain,
B. Paredes belen@kim.fmc.uam.es
Universidad Autónoma de Madrid, Spain,
A. Perales alvis@math.uc3m.es
Universidad Carlos III, Spain,
C. Pérez immcp61@pinar2.cti.csic.es
IMAFF - CSIC, Spain,
A. Pérez-Madrid agusti@ffn.ub.es
Universitat de Barcelona, Spain,
V. Piazza piazza@sns.it
Scuola Normale Superiore, Italy,
G. Platero gloria@marenostrum.icmm.csic.es
Instituto de Ciencia de Materiales CSIC, Spain,
D. Porras diego@kim.fmc.uam.es
Universidad Autónoma de Madrid, Spain,
H. Postma henk@qt.tn.tudelft.nl
Delft University of Technology, The Netherlands,
A. Pouydebasque pouydeb@polycnrs-gre.fr
Institut National des Sciences Appliquées, France,
I.R. Cantalapiedra inma@hal9000.upc.es
Universitat Politècnica de Catalunya, Spain,
D. Reguera davidr@precario.ffn.ub.es
Universitat de Barcelona, Spain,
J. Reijniers jonasr@uia.ua.ac.be
Universiteit Antwerpen, Belgium,
B. Reusch reusch@luna.physik.uni-freiburg.de
University of Freiburg, Germany,
F. Ritort ritort@afrau.ffn.ub.es
Universitat de Barcelona, Spain,
J. Rollbuehler joerg@tfp1.physik.uni-freiburg.de
University of Freiburg, Germany,
J.M. Rubí miguel@ffn.ub.es
Universitat de Barcelona, Spain,
G. Rubio gabino.rubio@uam.es
Universidad Autónoma de Madrid, Spain,
M. Sales msales@ffn.ub.es
Universitat de Barcelona, Spain,
D. Sánchez david@quijote.icmm.csic.es
Instituto de Ciencia de Materiales CSIC, Spain,
J. Sánchez-Ruiz jsanchez@math.uc3m.es
Universidad Carlos III, Spain,

I. Santamaría ivan@ffn.ub.es
Universitat de Barcelona, Spain,
S. Sasaki satoshi@will.brl.ntt.co.jp
NTT Basic Research Laboratories, Japan,
H. Schomerus henning@lorentz.leidenuniv.nl
Universiteit Leiden, The Netherlands,
G. Schön schoen@tfp.physik.uni-karlsruhe.de
Universität Karlsruhe, Germany,
F. Sols fernando.sols@uam.es
Universidad Autónoma de Madrid, Spain,
E. Soukhoroukov sukhorukov@ubaclu.unibas.ch
University of Basel, Switzerland,
P. Stamp stamp@physics.ubc.ca
University of British Columbia, Canada,
G. Suderow hermann.suderow@uam.es
Universidad Autónoma de Madrid, Spain,
S. Teitsworth teitso@phy.duke.edu
Duke University, USA,
C. Tejedor carlos.tejedor@uam.es
Universidad Autónoma de Madrid, Spain,
J.M Torres 183486@cienz.unizar.es
Universidad de Zaragoza, Spain,
H.E. Van den Brom vdbrom@rulkol.leidenuniv.nl
Leiden University, The Netherlands,
G. Van Soest vansoest@phys.uva.nl
Van der Waals-Zeeman Institute, The Netherlands,
C. Velasco cvelasco@math.uc3m.es
Universidad Carlos III, Spain,
J.A. Verges jav@venus.icmm.csic.es
Instituto de Ciencia de Materiales CSIC, Spain,
M. Wagner wagner@phy.cam.ac.uk
Hitachi Cambridge Laboratory, UK,
M. Wegewijs maarten@duttnto.tn.tudelft.nl
Delft University of Technology, The Netherlands,
J. Weis weis@klizix.mpi-stuttgart.mpg.de
Max-Planck-Institut fuer Festkoerperforschung, Germany,
D.S Wiersma wiersma@lens.unifi.it
European Laboratory for Non-linear Spectroscopy, Italy,
A. Yanson ayanson@rulkol.LeidenUniv.nl
Leiden University, The Netherlands

Lecture Notes in Physics

For information about Vols. 1–514
please contact your bookseller or Springer-Verlag

Vol. 515: C.-H. Bruneau (Ed.), Sixteenth International Conference on Numerical Methods in Fluid Dynamics. Proceedings. XV, 568 pages. 1998.

Vol. 516: J. Cleymans, H. B. Geyer, F. G. Scholtz (Eds.), Hadrons in Dense Matter and Hadrosynthesis. Proceedings, 1998. XII, 253 pages. 1999.

Vol. 517: Ph. Blanchard, A. Jadczyk (Eds.), Quantum Future. Proceedings, 1997. X, 244 pages. 1999.

Vol. 518: P. G. L. Leach, S. E. Bouquet, J.-L. Rouet, E. Fijalkow (Eds.), Dynamical Systems, Plasmas and Gravitation. Proceedings, 1997. XII, 397 pages. 1999.

Vol. 519: R. Kutner, A. Pękalski, K. Sznajd-Weron (Eds.), Anomalous Diffusion. From Basics to Applications. Proceedings, 1998. XVIII, 378 pages. 1999.

Vol. 520: J. A. van Paradijs, J. A. M. Bleeker (Eds.), X-Ray Spectroscopy in Astrophysics. EADN School X. Proceedings, 1997. XV, 530 pages. 1999.

Vol. 521: L. Mathelitsch, W. Plessas (Eds.), Broken Symmetries. Proceedings, 1998. VII, 299 pages. 1999.

Vol. 522: J. W. Clark, T. Lindenau, M. L. Ristig (Eds.), Scientific Applications of Neural Nets. Proceedings, 1998. XIII, 288 pages. 1999.

Vol. 523: B. Wolf, O. Stahl, A. W. Fullerton (Eds.), Variable and Non-spherical Stellar Winds in Luminous Hot Stars. Proceedings, 1998. XX, 424 pages. 1999.

Vol. 524: J. Wess, E. A. Ivanov (Eds.), Supersymmetries and Quantum Symmetries. Proceedings, 1997. XX, 442 pages. 1999.

Vol. 525: A. Ceresole, C. Kounnas, D. Lüst, S. Theisen (Eds.), Quantum Aspects of Gauge Theories, Supersymmetry and Unification. Proceedings, 1998. X, 511 pages. 1999.

Vol. 526: H.-P. Breuer, F. Petruccione (Eds.), Open Systems and Measurement in Relativistic Quantum Theory. Proceedings, 1998. VIII, 240 pages. 1999.

Vol. 527: D. Reguera, J. M. G. Vilar, J. M. Rubí (Eds.), Statistical Mechanics of Biocomplexity. Proceedings, 1998. XI, 318 pages. 1999.

Vol. 528: I. Peschel, X. Wang, M. Kaulke, K. Hallberg (Eds.), Density-Matrix Renormalization. Proceedings, 1998. XVI, 355 pages. 1999.

Vol. 529: S. Biringen, H. Örs, A. Tezel, J.H. Ferziger (Eds.), Industrial and Environmental Applications of Direct and Large-Eddy Simulation. Proceedings, 1998. XVI, 301 pages. 1999.

Vol. 530: H.-J. Röser, K. Meisenheimer (Eds.), The Radio Galaxy Messier 87. Proceedings, 1997. XIII, 342 pages. 1999.

Vol. 531: H. Benisty, J.-M. Gérard, R. Houdré, J. Rarity, C. Weisbuch (Eds.), Confined Photon Systems. Proceedings, 1998. X, 496 pages. 1999.

Vol. 532: S. C. Müller, J. Parisi, W. Zimmermann (Eds.), Transport and Structure. Their Competitive Roles in Biophysics and Chemistry. XII, 400 pages. 1999.

Vol. 533: K. Hutter, Y. Wang, H. Beer (Eds.), Advances in Cold-Region Thermal Engineering and Sciences. Proceedings, 1999. XIV, 608 pages. 1999.

Vol. 534: F. Moreno, F. González (Eds.), Light Scattering from Microstructures. Proceedings, 1998. XII, 300 pages. 2000

Vol. 535: H. Dreyssé (Ed.), Electronic Structure and Physical Properties of Solids: The Uses of the LMTO Method. Proceedings, 1998. XIV, 458 pages. 2000.

Vol. 536: T. Passot, P.-L. Sulem (Eds.), Nonlinear MHD Waves and Turbulence. Proceedings, 1998. X, 385 pages. 1999.

Vol. 537: S. Cotsakis, G. W. Gibbons (Eds.), Mathematical and Quantum Aspects of Relativity and Cosmology. Proceedings, 1998. XII, 251 pages. 1999.

Vol. 538: Ph. Blanchard, D. Giulini, E. Joos, C. Kiefer, I.-O. Stamatescu (Eds.), Decoherence: Theoretical, Experimental, and Conceptual Problems. Proceedings, 1998. XII, 345 pages. 2000.

Vol. 539: A. Borowiec, W. Cegła, B. Jancewicz, W. Karwowski (Eds.), Theoretical Physics. Fin de Siècle. Proceedings, 1998. XX, 319 pages. 2000.

Vol. 540: B. G. Schmidt (Ed.), Einstein's Field Equations and Their Physical Implications. Selected Essays. 1999. XIII, 429 pages. 2000

Vol. 541: J. Kowalski-Glikman (Ed.), Towards Quantum Gravity. Proceedings, 1999. XII, 376 pages. 2000.

Vol. 542: P. L. Christiansen, M. P. Sørensen, A. C. Scott (Eds.), Nonlinear Science at the Dawn of the 21st Century. Proceedings, 1998. XXVI, 458 pages. 2000.

Vol. 543: H. Gausterer, H. Grosse, L. Pittner (Eds.), Geometry and Quantum Physics. Proceedings, 1999. VIII, 408 pages. 2000.

Vol. 544: T. Brandes (Ed.), Low-Dimensional Systems. Interactions and Transport Properties. Proceedings, 1999. VIII, 219 pages. 2000

Vol. 545: J. Klamut, B. W. Veal, B. M. Dabrowski, P. W. Klamut, M. Kazimierski (Eds.), New Developments in High-Temperature Superconductivity. Proceedings, 1998. VIII, 275 pages. 2000.

Vol. 546: G. Grindhammer, B. A. Kniehl, G. Kramer (Eds.), New Trends in HERA Physics 1999. Proceedings, 1999. XIV, 460 pages. 2000.

Vol. 547: D. Reguera, G. Platero, L.L. Bonilla, J.M. Rubí(Eds.), Statistical and Dynamical Aspects of Mesoscopic Systems. Proceedings, 1999. XII, 357 pages. 2000.

Vol. 548: D. Lemke, M. Stickel, K. Wilke (Eds.), ISO Surveys of a Dusty Universe. Proceedings, 1999. XII, 432 pages. 2000.

Vol. 549: C. Egbers, G. Pfister (Eds.), Physics of Rotating Fluids. Proceedings, 1999. XIV, 439 pages. 2000.

Vol. 550: M. Planat (Ed.), Noise, Oscillators and Algebraic Randomness. Proceedings, 1999. XII, 417 pages. 2000.

Monographs

For information about Vols. 1–20
please contact your bookseller or Springer-Verlag

Vol. m 21: G. P. Berman, E. N. Bulgakov, D. D. Holm, Crossover-Time in Quantum Boson and Spin Systems. XI, 268 pages. 1994.

Vol. m 22: M.-O. Hongler, Chaotic and Stochastic Behaviour in Automatic Production Lines. V, 85 pages. 1994.

Vol. m 23: V. S. Viswanath, G. Müller, The Recursion Method. X, 259 pages. 1994.

Vol. m 24: A. Ern, V. Giovangigli, Multicomponent Transport Algorithms. XIV, 427 pages. 1994.

Vol. m 25: A. V. Bogdanov, G. V. Dubrovskiy, M. P. Krutikov, D. V. Kulginov, V. M. Strelchenya, Interaction of Gases with Surfaces. XIV, 132 pages. 1995.

Vol. m 26: M. Dineykhan, G. V. Efimov, G. Ganbold, S. N. Nedelko, Oscillator Representation in Quantum Physics. IX, 279 pages. 1995.

Vol. m 27: J. T. Ottesen, Infinite Dimensional Groups and Algebras in Quantum Physics. IX, 218 pages. 1995.

Vol. m 28: O. Piguet, S. P. Sorella, Algebraic Renormalization. IX, 134 pages. 1995.

Vol. m 29: C. Bendjaballah, Introduction to Photon Communication. VII, 193 pages. 1995.

Vol. m 30: A. J. Greer, W. J. Kossler, Low Magnetic Fields in Anisotropic Superconductors. VII, 161 pages. 1995.

Vol. m 31 (Corr. Second Printing): P. Busch, M. Grabowski, P.J. Lahti, Operational Quantum Physics. XII, 230 pages. 1997.

Vol. m 32: L. de Broglie, Diverses questions de mécanique et de thermodynamique classiques et relativistes. XII, 198 pages. 1995.

Vol. m 33: R. Alkofer, H. Reinhardt, Chiral Quark Dynamics. VIII, 115 pages. 1995.

Vol. m 34: R. Jost, Das Märchen vom Elfenbeinernen Turm. VIII, 286 pages. 1995.

Vol. m 35: E. Elizalde, Ten Physical Applications of Spectral Zeta Functions. XIV, 224 pages. 1995.

Vol. m 36: G. Dunne, Self-Dual Chern-Simons Theories. X, 217 pages. 1995.

Vol. m 37: S. Childress, A.D. Gilbert, Stretch, Twist, Fold: The Fast Dynamo. XI, 406 pages. 1995.

Vol. m 38: J. González, M. A. Martín-Delgado, G. Sierra, A. H. Vozmediano, Quantum Electron Liquids and High-Tc Superconductivity. X, 299 pages. 1995.

Vol. m 39: L. Pittner, Algebraic Foundations of Non-Com-mutative Differential Geometry and Quantum Groups. XII, 469 pages. 1996.

Vol. m 40: H.-J. Borchers, Translation Group and Particle Representations in Quantum Field Theory. VII, 131 pages. 1996.

Vol. m 41: B. K. Chakrabarti, A. Dutta, P. Sen, Quantum Ising Phases and Transitions in Transverse Ising Models. X, 204 pages. 1996.

Vol. m 42: P. Bouwknegt, J. McCarthy, K. Pilch, The W3 Algebra. Modules, Semi-infinite Cohomology and BV Algebras. XI, 204 pages. 1996.

Vol. m 43: M. Schottenloher, A Mathematical Introduction to Conformal Field Theory. VIII, 142 pages. 1997.

Vol. m 44: A. Bach, Indistinguishable Classical Particles. VIII, 157 pages. 1997.

Vol. m 45: M. Ferrari, V. T. Granik, A. Imam, J. C. Nadeau (Eds.), Advances in Doublet Mechanics. XVI, 214 pages. 1997.

Vol. m 46: M. Camenzind, Les noyaux actifs de galaxies. XVIII, 218 pages. 1997.

Vol. m 47: L. M. Zubov, Nonlinear Theory of Dislocations and Disclinations in Elastic Body. VI, 205 pages. 1997.

Vol. m 48: P. Kopietz, Bosonization of Interacting Fermions in Arbitrary Dimensions. XII, 259 pages. 1997.

Vol. m 49: M. Zak, J. B. Zbilut, R. E. Meyers, From Instability to Intelligence. Complexity and Predictability in Nonlinear Dynamics. XIV, 552 pages. 1997.

Vol. m 50: J. Ambjørn, M. Carfora, A. Marzuoli, The Geometry of Dynamical Triangulations. VI, 197 pages. 1997.

Vol. m 51: G. Landi, An Introduction to Noncommutative Spaces and Their Geometries. XI, 200 pages. 1997.

Vol. m 52: M. Hénon, Generating Families in the Restricted Three-Body Problem. XI, 278 pages. 1997.

Vol. m 53: M. Gad-el-Hak, A. Pollard, J.-P. Bonnet (Eds.), Flow Control. Fundamentals and Practices. XII, 527 pages. 1998.

Vol. m 54: Y. Suzuki, K. Varga, Stochastic Variational Approach to Quantum-Mechanical Few-Body Problems. XIV, 324 pages. 1998.

Vol. m 55: F. Busse, S. C. Müller, Evolution of Spontaneous Structures in Dissipative Continuous Systems. X, 559 pages. 1998.

Vol. m 56: R. Haussmann, Self-consistent Quantum Field Theory and Bosonization for Strongly Correlated Electron Systems. VIII, 173 pages. 1999.

Vol. m 57: G. Cicogna, G. Gaeta, Symmetry and Perturbation Theory in Nonlinear Dynamics. XI, 208 pages. 1999.

Vol. m 58: J. Daillant, A. Gibaud (Eds.), X-Ray and Neutron Reflectivity: Principles and Applications. XVIII, 331 pages. 1999.

Vol. m 59: M. Kriele, Spacetime. Foundations of General Relativity and Differential Geometry. XV, 432 pages. 1999.

Vol. m 60: J. T. Londergan, J. P. Carini, D. P. Murdock, Binding and Scattering in Two-Dimensional Systems. Applications to Quantum Wires, Waveguides and Photonic Crystals. X, 222 pages. 1999.

Vol. m 61: V. Perlick, Ray Optics, Fermat's Principle, and Applications to General Relativity. X, 220 pages. 2000.

Vol. m 63: R. J. Szabo, Ray Optics, Equivariant Cohomology and Localization of Path Integrals. XII, 315 pages. 2000.

Vol. m 64: I. G. Avramidi, Heat Kernel and Quantum Gravity. X, 143 pages. 2000.

Kolibri 360 GmbH
Ferdinand-Jühlke-Straße 7
99095 Erfurt, Deutschland
produktsicherheit@kolibri360.de